CENTRIFUGAL PUMPS:

Selection, operation and maintenance

by Igor Karassik and Roy Carter

Here is a comprehensive reference book for users and designers of centrifugal pumps throughout industry. Component parts, pump drives, performance characteristics, system-head curves, controls, and priming are all discussed from the point of view of the buyer and the user of pumping equipment.

CENTRIFUGAL PUMPS gathers together and synthesizes information covering the entire field of centrifugal pumps, their appurtenances, and control. It describes and thoroughly illustrates all types of pumps, including vertical, self-priming, and regenerative pumps; it also discusses pumps for various areas of industrial service. One chapter details what the buyer should know about the conditions of service before he orders a pump. Other chapters discuss trouble shooting, installation, and general maintenance. A useful data section completes the book.

The key word which best describes this work is practicality. Theory is only introduced where it is essential to the understanding of specific problems in planning, selecting, operating, and maintaining pump systems. Because of its practicality, this book will be of everyday use to anyone concerned with the moving of liquids or gases in bulk.

IGOR J. KARASSIK is consulting engineer and manager of planning at the Harrison Division of Worthington Corporation.

For the past 27 years, Karassik has been engaged in research and design work on single and multistage pumps. He has specialized in the application of multistage high pressure pumps, especially for steam power stations. He is noted for his work in developing and designing improved boiler feed cycles for central power station plants, and holds numerous patents in this area.

Karassik has also achieved prominence as a technical author. He has written almost 300 articles on centrifugal pumps and steam power plants for technical journals. Including reprints and translations, his articles have appeared in 620 issues of 56 different magazines, in a number of languages. He is co-author of the book *Pump Questions and Answers,* and is a three-time winner of the first prize in the annual Hydraulic Institute engineering essay contest.

Born in Russia, Karassik emigrated to Turkey in 1919, to France in 1924, and to the United States in 1928. He has received B. S. and M. S. degrees from the Carnegie Institute of Technology.

The late ROY CARTER was public works consultant and manager of the Volute Section at the Worthington Corporation, until his untimely death in 1958. Although Karassik was solely responsible for the execution of *Centrifugal Pumps,* he has insisted that the book carry Carter's name as co-author, "Because its writing reflects our many discussions, and its chapters contain much that we had conceived together."

Centrifugal Pumps

Centrifugal Pumps

SELECTION,
OPERATION,
and
MAINTENANCE

Igor J. Karassik
CONSULTING ENGINEER, WORTHINGTON CORPORATION

and Roy Carter

McGRAW-HILL BOOK COMPANY, INC.

New York   Toronto   London

Foreword

The subject of centrifugal pumps has received much attention in technical literature both here and abroad. However, the authors felt that most of this literature placed greatest emphasis on centrifugal pump theory, with insufficient stress on the more practical side of the problem. This practical side is more important to most engineers and users, as these people put centrifugal pumps to use while only a small minority actually design the equipment.

One aim of this book is to guide the centrifugal pump user in system design and equipment selection for the most satisfactory combination of the two. It is also intended to provide useful information about equipment already installed as a guide to maximum service with minimum maintenance and unscheduled outage. The structural details and component parts of centrifugal pumps are described and methods are recommended for restoring each component to its initial condition after deterioration in service. In addition, special chapters are devoted to vertical pumps, self-priming units, and the so-called regenerative pumps. These are followed by a discussion of construction materials. A detailed presentation is given on the concept of "heads," conditions of service, and performance characteristics of various types of centrifugal pumps. System-head curves and their effect on pump output and selection are also fully discussed.

An important factor in centrifugal pump application and operation that has often been neglected is the controls. This subject has been given special attention.

Because successful pump application also depends on a harmonious combination of pump and driver, a chapter is included on pump drivers. Another important subject, priming, is discussed in great detail.

Nearly all centrifugal pumping services have their individual problems and requirements. These services range from general water supply, sewage, drainage, and irrigation to power plant, process work, and other specialized applications. Growth and change in processes and industries have contributed to the development of new designs for the ever-increasing number of pumps. Many special designs are therefore available today that may be severely limited in application flexibility. Centrifugal pump users should have a general knowledge of specialized designs to help assure proper application. The chapter on "Services" covers these special types and presents related operational information. It is supplemented by a chapter on the preparation of inquiries and ordering procedures.

One important section of the book is devoted to the installation, operation, and maintenance of centrifugal pumps. Finally, to make this book as useful as possible, a general Data Section contains valuable data required for engineering pumping installations and analyzing the performance of existing units.

The authors have attempted to avoid complex technical explanations and involved theoretical discussions having little practical value to centrifugal pump users. Theoretical design data would only suggest that the user is expected to judge the excellence of the designer. This aim is not part of, nor compatible with, the objectives of the book, which are to provide practical and useful knowledge of centrifugal pump construction, application, control, installation, operation, maintenance, and trouble-shooting.

The data in this book apply to all makes and types of centrifugal pumps. Wherever possible, therefore, illustrations have been selected from a wide group of manufacturers. For obvious reasons, however, we had greatest access to the extensive files of the Worthington Corporation. For many subjects—for example, individual pump parts—the illustrations would be similar, regardless of source, and therefore most of these were selected from the Worthington files. Wherever photographs of complete pumps or sectional drawings that are not from Worthington are reproduced, the captions give credit to the pump manufacturer responsible for the design. The authors wish to extend their thanks to the Worthington Corporation, the Allis-Chalmers Mfg. Co., Byron-Jackson Co., the DeLaval Steam Turbine Co., Ingersoll-Rand Co., Pacific Pump Co., and many others who very graciously granted the right to reproduce equipment photographs and drawings.

The authors also wish to thank numerous magazines including *Air Conditioning,* *Heating and Ventilating, Power, Power Engineering, Southern Power and Industry, Water and Sewage Works,* and many others for their kind permission to utilize material from articles by the authors that had originally appeared in their pages.

We are indebted to the Hydraulic Institute for the permission to reproduce a number of charts and data from its Standards.

Finally, the authors wish to express their thanks to Messrs. A. H. Borchardt, G. F. Habach, L. H. Garnar, W. C. Krutzsch, C. J. Tullo, and many other associates at Worthington Corporation for providing valuable advice and constructive criticism.

Roy Carter and I decided to undertake this book a number of years ago. Unfortunately, Mr. Carter did not live to see it completed. He passed away unexpectedly in September 1958. I decided that the book *should* be completed and therefore continued the task alone. I hope it measures up to our mutual expectations. Because its writing reflects our many discussions, and its chapters contain much that we had conceived together as articles for technical magazines, this book carries both our names as co-authors.

Igor J. Karassik

Contents

I
PUMP TYPES
and
CONSTRUCTION

1 *Classification and Nomenclature*

Pumping may be defined as the addition of energy to a fluid to move it from one point to another. It is not, as frequently thought, the addition of pressure. Because energy is capacity to do work, adding it to a fluid causes the fluid to do work, normally flowing through a pipe or rising to a higher level.

A centrifugal pump is a machine consisting of a set of rotating vanes enclosed within a housing or casing. The vanes impart energy to a fluid through centrifugal force. Thus, stripped of all refinements, a centrifugal pump has two main parts: (1) A rotating element, including an impeller and a shaft, and (2) a stationary element, made up of a casing, stuffing box, and bearings. The structural details of these elements and all refinements applied in modern pump construction are covered by Chap. 2 through 13.

One of the most important factors contributing to the increasing use of centrifugal pumps has been the universal development of electric power. This century has seen electricity replacing small steam plants as the main industrial power source. Although reciprocating pumps were ideal for steam drive, the development of the electric motor permitted use of the much lighter and cheaper direct-connected centrifugal pump. Even though early centrifugal pumps would be considered inefficient by modern performance standards, their lower first cost more than compensated for this shortcoming. The centrifugal pump also immediately demonstrated other important advantages over the reciprocating pump. For example, the centrifugal pump gives steady flow at uniform pressures without pressure surges. It provides the greatest possible flexibility, developing a specific maximum discharge pressure under any operating condition with delivery controlled by either speed variation or throttling.

Naturally, manufacturers working to widen the field of centrifugal pump applications through experience and research have greatly improved the operating range of pressures, the efficiency, and the mechanical and hydraulic design of their product. Concurrently, electric motor builders improved their designs, permitting pump manufacturers to use higher rotative speeds and develop pumps suitable for higher heads. So, over the last half-century, the application of centrifugal pumps has been greatly extended in both pressure and capacity. Centrifugal pumps have been built in sizes ranging from a few gallons per minute to the mammoth 144-in. pumps at Grand Coulee, which handle 1,350 cfs (605,000 gpm) against 310 ft total head, driven by 65,000-hp motors. As far as pressures are concerned, centrifugal pumps may range from the single-stage cellar drainer, which develops 10 to 15 ft head, to the multi-stage boiler feed units for super-pressure power plants that develop over 6,000 psi discharge pressure. And, centrifugal pumps have been built to operate at speeds as high as 10,000 rpm.

Modern trends in small- and medium-capacity range

An interesting trend in small- and medium-size centrifugal pumps is extensive standardization promoted by the fact that about 60 per cent of all pumps in use are of centrifugal design and over 75 per cent of these are in a head and capacity range that can be met by standardized end-suction pumps.

A typical example of this standardization trend is a line of pumps consisting of a number of liquid ends, all suitable for mounting (1) on a motor for close-coupled connection, (2) on a bearing frame for coupled or belt drive, or (3) on a close-coupled turbine. Many standard alternatives are then incorporated into the basic plan, both in materials selected and mechanical construction, thereby eliminating "specials." The use of such an integrated line can result in better delivery service, a wider selection of standardized units, and dollar savings through the maximum use of interchangeable parts. In one typical case, some 100 sizes of pumps, using interchangeable parts, can produce over 60,000 different, standard combinations to suit almost any user's needs (Fig. 1.1).

Centrifugal pump nomenclature

In a centrifugal pump the liquid is forced, by either atmospheric or other pressure, into a set of rotating vanes. These vanes constitute an impeller that discharges the liquid at a higher velocity at its periphery. The velocity is then converted into pressure energy by means of a volute (Fig. 1.2) or by a set of stationary diffusion vanes (Fig. 1.3) surrounding the impeller periphery. Pumps with volute casings are generally called volute pumps, whereas those with diffusion vanes are called diffuser pumps. Diffuser pumps were once quite commonly called turbine pumps, but this term has recently been more selectively applied to vertical deep-well centrifugal diffuser pumps, now called vertical turbine pumps.

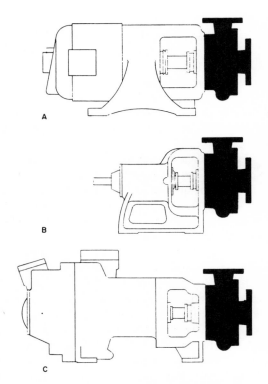

Fig. 1.1 Typical interchangeable standardized line of single-stage end-suction centrifugal pumps

A. Liquid end mounted close-coupled on motor

B. Same liquid end mounted on bearing frame

C. Same liquid end mounted close-coupled on steam turbine.

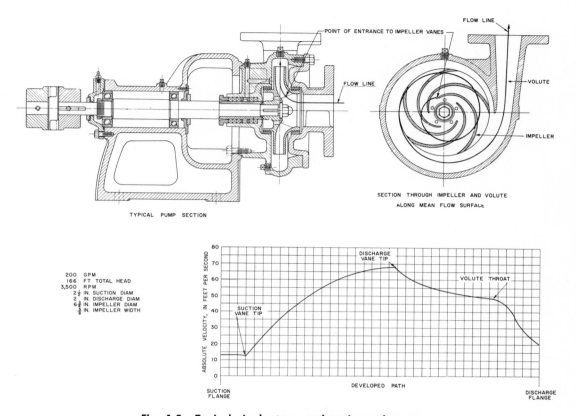

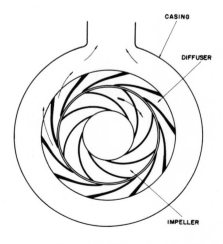

Fig. 1.2 Typical single-stage end-suction volute pump

Most people find it difficult to visualize the path of the liquid passing through a centrifugal pump. This flow path is shown in Fig. 1.2 for a modern end-suction volute pump operating at rated capacity (capacity at which best efficiency is obtained).

Besides being classified in terms of energy conversion, centrifugal pumps are divided into other categories, several of which relate to the impeller. First, impellers are classified according to the major direction of flow in reference to the axis of rotation. Thus, centrifugal pumps may have:

1. Radial-flow impellers (see Chap. 4, Fig. 4.1, 4.10, 4.11)

2. Axial-flow impellers, now usually called propeller type (Fig. 4.6)

3. Mixed-flow impellers, which combine radial- and axial-flow principles (Fig. 4.3 and 4.5).

Fig. 1.3 Typical diffuser pump

Fig. 1.4 Birotor pump

These impellers are further classified according to the flow into the suction edges of the vanes:

1. Single-suction with a single inlet on one side (see Chap. 4, Fig. 4.1, 4.4, 4.5, 4.11)
2. Double-suction, with water flowing to the impeller symmetrically from both sides (Fig. 4.2, 4.3, 4.14).

Impellers are categorized according to their mechanical construction as follows:

1. Enclosed, with shrouds or sidewalls enclosing the waterways (see Chap. 4, Fig. 4.1, 4.2, 4.3, 4.4)
2. Open, with no shrouds (Fig. 4.6, 4.9, 4.10)
3. Semiopen, or semiclosed (Fig. 4.12).

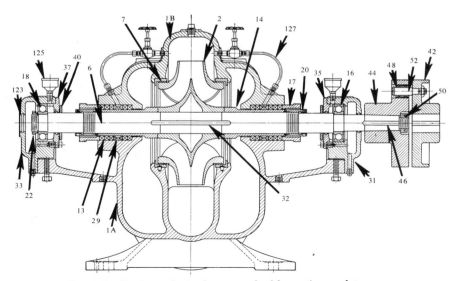

Fig. 1.5 Horizontal single-stage double-suction volute pump

Numbers refer to parts listed in Table 1.1.

If the pump is one in which head is developed by a single impeller, it is called a single-stage pump. Often the total head to be developed requires the use of two or more impellers operating in series, each taking its suction from the discharge of the preceding impeller. For this purpose, two or more single-stage pumps may be connected in series, or all the impellers may be incorporated in a single casing. The unit is then called a multistage pump.

In the early development of centrifugal pumps, birotor pumps (Fig. 1.4) and even trirotor pumps were fairly common. In effect, these were two one-half capacity or three one-third capacity pumps built into one casing and operated in parallel.

The mechanical design of the casing provides the added pump classification of axially split or radially split, and the axis

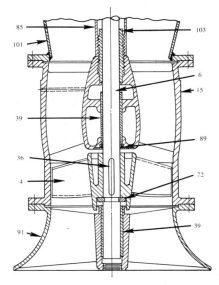

Fig. 1.6 Vertical wet-pit diffuser pump bowl

Numbers refer to parts listed in Table 1.1.

TABLE 1.1   RECOMMENDED NAMES OF CENTRIFUGAL PUMP PARTS

These parts are called out in Fig. 1.5, 1.6, and 2.7.

Item no.	Name of part	Item no.	Name of part
1	Casing	33	Bearing housing (outboard)
1A	Casing (lower half)	35	Bearing cover (inboard)
1B	Casing (upper half)	36	Propeller key
2	Impeller	37	Bearing cover (outboard)
4	Propeller	39	Bearing bushing
6	Pump shaft	40	Deflector
7	Casing ring	42	Coupling (driver half)
8	Impeller ring	44	Coupling (pump half)
9	Suction cover	46	Coupling key
11	Stuffing box cover	48	Coupling bushing
13	Packing	50	Coupling lock nut
14	Shaft sleeve	52	Coupling pin
15	Discharge bowl	59	Handhole cover
16	Bearing (inboard)	68	Shaft collar
17	Gland	72	Thrust collar
18	Bearing (outboard)	78	Bearing spacer
19	Frame	85	Shaft enclosing tube
20	Shaft sleeve nut	89	Seal
22	Bearing lock nut	91	Suction bowl
24	Impeller nut	101	Column pipe
25	Suction head ring	103	Connector bearing
27	Stuffing box cover ring	123	Bearing end cover
29	Seal cage	125	Grease (oil) cup
31	Bearing housing (inboard)	127	Seal piping (tubing)
32	Impeller key		

Fig. 1.7 Old double-suction pump with separately cast suction elbows

Fig. 1.8   Double-suction pump evolved from that in Fig. 1.7

Features integrally cast suction and discharge passages, separate casing heads, and radially split casing.

Fig. 1.9   Double-suction pump with axially split casing for convenience of dismantling

The construction shown is still somewhat crude, having old-style sleeve bearings.

Fig. 1.10 Modern double-suction single-stage pump

Fig. 1.11 Old style multistage pump with radially split casing

Fig. 1.12 Modern high-pressure multistage pumps (for over 1,500 psi) with radially split casings

Fig. 1.13   Early motor-mounted pump (around 1905)

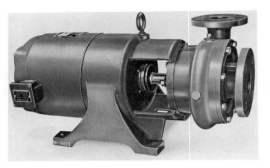

Fig. 1.14   Modern motor-mounted pump

of rotation determines whether the pump is horizontal-shaft, vertical-shaft, or (occasionally) inclined-shaft. Usually these are referred to simply as horizontal or vertical units.

Horizontal centrifugal pumps are classified still further according to suction nozzle location:

1. End-suction (Fig. 1.2, 1.14, 2.9)
2. Side-suction (Fig. 1.10, 2.10, 2.12, 3.5)
3. Bottom-suction (Fig. 2.19)
4. Top-suction (Fig. 22.45).

Some pumps operate with the total liquid flow conducted to and from the unit by piping. Other pumps, most often vertical types, are submerged in their suction supply. Vertical-shaft pumps are therefore called either dry-pit or wet-pit types. If the wet-pit pumps are axial-flow, mixed-flow, or vertical-turbine types, the liquid is discharged up through the supporting drop or column pipe to a discharge point either above or below the supporting floor. These

pumps are consequently designated as above-ground discharge or below-ground discharge units.

The basic elements of a centrifugal pump are its impeller, casing, shaft, and bearings, but there are other necessary parts. Various names have been given these parts by different manufacturers, often leading to confusion. Figures 1.5 and 1.6 show typical constructions of a horizontal double-suction volute pump and the bowl section of a single-stage axial-flow propeller pump. A vertical dry-pit single-suction volute pump is shown in Chap. 2, Fig. 2.7. Names recommended for various parts by the Hydraulic Institute are given in Table 1.1.

The reader may be interested in comparing centrifugal pumps of the 1900's with their modern counterparts. These are illustrated in Fig. 1.7 through 1.14, which clearly show many of the changes in mechanical construction that were necessary for improved service and maintenance. Other changes simply reflect refinements in design, foundry, or machine shop practice.

2 *Casings and Diffusers*

The impeller of a centrifugal pump discharges liquid at a high velocity. One function of the pump casing is to reduce this velocity and convert kinetic energy into pressure energy, either by means of a volute or a set of diffusion vanes.

The volute-casing pump (see Fig. 1.2) derives its name from the spiral-shaped casing surrounding the impeller. This casing section collects the liquid discharged by the impeller and converts velocity energy into potential energy. A centrifugal pump volute increases in area from its initial point until it encompasses the full 360 degrees around the impeller and then flares out to the final discharge opening. The wall dividing the initial section and the discharge nozzle portion of the casing is called the tongue of the volute, or the "cut-water." The diffusion vanes and concentric casing of a diffuser pump fill the same function as the volute casing in energy conversion.

In propeller and other pumps in which axial-flow impellers are used, it is not practical to use a volute casing; rather, the impeller is enclosed in a pipe-like casing. Normally, diffusion vanes are used behind the impeller proper, but in certain extremely low-head units, these vanes may be omitted.

A diffuser-type centrifugal pump was illustrated in Fig. 1.3. The development of the diffuser appreciably improved the efficiency of the rather crude volute forms characteristic of the early days of centrifugal pump construction. Later improvements in the hydraulic design of impellers and volute casings made the diffuser of little, if any, value in increasing pump efficiency. It is therefore seldom applied to a single-stage volute pump, although it possesses structural as well as hydraulic advantages that may sometimes be useful. For example, one advantage is that such a construction balances radial reactions on the rotor.

At present, except for certain high-pressure multistage pump designs, the major application of diffusion vane pumps is in vertical turbine pumps and in single-stage, low-head propeller pumps (Fig. 2.1).

Unfortunately, the use of diffusers may impair the hydraulic characteristics of the pump. The fast-moving liquid from the impeller can meet the fixed vanes of the diffuser without shock only when the pump is operating at rated capacity, for only then

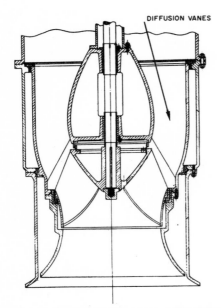

DIFFUSION VANES

Fig. 2.1 Vertical propeller pump with diffusion vanes

does the angle of the vanes correspond to the angle at which the liquid leaves the impeller. At all other rates of flow, the multiple vanes cause shock and turbulence, so that the pump may operate in an unstable condition. As a matter of fact, when flow is restricted to as low as 5–10 per cent of normal capacity, the shock and turbulence may become sufficiently severe to reduce the total head generated. As a result, the head-capacity curve of diffuser-type pumps could easily acquire a "droop" in the shut-off capacity area, making the pump unsatisfactory for parallel operation. Do not interpret this to mean that a diffuser pump *always* produces a drooping characteristic. However, such a curve can result from this design unless extreme care is taken in layout of the impeller and diffuser combination. Pump manufacturers have long tried to stabilize diffuser pump head-capacity curves, and various solutions are available, based on proper selection of impeller vane angles, curvature of the impeller blades, and careful design of diffuser passageways.

Another problem arising from use of diffusers is the potential pump flexibility. Obviously, pump manufacturers try to obtain as much coverage from a single pump pattern as possible, to keep the number of patterns comprising a complete line of pumps at a minimum and to reduce the necessary number of parts in stock. With a volute pump, the impeller diameter may be decreased as much as 20 per cent from its maximum value without appreciably reducing the pump efficiency caused by increased hydraulic losses. On the other hand, similar reduction in diameter of a diffuser-type pump impeller would produce unacceptable performance. The increased gap between the impeller periphery and the diffuser inlet vanes would result in excessive hydraulic losses. For this reason, a maximum-diameter impeller can be cut only from 5 to 10 per cent. Further reduction requires a different diffuser pattern with a smaller inlet vane diameter.

Impeller cutdown restrictions necessitate an increased parts inventory. Also, the flexibility of constructed units is limited because a change in conditions of service, otherwise taken care of by an impeller cutdown, may also require a new diffuser for satisfactory performance.

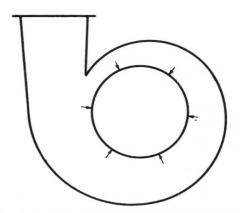

Fig. 2.2 Zero radial reaction in single-volute casing

Uniform pressures exist at design capacity.

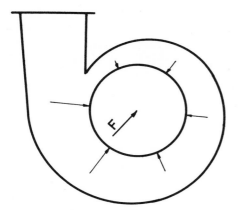

Fig. 2.3 Radial reaction in a single-volute casing

Uniform pressures do not exist at reduced capacities.

Radial thrust

In a *single-volute pump casing design* (Fig. 2.2), uniform or near-uniform pressures act on the impeller when the pump is operated at design capacity (which coincides with the best efficiency). At other capacities (Fig. 2.3), the pressures around the impeller are not uniform, and there is a resultant radial reaction (F). A graphical representation of the typical change in this force with pump capacity is shown in Fig. 2.4; note that the force is greatest at shut-off.[1]

For any percentage of capacity, radial reaction is a function of total head, and of the width and diameter of the impeller. Thus, a high-head pump with a large-diameter impeller will have a much greater radial reaction force at partial capacities than a low-head pump with a small-diameter impeller. A zero radial reaction is not often realized; the minimum reaction occurs at design capacity. In a diffuser-type pump, which has the same tendency for

[1] In Fig. 2.2, 2.3, 2.4, and 2.5, no attempt has been made to show correct quantitative force values for a specific example nor to locate the exact resultant force. The magnitude and direction of forces vary with the type of pump, casing design, and many other factors.

overcapacity unbalance as a single-volute pump, the reaction is limited to a small arc repeated all around the impeller, with the individual forces cancelling each other.

In a centrifugal pump design, shaft diameter and bearing size can be affected by allowable deflection as determined by shaft span, impeller weight, radial reaction forces, and the torque to be transmitted. Formerly, standard designs compensated for reaction forces if maximum-diameter pump impellers were used only for operations exceeding 50 per cent of design capacity. For sustained operations at lower capacities, the pump manufacturer, if properly advised, would supply a heavier shaft, usually at a much higher cost. Sustained operation at extremely low flows without informing the manufacturer at the time of purchase is a much more common practice today. The result—broken shafts, especially on high-head units.

Because of the increasing operation of pumps at reduced capacities, it has become desirable to design standard units to accommodate such conditions. One solution is to use heavier shafts and bearings. Except for low-head pumps in which only a small additional load is involved this solution is not economical. The only practical answer

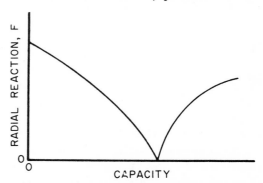

Fig. 2.4 Magnitude of radial reaction in single-volute casing

F decreases from shut-off to design capacity and then increases with over-capacity. With over-capacity, the reaction is roughly in the opposite direction from that with partial capacity.

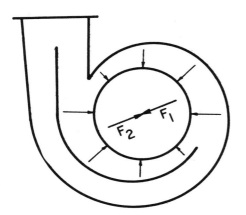

Fig. 2.5   Radial reactions in double-volute pump

is a casing design that develops a much smaller radial reaction force at partial capacities. One of these is the double-volute casing design, also called twin-volute or dual-volute.

The application of the double-volute design principle to neutralize reaction forces at reduced capacity is illustrated in Fig. 2.5. Basically, this design consists of two 180-deg volutes; a passage external to the second joins the two into a common discharge. Although a pressure unbalance exists at partial capacity through each 180-deg arc, forces F_1 and F_2 are approximately equal and opposite, thereby producing little, if any, radial force on the shaft and bearings. Although the double-volute casing design principle has been known for some time, it was once a serious manufacturing problem, especially with horizontal-discharge axially split double-suction pumps. This was because the division wall spanned the split (Fig. 2.6). In the past few years, improved foundry techniques have permitted more accurate castings and use of the double-volute casing design has become feasible in a commercial line of intermediate and large-size double-suction pumps.

The double-volute design has many "hidden" advantages. For example, in large-capacity medium- and high-head single-

stage vertical pump applications, the rib forming the second volute and separating it from the discharge waterway of the first volute strengthens the casing (Fig. 2.7).

When the principle of the double volute is applied to individual stages of a multistage pump, it becomes a twin-volute. The question has been broached whether this design should be called twin-volute or a two-vane diffuser, but the first has become the accepted form. A typical twin-volute is illustrated in Fig. 2.8. The kinetic energy of the water discharged from the impeller must be transformed into pressure energy, then turned back 180 deg to enter the impeller of the next stage. The twin-volute, therefore, also acts as a return channel. The back view in Fig. 2.8 shows this, as well as the guide vanes used to straighten the flow into the next stage.

Solid and split casings

Solid casing is a design in which the discharge waterways that lead to the discharge nozzle are all in one casting or fabricated piece. It must have one side open so that the impeller may be introduced into the casing; however, it cannot be completely solid, and designs normally called solid casing are really radially split (Fig. 2.9, 2.12, 2.13, and 2.14).

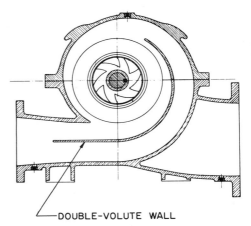

—DOUBLE-VOLUTE WALL

Fig. 2.6   Transverse view of double-volute casing pump

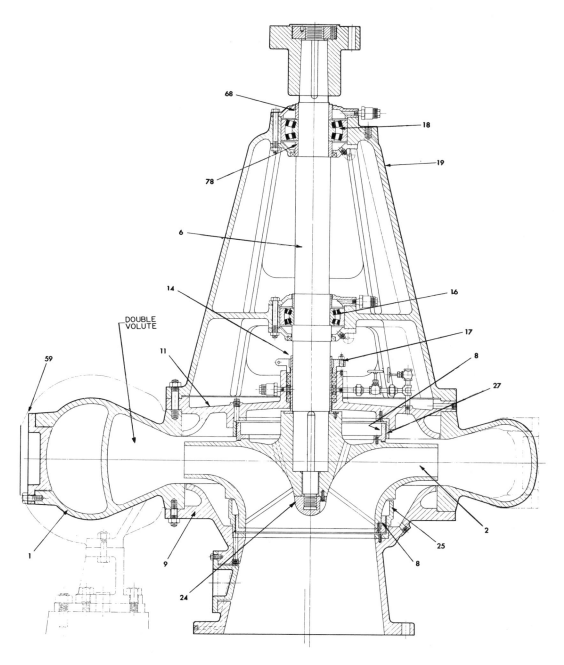

DOUBLE
VOLUTE

Fig. 2.7 Sectional view of vertical-shaft end-suction pump with a double-volute casing

Numbers refer to parts listed in Table 1.1.

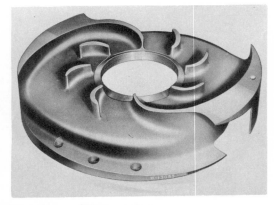

Fig. 2.8   Twin-volute of a multistage pump

Front view (left) and back view (right).

Split casing is a casing made of two or more parts. The term "horizontally split" had regularly been used to describe horizontal double-suction pumps, indicating that the casing was divided by a horizontal plane through the shaft centerline or axis (Fig. 2.10). That designation was an unfortunate choice because applications of the same pump design for vertical use or with the nozzle position rotated caused confusion. The term "axially split" is now preferred. Since both the suction and discharge nozzles of axially split pumps are usually in the same half of the casing, the other half may be removed (upper half in the case of horizontal pumps) for inspection without disturbing the bearings or piping.

Like its counterpart, "horizontally split," the term "vertically split" is unfortunate. It refers to a casing split in a plane perpendicular to the axis of rotation. The term "radially split" is now preferred.

Fig. 2.9   Radially split casing, end-suction single-stage pump with overhung impeller

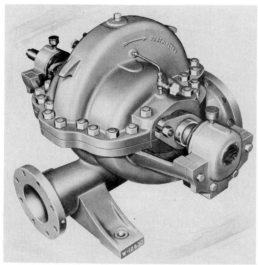

Fig. 2.10 Axially split casing, horizontal double-suction volute pump

In some special horizontal pump applications, it is desirable to use a casing split along a plane passing through the pump axis but inclined to the horizontal.

This construction is primarily used when the need for a vertical discharge is combined with the desire for the convenience of an axially split casing (Fig. 2.11).

End-suction pumps

Most end-suction single-stage pumps are made of one-piece solid casings. At least one side of the casing must have an opening with cover, so the impeller can be assembled in the pump. If the cover is on the suction side, it becomes the casing sidewall and contains the suction opening (Fig. 2.12). This is called the suction cover or casing suction head. Other designs are made with stuffing box covers (Fig. 2.13), whereas still others have both casing suction covers and stuffing box covers (Fig. 2.7 and 2.14).

Fig. 2.11 Horizontal end-suction pump with casing split along plane through its axis but inclined from horizontal

(Courtesy Allis-Chalmers.)

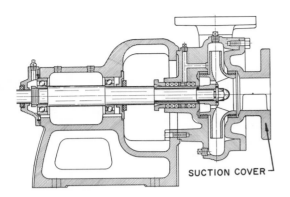

Fig. 2.12   Sectional view of end-suction pump with radially split casing

Note the suction cover.

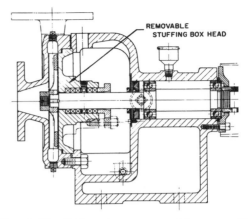

Fig. 2.13   End-suction pump with removable stuffing box head

For general service, the end-suction single-stage pump design is extensively used for both motor-mounted and coupled-type small pumps up to 4- and 5-in. discharge size. In all of these the small size makes it feasible to cast the volute and one side integrally. Whether the stuffing box or suction side is made integral with the casing is usually determined by the most eco-

nomical pump design. For larger pumps, especially those for special service, such as sewage handling, there is a demand for pumps of both rotations. A design with separate suction and stuffing box heads permits use of the same casing for either rotation if the flanges on the two sides are made identical. There is also a demand for vertical pumps that can be disassembled by

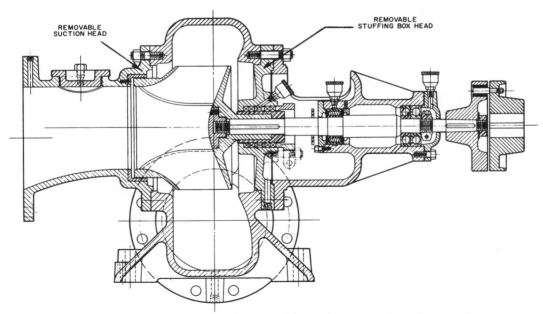

Fig. 2.14   End-suction pump with removable suction and stuffing box heads

removing the rotor and bearing assembly from the top of the casing. Many horizontal applications of the same pumps, however, require partial dismantling from the suction side. Such lines are therefore most adaptable when they have separate suction and stuffing box covers.

Casing construction for open-impeller pumps

In the inexpensive open-impeller pump, the impeller rotates within close clearance of the pump casing (Fig. 2.13). If the intended service is more severe, a sideplate is mounted within the casing to provide a renewable close clearance guide to the water flowing through the open impeller. One of the advantages of using sideplates is that abrasion-resistant material, like stainless steel, may be used for the impeller and sideplate whereas the casing itself may be made of a less costly material (Fig. 2.15). Another special example, illustrated in Fig. 2.16, shows a pump used to handle paper pulp stock containing abrasive material.

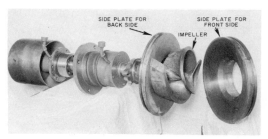

Fig. 2.16 Rotor and casing sideplates of paper-pulp pump handling abrasive stock

Although double-suction, open-impeller pumps are seldom used today, they were common in the past and were generally made with sideplates (Fig. 2.17).

Prerotation and stop-pieces

Improper entrance conditions and incorrect suction approach shapes may cause the liquid column in the suction pipe to spiral for some distance ahead of the actual impeller entrance. This phenomenon is called "prerotation." Many operational and design factors may cause it in either vertical or horizontal pumps.

Prerotation is usually harmful to pump operation because the liquid enters between the impeller vanes at an angle not predicted by the designer. This entrance frequently lowers the net effective suction head and pump efficiency. Various means

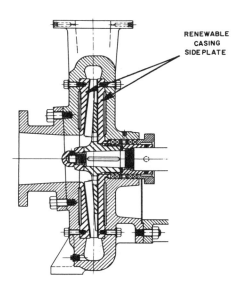

Fig. 2.15 Renewable casing sideplates in end-suction pump to provide close clearance for open-type impeller

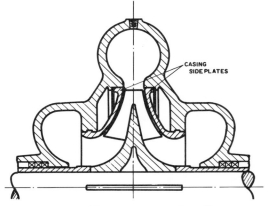

Fig. 2.17 Double-suction open-impeller pump with casing sideplates

are used to avoid it both in the construction of the pump and design of the suction approaches.

Practically all horizontal single-stage double-suction pumps and most multistage pumps have a suction volute that guides the liquid in streamline flow to the impeller eye. The flow comes to the eye at right angles to the shaft and separates unequally on both sides of the shaft. Moving from the suction nozzle to the impeller eye, the suction waterways reduce in area, meeting in a projecting section of the sidewall dividing the two sections. This dividing projection is called a stop-piece.

Practically all double-suction axially split casing pumps have a side-discharge nozzle, and either a side or a bottom-suction nozzle. If the suction nozzle is placed on the side of the pump casing with its axial centerline at right angles to the vertical centerline (see Fig. 2.10), the pump is classified as a side-suction pump. If its suction nozzle points vertically downwards (Fig. 2.19), the pump is called a bottom-suction pump. Single-stage bottom-suction pumps are rarely made in sizes below 10 in. discharge nozzle diameter.

Special nozzle positions can sometimes be provided for double-suction axially split

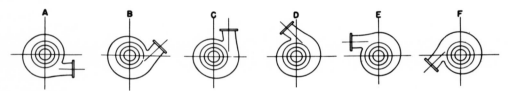

Fig. 2.18 Possible positions of discharge nozzles for horizontal end-suction solid-casing pump
Rotation illustrated is counter clockwise from suction end.

To prevent prerotation in end-suction pumps, a radial-fin stop-piece projecting toward the center may be cast into the suction-nozzle wall.

Nozzle locations

The discharge nozzle of end-suction single-stage horizontal pumps is usually in a top-vertical position (see Fig. 2.9, 2.12, and 2.13). However, other nozzle positions may be obtained, such as top-horizontal, bottom-horizontal or bottom-vertical discharge. Figure 2.18 illustrates the flexibility available in discharge nozzle location. However, sometimes the horizontal pump frame, bearing bracket, or bedplate interfere with the discharge flange, prohibiting a bottom-horizontal or bottom-vertical discharge nozzle position. In other instances, solid casings cannot be rotated for various nozzle positions because the stuffing box seal connection would become inaccessible.

casing pumps to meet special piping arrangements, for example, a vertically split casing with bottom suction and top discharge in one half of the casing. As these special designs are usually costly, they should be avoided.

Centrifugal pump rotation

Because suction and discharge nozzle location is affected by pump rotation, it is very important to understand the means used to define the direction of rotation.

According to Hydraulic Institute Standards, rotation is defined as clockwise or counterclockwise by looking at the driven end of a horizontal pump or looking down on a vertical unit. Some manufacturers still designate rotation of a horizontal pump from its outboard end. Therefore, to avoid misunderstanding, clockwise or counterclockwise rotation should always be clarified by including the direction from which one looks at the pump.

The terms "inboard end" and "outboard end" are used only with horizontal pumps. Inboard end is the one closest to the driver, whereas the outboard end is the one farthest away. The terms lose their significance with dual-driven pumps and are not then used. Any centrifugal pump casing pattern may be arranged for either clockwise or counterclockwise rotation, except for end-suction pumps that have integral heads on one side; these require separate directional casing patterns.

Casing hand holes

Casing hand holes are furnished primarily on pumps handling sewage or stringy materials that may become lodged on the impeller suction vane edges or on the tongue of the volute. They permit removal of this material without dismantling the complete pump. End-suction pumps used for handling such liquids are provided with hand holes or access to the suction side of the impellers. These are located on the suction head or in the suction elbow. Hand holes are also included in drainage, irrigation, circulating, and supply pumps if foreign matter may become lodged in the waterways. On the very large pumps, manholes provide access to the interior for both cleaning and inspection.

Mechanical features of casings

Most single-stage centrifugal pumps are intended for service with moderate pressures and temperatures. As a result, pump manufacturers usually design a special line or lines of pumps for high operating pressures and temperatures rather than make their standard line unduly expensive by having it cover too wide a range of operating conditions.

If axially split casings are subject to high pressure, they tend to "breathe" at the split joint.[1] This leads to misalignment of the

[1] "Breathe" is a word commonly used to describe the separation of two parts of an axially split pump under pressure.

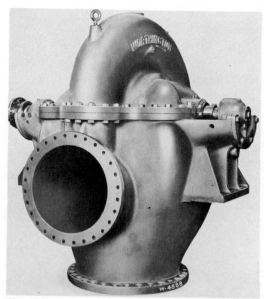

Fig. 2.19   Bottom-suction axially split casing, single-stage pump

rotor and, even worse, to leakage. For such conditions, internal and external ribbing is applied to casings at the points subject to greatest stress. In addition, whereas most pumps are supported by feet at the bottom of the casing, high-temperature conditions require centerline support so that expansion as the pump becomes heated will not cause misalignment.

Series units

For large-capacity, medium–high-head service conditions that require such an arrangement, two single-stage double-suction pumps may be connected in series on one baseplate with a single driver. Such an arrangement is very common in waterworks applications for heads of 250 to 400 ft. One series arrangement uses a double-extended shaft motor in the middle, driving two pumps connected in series by piping (Fig. 2.20). In a second type, a standard motor is used with one pump having a double-extended shaft (Fig. 2.21). This latter arrangement may be limited, because the shaft of the pump next to the motor must be strong

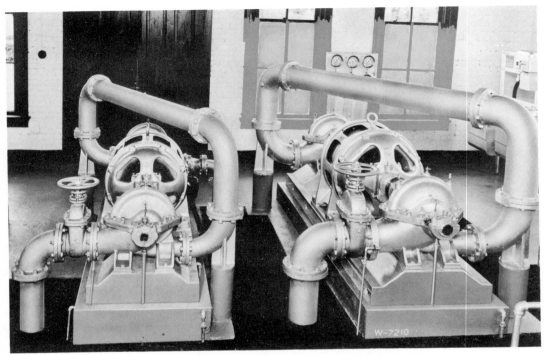

Fig. 2.20   Series unit (motor in middle)

enough to transmit the total pumping horsepower.

If the total pressure generated by such a series unit is relatively high, the casing of the second pump may require ribbing.

Casing maintenance

Pumps that handle noncorrosive water or liquids are not usually subject to extensive casing wear. However, the casing waterways should always be thoroughly cleaned and repainted during a complete overhaul. A suitable paint should be used that firmly adheres to the metal so that the water velocity will not wash or jet it off.

An enamel-like finish is the most efficient. A program of casing cleaning and repainting should be established on the basis of local conditions. This will prevent the protective coat from ever fully eroding before replacement, thereby preventing corrosion.

Pumps handling gritty or sandy water naturally are more subject to casing

troubles. Erosion or wear can be reduced by selecting pumps for low water velocities, and with fine grain metal casings. The use of a 1 to 2 per cent nickel cast iron usually gives sufficient added resistance, but if a high concentration of sand is evident, a more resistant and expensive material might be warranted.

Progress has recently been made in processes of rubber coating pump waterways, and this technique may be desirable in some applications. If the casings of pumps handling sandy or gritty water are to be protected primarily by periodic painting, a suitable type pump should be carefully selected for local water conditions and a special maintenance schedule established.

In these difficult pumping applications, the casing should be regularly examined for corrosion, which will be indicated by cast-iron graphitization. This occurs when the ferrous particles are washed out by electrolytic action and deposited on bronze pump parts. If severe graphitization takes

Fig. 2.21   Series unit (motor at end)

place, the manufacturer should be consulted on the possibility of substituting materials more impervious to the pumped liquid.

If the casing is pitted or eroded in places, it can be restored by welding, brazing, silver soldering, or metal spraying, depending upon the construction material and the facilities available. The authors know of several large centrifugal pumps in waterworks service in which corroded areas, located where water velocities are low, are actually filled with properly anchored concrete.

Special care must be taken to examine and recondition metal-to-metal fits where stationary parts such as casing rings, diffusers, or stage-pieces seat in the casing. If the casing is steel, and these fits show signs of erosion, it might be advantageous to face them with 18-8 stainless steel and refinish.

Frequently, the cut-water, or volute tongue as it is also called, becomes eroded, for example when a pump handles water with some sand in suspension or when the periphery of the impeller is located too close

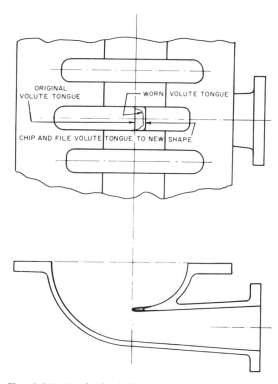

Fig. 2.22   Method of filing worn volute tongue

to the tongue. Another fairly common cause of erosion in this area is galvanic action between a cast-iron casing and bronze fittings. The cast iron graphitizes and wears away most perceptibly in areas of high velocity, like those near the volute tongue.

The best way to correct this condition is to cut back the tongue so that it is straight across and then file it to a smooth rounded edge (Fig. 2.22). This cut back does not affect pump capacity unfavorably; on the contrary, it is often used to squeeze out an extra small percentage of capacity without putting in an impeller of larger diameter. The added capacity comes from the increase in the casing throat area, which causes an increase for a given casing velocity.

Care should be taken not to distort or warp the casing during overhaul. After repairs are completed, the horizontal flanges of an axially split casing should be finished to a flat surface with hand tools. Of course, if the repairs are very serious, the pump is serviced at the manufacturer's shop. The casing flanges may have to be refinished at that time and the casing rebored.

Except for some special designs, every pump has gaskets that are subject to damage when the pump is opened. If the old gasket adheres to the lower half of the casing and is in good condition, it is not necessary to replace it. However, it should be replaced if it is damaged in any way, and

for this reason a new gasket should always be available.

The new gasket should be of the same thickness as the original and, if possible, of the same type of material so that it will have the same compression characteristics. Too thick a gasket usually leads to leakage. If the gasket is thinner than the original, tightening of the two casing halves may exert undue force on casing wearing rings and distort them.

In installing a new gasket, the inner edge must be accurately trimmed along the edge of the stuffing box bore. At all points where the gasket abuts on the outer diameter and the sides of stationary parts, the edges must be trimmed squarely and neatly, allowing sufficient gasket overlap. Tightening the upper half of the casing will effectively press the gasket edges against the stator parts, insuring proper sealing. This trimming operation is best accomplished by first cementing the gasket to the lower casing half with shellac and then cutting all edges square with a razor blade. Of course, all foreign matter must be removed from the casing flanges before the gasket is applied to the lower casing half.

In reassembling the pump, it is recommended that powdered graphite be rubbed into the gasket before the top casing half is replaced. This action will prevent the gasket from sticking to the upper half when it is again removed.

3. The impeller hubs must not extend through the interstage portion of the casing separating adjacent stages.

Except for some special pumps that have an internal and enclosed bearing at one end, and therefore only one stuffing box, most multistage pumps fulfill the first condition. But because of structural requirements, the last two conditions are not practical. A slight residual thrust is usually present in multistage opposed-impeller pumps, unless impeller hubs or wearing rings are located on different diameters for various stages. Because such a construction would eliminate axial thrust only at the expense of reduced interchangeability and increased manufacturing cost, this residual thrust, being relatively small, is usually carried on the thrust bearing.

Arrangement of stages in opposed-impeller multistage pumps

Once a multistage pump is balanced by an opposed-impeller design, the best sequence in which the individual stages are to be arranged within the pump casing must be determined. This problem is not simple, as illustrated by analyzing the best arrangement for a six-stage pump. The total number of possible arrangements is the permutation of 6, that is, $6 \times 5 \times 4 \times 3 \times 2 \times 1$, or 720. Of course, half this number are duplicates, because they describe the same arrangement as viewed in a mirror. The problem is actually simplified by the fact that most potential sequences are rather complicated and can be eliminated entirely. Figure 5.10 illustrates four of the most logical possibilities for a six-stage pump.

Three factors enter into the analysis of a satisfactory stage arrangement, and the final solution must be a compromise between the "best" individual solutions, each satisfying the following requirements:

1. The arrangement of stages provides the minimum possible leakage at the run-

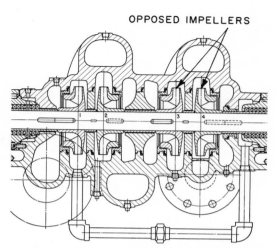

OPPOSED IMPELLERS

Fig. 5.9 Four-stage pump with opposed impellers

ning joints and maintains this minimum over a long period of time.

2. The various stage impellers are arranged so that the stuffing boxes are subject to the lowest pressure in the pump.

3. The sequence of stages precludes excessive complications in the forming of interstage passages.

Running joint leakage

To combat internal leakage, the pump designer must determine whether it is preferable to keep a single running joint under relatively high pressure (as in Fig. 5.10, in which joint A is subject to a three-stage pressure differential) or to have many joints under a moderate pressure differential (such as the five interstage joints in Fig. 5.11). Experimental data indicate that the latter alternative is best. If the running clearances remain undisturbed throughout the life of the pump, the single running joint under a high pressure differential would no doubt prove satisfactory. However, the original clearances will not be maintained throughout the useful life of the pump, and the increase in the clearances will be considerably greater with the higher pressure differential *unless a relatively long running joint is provided.* As a

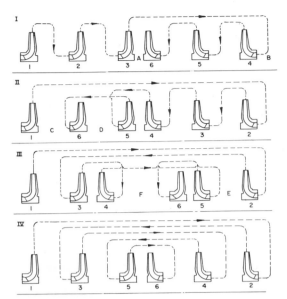

Fig. 5.10 Stage arrangements for six-stage axially balanced pump

I. Joint A subject to three-stage pressure differential; one stuffing box under high pressure at B. II. Arrangement with two high-pressure joints, including four-stage pressure differential at C and two-stage differential at D. III. Joints E and F under two-stage pressure differential. IV. All running joints subject to only one-stage pressure differential.

result, the total internal leakage of a pump with a single joint under high pressure differential will rapidly increase beyond that in a pump with several low-pressure-drop running joints. It follows that the renewal of parts subject to wear will necessarily be more frequent with the single joint.

High pressure differences not only lead to increased leakage and wear, but they can also seriously affect axial balance. It must be remembered that theoretical axial thrust balance depends upon the equality of forces acting on the two sets of opposed impellers. If this equality is destroyed by unequal leakage from excessive and uneven wear on the several running joints, there would be unbalance and the pump itself would ultimately fail. Therefore, reducing pressure differences between adjoining stage chambers is a key factor in maintaining axial balance in an opposed-impeller multistage pump.

These considerations can now be used to restate the first stage-arrangement factor, as follows:

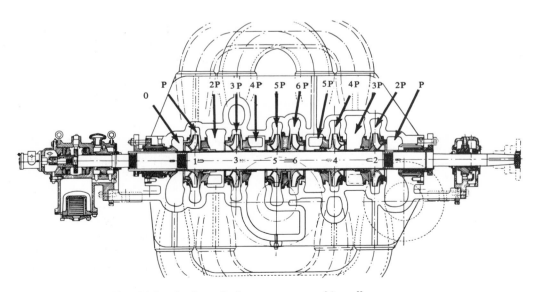

Fig. 5.11 Section of six-stage opposed-impeller pump

Suction pressure equals zero; pressure generated by each impeller is indicated by P.

1. The pressure difference between any two adjoining stage chambers is kept to a minimum, and several low-pressure running joints are to be preferred to a fewer number of joints under relatively high pressure differentials.

By referring to Fig. 5.10, it can be seen that arrangements I, II, and III have greater pressure differences at the running joints than arrangement IV. Because of the sequence of stages used in I, there is a three-stage pressure differential across running joint A, separating the third and sixth stages. Arrangement II has two high-pressure joints; joint C is under a four-stage and joint D is under a two-stage pressure differential. Likewise, arrangement III has two leakage joints, E and F, that are subject to a two-stage differential.

Arrangement IV, on the other hand, is such that the pressure difference between any two adjoining stage chambers does not exceed the pressure generated by a single stage. This arrangement is illustrated in the six-stage pump in Fig. 5.11. For purposes of simplicity, it has been assumed that the suction pressure is zero, the pressure generated by each impeller being represented by P. The sequence of the stages and the connections between successive stages are clearly illustrated. Reading from left to right, the pressures in the individual chambers are, 0, P, $2P$, $3P$, $4P$, $5P$, $6P$, $5P$, $4P$, $3P$, $2P$, and finally P.

Stuffing box pressures

In all the sequences illustrated in Fig. 5.10 except arrangement I, the effect of the sequence on the stuffing box pressures is satisfactory. By placing the two lowest pressure stage impellers at the two ends of the casing, the stuffing box packings are subjected to only the lowest pressures in the pump, namely, the suction and first-stage discharge pressures.

Simplicity of interstage passage structure

The third factor affecting staging sequence is not easily reconciled with the first two. Its importance, however, is secondary, and it should not take precedence over the reduction of interstage pressure differences. If the individual stage pressures are relatively low, of course, a compromise may be most satisfactory. For example, Fig. 5.9 shows a medium-pressure four-stage pump in which the advantage of locating stages in ascending order (from left to right) far outweighs the disadvantage of having to provide a pressure breakdown ahead of the stuffing box at the right side of the suction.

6 *Hydraulic Balancing Devices*

A single-suction impeller is subject to axial hydraulic thrust caused by the pressure differential between its two faces. If all the single-suction impellers of a multistage pump face in the same direction, the total theoretical hydraulic axial thrust acting towards the suction end of the pump will be the sum of the individual impeller thrusts. The thrust magnitude (in pounds) will be approximately equal to the product of the net pump pressure (in pounds per square inch) and the annular unbalanced area (in square inches). Actually the axial thrust turns out to be about 70 to 80 per cent of this theoretical value.

Some form of hydraulic balancing device must be used to balance the axial thrust and to reduce the pressure on the stuffing box adjacent to the last-stage impeller. This hydraulic balancing device may be a balancing drum, a balancing disk, or a combination of the two.

Balancing drums

The balancing drum is illustrated in Fig. 6.1. The balancing chamber at the back of the last-stage impeller is separated from the pump interior by a drum that is either keyed or screwed to the shaft and therefore rotates with the shaft. The drum is separated by a small radial clearance from the stationary portion of the balancing device, called the "balancing drum head," that is fixed to the pump casing.

The balancing chamber is connected either to the pump suction or to the vessel from which the pump takes its suction. Thus, the back pressure in the balancing chamber is only slightly higher than the suction pressure, the difference between the two being equal to the friction losses between this chamber and the point of return. The leakage between the drum head is, of course, a function of the differential pressure across the drum and of the clearance area.

The forces acting on the balancing drum in Fig. 6.1 are the following:

1. Towards the discharge end—the discharge pressure multiplied by the front balancing area (area "B") of the drum

2. Towards the suction end—the back pressure in the balancing chamber multiplied by the back balancing area (area "C") of the drum.

The first force is greater than the second, thereby counterbalancing the axial thrust exerted upon the single-suction impellers.

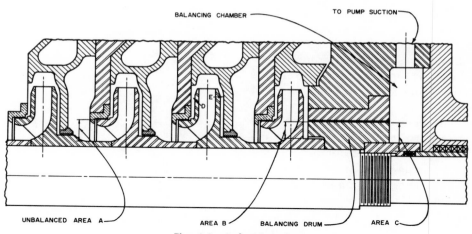

Fig. 6.1 Balancing drum

Actually, the drum diameter can be selected to balance axial thrust completely or within 90 to 95 per cent, depending on the desirability of carrying any thrust bearing loads.

It has been assumed in the preceding simplified description that the pressure acting on the impeller walls is constant over their entire surface and that the axial thrust is equal to the product of the total net pressure generated and the unbalanced area. Actually, this pressure varies somewhat in the radial direction because of the centrifugal force exerted upon the water by the outer impeller shroud (see Fig. 5.2). Furthermore, the pressures at two corresponding points on the opposite impeller faces (D and E, Fig. 6.1) may not be equal because of variation in clearance between the impeller wall and the casing section separating successive stages. Finally, pressure distribution over the impeller wall surface may vary with head and capacity operating conditions.

This pressure distribution and design data can be determined by test quite accurately for any one fixed operating condition, and an effective balancing drum could be designed on the basis of the forces resulting from this pressure distribution. Unfortunately, varying head and capacity conditions change the pressure distribution, and as the area of the balancing drum is necessarily fixed, can destroy the equilibrium of the axial forces. The objection to this is not primarily the amount of the thrust, but rather that the direction of the thrust cannot be predetermined because of the uncertainty about internal pressures. Still, it is advisable to predetermine normal thrust direction, as this can influence external mechanical thrust bearing design. Because 100 per cent balance is unattainable in practice and the slight but predictable unbalance can be carried on a thrust bearing, the balancing drum is often designed to balance only 90 to 95 per cent of total impeller thrust.

Balancing drum modifications

To reduce the balancing drum leakage, a series of steps along the drum with small relief chambers at each step is sometimes provided. The drum surface is also frequently serrated.

Experience indicates that the most successful simple balancing drum designs are of relatively long length. The length reduces the pressure drop per linear unit and thus decreases the rate of wear. This design, however, suffers from the fact that it substantially increases the pump shaft span.

Figure 6.2 illustrates a modification of the balancing drum that incorporates a "labyrinth" construction with concentric pressure-reducing passages. The example illustrated provides approximately 18 inches of effective drum length in only 7½ in. of axial length.

The balancing drum satisfactorily balances the axial thrust of single-suction impellers and reduces pressure on the discharge side stuffing box. It lacks, however, the virtue of automatic compensation for any changes in axial thrust caused by varying impeller reaction characteristics. In effect, if the axial thrust and balancing drum forces become unequal, the rotating element will tend to move in the direction of the greater force. The thrust bearing must then prevent excessive movement of the rotating element. The balancing drum performs no restoring function until such time as the drum force again equals the axial thrust. This automatic compensation is the major feature that differentiates the balancing disk from the balancing drum.

Balancing disks

The operation of the simple balancing disk is illustrated in Fig. 6.3. The disk is fixed to and rotates with the shaft. It is separated from the balancing disk head installed as a casing part, by a small axial clearance. The leakage through this clearance flows into the balancing chamber and from there either to the pump suction or

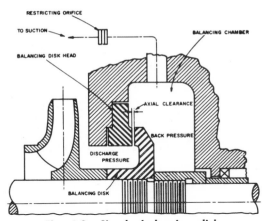

Fig. 6.3 Simple balancing disk

to the vessel from which the pump takes its suction. The back of the balancing disk is subject to the balancing chamber back pressure whereas the disk face experiences a range of pressures. These vary from discharge pressure at its smallest diameter to back pressure at its periphery. The inner and outer disk diameters are chosen so that the difference between the total force acting on the disk face and that acting on its back will balance the impeller axial thrust.

If the axial thrust of the impellers should exceed the thrust acting on the disk during operation, the latter is moved towards the disk head, reducing the axial clearance between the disk and the disk head. The amount of leakage through the clearance is reduced so that the friction losses in the leakage return line are also reduced, lowering the back pressure in the balancing chamber. This automatically increases the pressure difference acting on the disk and moves it away from the disk head, increasing the clearance. Now, the pressure builds up in the balancing chamber, and the disk is again moved towards the disk head until an equilibrium is reached.

To assure proper balancing disk operation, the change in back pressure in the balancing chamber must be of an appreciable magnitude. Thus, with the balancing disk wide open with respect to the disk head, the back pressure must be substan-

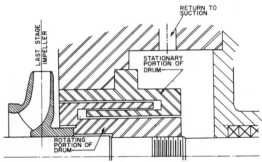

Fig. 6.2 Labyrinth balancing drum

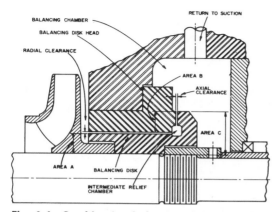

Fig. 6.4 Combination balancing disk and drum

1. Towards the discharge end—the sum of the discharge pressure multiplied by area *A*, plus the average intermediate pressure multiplied by area *B*

2. Towards the suction end—the back pressure multiplied by area *C*.

Whereas the "position-restoring" feature of the simple balancing disk required an undesirably wide variation of the back pressure, it is now possible to depend upon a variation of the intermediate pressure to achieve the same effect. Here is how it works. When the pump rotor moves towards the suction end (to the left, in Fig. 6.4) because of increased axial thrust, the axial clearance is reduced, and pressure builds up in the intermediate relief chamber, increasing the average value of the intermediate pressure acting on area *B*. In other words, with reduced leakage, the pressure drop across the radial clearance decreases, increasing the pressure drop across the axial clearance. The increase in intermediate pressure forces the balancing disk towards the discharge end until equilibrium is reached. Movement of the pump rotor towards the discharge end would have the opposite effect of increasing the axial clearance and the leakage and decreasing the intermediate pressure acting on area *B*.

Figure 6.5 illustrates the pressure distribution in a combination balancing disk and drum. No attempt is made to describe the exact manner in which the pressure decreases between any two points, although this curve is not necessarily a straight line. Also, this illustration is not quantitatively correct. It only serves to show that changes in the balancing device position vary the internal pressure distribution without altering the back pressure. The only possible variation may be caused by pressure changes at the point where the balancing device leakage is returned to the system. An orifice may still be located in the return line. Its function now, however, is not that of changing back pressure but rather of gaging the volume of leakage flow. This flow

tially higher than the suction pressure to give a resultant force that restores the normal disk position. This can be accomplished by introducing a restricting orifice in the leakage return line that increases back pressure when leakage past the disk increases beyond normal. The disadvantage of this arrangement is that the pressure on the stuffing box packing is variable—a condition that is injurious to the life of the packing and therefore to be avoided. The higher pressure that can occur at the packing is also undesirable.

Combination disk and drum

For the reasons just described, the simple balancing disk is seldom used. The combination balancing disk and balancing drum (Fig. 6.4) was developed to obviate the shortcomings of the disk while retaining the advantage of automatic compensation for axial thrust changes.

The rotating portion of this balancing device consists of a long cylindrical body that turns within a drum portion of the disk head. This rotating part incorporates a disk similar to the one previously described. In this design, radial clearance remains constant regardless of disk position, whereas the axial clearance varies with the pump rotor position. The following forces act on this device:

should not be throttled outside the balancing device; the orifice pressure drop is negligible, ranging from about 2 to 20 psi.

Balancing device modifications

There are now in use numerous hydraulic balancing device modifications. One typical design separates the drum portion of a combination device into two halves, one preceding and the second following the disk (Fig. 6.6). The virtue of this arrangement is a definite cushioning effect at the intermediate relief chamber, thus avoiding too positive a restoring action, which might result in the contacting and scoring of the disk faces.

Materials selection for balancing devices will be treated in a subsequent section. However, it is imperative to remember that both the material and the design are extremely important. If the balancing device wears appreciably, the pump rotor will automatically shift, setting up new axial reactions that may compound the condition.

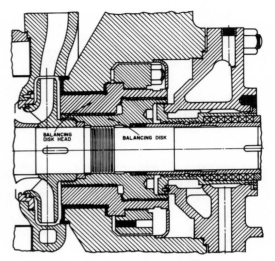

Fig. 6.6 Combination balancing disk and drum with disk located in center portion of drum

There is also the danger that rotor shift may cause the impellers to contact some internal pump part, such as a stage-piece, resulting in the complete destruction of the unit.

Individual axial thrust balancing

A design sometimes used to balance the axial thrust of single-suction impellers without the use of a hydraulic balancing device is illustrated in Fig. 6.7. It provides

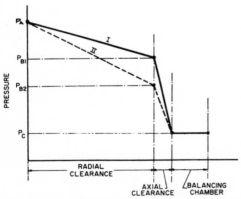

Fig. 6.5 Pressure distribution in combination balancing disk and drum

KEY:
P_a = *discharge pressure*
P_b = *pressure at intermediate relief chamber*
P_c = *back pressure*
I = *normal pressure distribution*
II = *pressure distribution after disk moves away from disk head*
$P_a - P_b$ = *pressure drop through drum portion*
$P_b - P_c$ = *pressure drop through disk portion.*

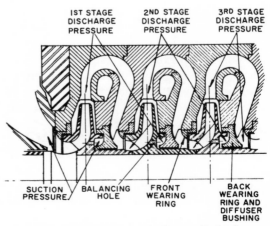

Fig. 6.7 Balancing axial thrust of single-suction impellers with back wearing rings

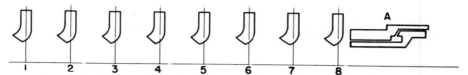

Fig. 6.8 Balancing axial thrust with balancing device in eight-stage pump

Joint A is subject to a differential pressure of eight stages.

the individual impellers with wearing rings both in front and back, the inner diameter of both rings being the same to equalize the thrust areas. Balancing holes are drilled through the impellers to equalize the pressures from front to back. This prevents the leakage water that flows across the back sealing surface from collecting in the annular space back of the impeller and building up the pressure at that point. For convenience, the back wearing ring may form an integral part with the diffuser or stage-piece bushing.

This design theoretically provides axial balance. Although the pressures on the two impeller sides may not be exactly equal in practice (because of unequal wearing ring leakage), the amount of unbalance is rather small and can usually be accommodated by the thrust bearing. Unfortunately, the use of a back wearing ring becomes less justifiable when one considers the effects of this construction on multistage pump internal leakage and mechanical design.

Normally, pumps with impellers arranged in ascending stage order enjoy almost negligible pressure difference and leakage across the stage-piece separating two consecutive stages. However, with a wearing ring at the back of the impeller, this difference becomes equivalent to the pressure generated by one stage. Thus, two additional clearance joints subject to a full one-stage pressure difference are now used, namely, the back wearing ring and the stage-piece joints.

A back wearing ring design is lacking in one other respect, compared to a hydraulic balancing device. That is, it does not reduce the pressure on the discharge-end stuffing box, which is now subjected to the suction pressure of the last stage. Unless the stuffing box is packed against this pressure (a pronounced improbability), it is necessary to provide some form of pressure-reducing mechanism ahead of the box. This requirement, in addition to the space needed for the back wearing rings, substantially increases the total shaft span of a multistage pump and makes the design even less desirable.

Comparison of balancing devices and opposed impellers

On the surface, it would appear that the choice between using a balancing device and arranging impellers in opposed sequence to balance axial thrust reflects a basic difference in design philosophy. Consequently, this choice has always been controversial among designers and users. The

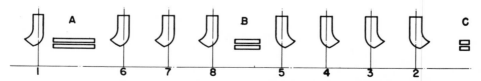

Fig. 6.9 Balancing axial thrust with opposed impellers in eight-stage pump

Joint A is subject to a differential pressure of four stages, joint B three stages, and joint C one stage, for a total of eight stages.

argument supporting each method centers on the presence or absence of a balancing device subject to a differential pressure equal to the total pressure generated by the pump.

Actually, the argument is strictly semantic. The total differential pressure has to be broken down in either case. Whether this is accomplished across a single running joint (Fig. 6.8) or the balancing device is split up into three separate portions distributed throughout the pump (Fig. 6.9) and given a different name is immaterial *so long as the running joints in each case are of proper length.* Wear is essentially a function of the pressure drop per inch of running joint length, and if the lengths of these joints are chosen to maintain the same pressure drop per inch, the wear will not be affected by the number of joints nor by the pressure differential across them.

The only significant difference between these hydraulic balancing designs is that the device located at the extreme end of all the impellers lends itself to leakage flow measurement whereas the three-part balancing device of Fig. 6.9 does not. It is more practical to test the progress of wear in a pump provided with a balancing device than in an opposed-impeller pump. This difference is not one, however, that all pump users hold in equal regard, and therefore personal preferences usually make the choice between the two arrangements.

7 *Shafts and Shaft Sleeves*

The basic function of a centrifugal pump shaft is to transmit the torques encountered in starting and during operation while supporting the impeller and other rotating parts. It must do this job with a deflection less than the minimum clearance between rotating and stationary parts. The loads involved are (1) the torques, (2) the weight of the parts, and (3) both radial and axial hydraulic forces. In designing a shaft, the maximum allowable deflection, the span or overhang, and the location of the loads all have to be considered, as does the critical speed of the resulting design.

Shafts are usually proportioned to withstand the stresses set up when a pump is started quickly, for example, when the driving motor is thrown directly across the line. If the pump handles hot liquids, the shaft is designed to withstand the stress set up when the unit is started cold without any preliminary warm-up.

Critical speeds

As critical speed is a key factor in the selection of shaft diameters, the centrifugal pump user ought to have a general knowledge of this subject.

Any object made of an elastic material has a natural period of vibration. When a pump rotor or shafting rotates at any speed corresponding to its natural frequency, minor unbalances will be magnified. These speeds are called the critical speeds.

In conventional pump designs, the rotating assembly is theoretically uniform around the shaft axis, and the center of mass should coincide with the axis of rotation. This theory will not hold for two reasons. First, there are always minor machining or casting irregularities, and second, there will be variations in metal density of each part. Thus, even in vertical shaft machines having no radial deflection caused by the weight of the parts, this eccentricity of the center of mass produces centrifugal force and therefore a deflection when the assembly rotates. At the speed at which the centrifugal force exceeds the elastic restoring force, the rotor will vibrate as though it were seriously unbalanced. If it is run at that speed without restraining forces, the deflection will increase until the shaft fails.

Rigid and flexible shaft designs

The lowest critical speed is called the first critical speed; the next higher is called the second, and so forth. In centrifugal pump nomenclature, a "rigid shaft" means one with an operating speed lower than its first critical speed, whereas a flexible shaft is one with an operating speed higher than its first critical speed.

Once an operating speed has been selected, the designer must still determine the relative shaft dimensions. In other words, he must decide whether the pump will operate above or below the first critical speed. Actually, the shaft critical speed can be reached and passed without danger because frictional forces tend to restrain the deflection. These forces are exerted by the surrounding liquid, the stuffing box packing, and the various internal leakage joints acting as internal liquid-lubricated bearings. Once the critical speed is passed, the pump will run smoothly again up to the second speed corresponding to the natural rotor frequency, and so on to the third, fourth, and all higher critical speeds.

Designs rated for 1,750 rpm (or lower) are usually of the rigid-shaft type. On the other hand, high-head, 3,600-rpm (or higher) multistage pumps like those in boiler feed service, are usually of the flexible shaft type. Although it may sound unsafe to bring a flexible shaft pump up to full speed through the first critical speed, with maximum vibration at that point, any deviation from exact critical speed immediately restores the elastic resistance of the shaft.

Shaft operation at exact critical speed will not necessarily cause pump failure. The amount of vibration involved depends primarily on the amount of unbalance in the shaft and rotation mass. Fine workmanship and careful balance can reduce vibrations to an imperceptible minimum. It is therefore possible to operate centrifugal pumps above their critical speeds for the following two reasons: (1) Very little time is required to attain full speed from rest (the time required to pass through the critical speed must therefore be extremely short),

and (2) the pumped liquid in the stuffing box packing and the internal leakage joints act as restraining forces on the vibration.

Experience has proved that although it was usually assumed necessary to use shafts of such rigidity that the first critical speed is at least 20 per cent above the operating speed, equally satisfactory results can be obtained with lighter shafts with a first critical speed of about 60 to 75 per cent of the operating speed. This, it is felt, is a sufficient margin to avoid any danger caused by operation close to the critical.

The influence of shaft deflection

To understand the effect of critical speed upon the selection of shaft size, consider the fact that the first critical speed of a shaft is linked to its static deflection by an immutable mathematical relation. Shaft deflection depends upon the weight of the rotating element (w), the shaft span (l) and the shaft diameter (d). The basic formula is:

$$f = \frac{wl^3}{mEI}$$

in which:

   f = deflection, in inches
   w = weight of the rotating element, in pounds
   l = shaft span, in inches
   m = coefficient depending on shaft support method and load distribution
   E = modulus of elasticity of shaft materials, in pounds per square inch
   I = moment of inertia ($d^4/64$), in inches^4.

This formula is given in its most simplified form, that is, for a shaft of constant diameter. If the shaft is of varying diameter (the usual situation), deflection calculations are much more complex. A graphical deflection analysis is then the most practical answer.

We must remember that the preceding formula solves only for static deflection, which alone affects critical speed calcula-

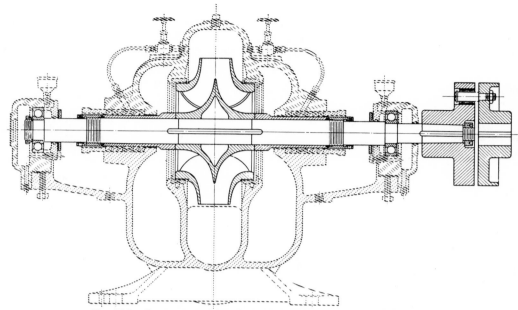

Fig. 7.1   Rotor assembly of a single-stage double-suction pump

tions. The actual shaft deflection—which must be determined to establish minimum permissible internal clearances—must take into account all transverse hydraulic reactions on the rotor, the actual weights of the rotating element, and other external loads, like belt pull.

It is not necessary to calculate exact deflection to make a relative shaft comparison. Instead, a factor can be developed that will be representative of relative shaft deflections. As the major portion of rotor weight is in the shaft, and as methods of bearing support and modulus of elasticity are common to similar designs, deflection (f) can be shown as follows:

$$f = \text{function of } \frac{(ld^2)(l^3)}{d^4}$$

or, $f = \text{function of } \dfrac{l^4}{d^2}$

In other words, pump deflection varies approximately as the fourth power of shaft span and inversely as the square of the shaft diameter. Therefore, the lower the (l^4/d^2) factor for a given pump, the lower

the unsupported shaft deflection, essentially in proportion to this factor.

For practical purposes, the first critical speed N_c (in rpm) can be calculated as:

$$N_c = \frac{187.7}{\sqrt{f}}$$

To maintain internal clearances at the wearing rings, it is usually desirable to limit shaft deflection under most adverse conditions to 0.005 to 0.006 in. It follows that a shaft design permitting a deflection of 0.005 to 0.006 will have a first critical speed of 2,400 to 2,650 rpm. This is the reason for using rigid shafts for pumps that operate at 1,750 rpm or lower. Multistage pumps operating at 3,600 rpm or higher use shafts of equal stiffness (for the same purpose of avoiding wearing ring contact). However, their corresponding critical speed is about 25 to 40 per cent less than their operating speed. This margin is sufficient to avoid any danger to the operation caused by critical speed effect.

So long as the operating speed is sufficiently different from the critical speed and the running clearances have been properly set to accommodate shaft clearances, it matters little whether the first critical speed is above or below the operating speed. A well-balanced, flexible-shaft pump is a safer piece of equipment than a poorly balanced, rigid-shaft pump.

Shaft sizing

We have stated that shaft diameters usually have larger dimensions than are actually needed to transmit the torque. A factor that assures this conservative design is the

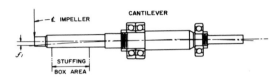

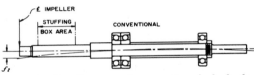

Fig. 7.2 Cantilever and conventional shaft designs for end-suction overhung-impeller pump

requirement for ease of rotor assemble. The shaft diameter must be stepped up several times from the end of the coupling to its center to facilitate impeller mounting (Fig. 7.1). Starting with the maximum diameter, at the impeller mounting in Fig. 7.1, there is a stepdown for the shaft sleeve, another for the external shaft nut, followed by several more for the bearings and the coupling. Therefore, the shaft diameter at the impellers exceeds that required for torsional strength at the coupling by at least an amount sufficient to provide all intervening stepdowns.

One frequent exception to shaft oversizing at the impeller occurs in units consisting of two double-suction single-stage pumps operating in series, one of them fitted with a double-extended shaft. As this pump must transmit the total horsepower

for the entire series unit, the shaft diameter at its inboard bearing may have to be greater than the normal diameter.

Shaft design of end-suction overhung-impeller pumps presents a somewhat different problem. One method for reducing shaft deflection at the impeller and stuffing box—where concentricity of running fits is extremely important—is to considerably increase shaft diameter between the bearings (Fig. 7.2). This design, incidentally, permits shortening the shaft span between the bearings, providing a more compact unit.

Except in certain smaller sizes, centrifugal pump shafts are protected against wear, erosion, and corrosion by renewable shaft sleeves. In very small pumps, however, shaft sleeves present a certain disadvantage. As the sleeves cannot appreciably contribute to shaft strength, the shaft itself must be designed for the full maximum stress. Shaft diameter is then materially increased by the addition of the sleeve, as the sleeve thickness cannot be decreased beyond a certain safe minimum. The impeller suction area may therefore become dangerously reduced, and if the eye diameter is increased to maintain a constant eye area, the liquid pick-up speed must be increased unfavorably. Other disadvantages accrue from greater hydraulic and stuffing box losses caused by increasing the effective shaft diameter out of proportion to the pump size.

To eliminate these shortcomings, very small pumps frequently use shafts of stainless steel or some other material that is sufficiently resistant to corrosion and wear not to need shaft sleeves. One such pump is illustrated in Fig. 7.3. Manufacturing costs, of course, are much less for this type of design, and the cost of replacing the shaft is about the same as the cost of new sleeves (including installation).

The question is sometimes asked: "Do impeller hubs and shaft sleeves make for a stiffer assembly?" They do. Stuffing boxes and internal leakage joints act as bearings to help support the rotor. Pumps designed to depend on these factors start to vibrate

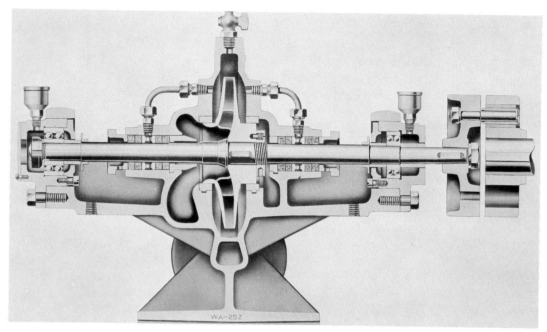

Fig. 7.3 Section of small centrifugal pump without shaft sleeves

when the leakage joints or packing become worn.

Pumps using mechanical seals instead of stuffing boxes should have reasonably stiff shafts to keep deflection to a minimum.

Shaft maintenance

Except in small pumps without shaft sleeves, it is unusual to replace a centrifugal pump shaft. Although the shaft may have to be replaced because of damage resulting from the failure of other parts, it will usually last the life of the pump.

During pump overhaul, the shaft should be carefully examined for any sign of wear or irregularities, especially at all the important fits, such as the impeller hub bores, under the shaft sleeves, and at the bearings.

The shaft may be damaged by rusting or pitting caused by leakage under the impellers or shaft sleeves. If the pump is fitted with ball bearings, the shaft may become damaged by turning in the inner race. If babbitt bearings are used, it may wear in the journals or become grooved or its fit at

the coupling may become loose. Small pumps without shaft sleeves may wear at the stuffing boxes.

It is also important to check the shaft condition at the keyways. Twisting of the shaft, excessive thermal stresses, corrosion, or even a poor original fit may have loosened the impellers, resulting in keyway wear. If the condition is not corrected, it will aggravate rapidly, producing a very noisy operation and possibly causing shaft failure. Finally, the shaft should be examined carefully for fatigue cracks, although these are rather rare.

After visual inspection, the shaft should be placed on centers and checked for concentricity. A bent or distorted shaft should not be corrected, as the process is difficult even in a shop specially equipped to handle the job. Nor should a shaft that has been damaged be welded, because distortion will always result. Bent or distorted shafts should be replaced.

If the cost of a new shaft is high, and if the proper facilities are available, a worn shaft may sometimes be repaired, by metal

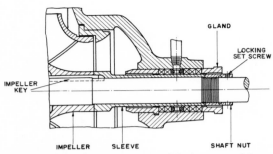

Fig. 7.4 Sleeve with external lock nut and impeller key

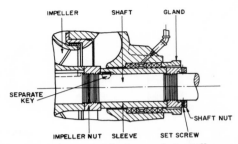

Fig. 7.5 Sleeve with internal impeller nut, external shaft sleeve nut, and separate key

spraying and remachining. Such repairs should not be undertaken without familiarity with the shaft material and appropriate metal spraying methods. After the shaft has been repaired, it must be checked for possible distortion and then rechecked after complete rotor assembly to make sure it has not been distorted by excessive tightening of the shaft nuts.

As previously mentioned a steel shaft can rust because of leakage between the shaft and impeller hub if the casting is porous at that point. Such leakage is hard to control. But leakage through the metal-to-metal joint of a sleeve and impeller hub can often be controlled by using packing. Corrosive seepage or leakage may often warrant use of a corrosion-resistant shaft material in place of regular steel. For example, monel shafts are commonly used in pumps

handling sea water even though shaft sleeves provide full protection against direct liquid contact.

Shaft sleeves

Pump shafts are usually protected from erosion, corrosion, and wear at stuffing boxes, leakage joints, internal bearings, and in the waterways by renewable sleeves. The most common shaft sleeve function is that of protecting the shaft from wear at a stuffing box. Thus shaft sleeves serving other functions are given specific names to indicate their purpose. For example, a shaft sleeve used between two multistage pump impellers in conjunction with the interstage bushing to form an interstage leakage joint is called an interstage or distance sleeve.

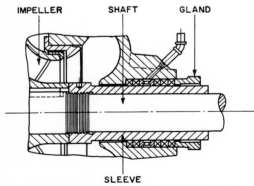

Fig. 7.6 Sleeve threaded onto shaft with no external lock nut

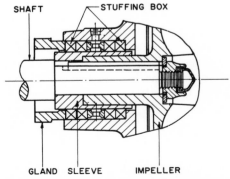

Fig. 7.7 Sleeve for pumps with overhung-impeller hubs extending into stuffing box

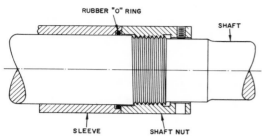

Fig. 7.8 Shaft sleeve seal to prevent leakage along shaft

In medium-size centrifugal pumps with two external bearings on opposite sides of the casing (the common double-suction and multistage varieties), the favored shaft sleeve construction uses an external shaft nut to hold the sleeve in axial position against the impeller hub. Sleeve rotation is prevented by a key, usually an extension of the impeller key (Fig. 7.4). If the axial thrust exceeds the frictional grip of the impeller on the shaft, it is transmitted through the sleeve to the external shaft nut.

In larger high-head pumps, a high axial load on the sleeve is possible, and a design like that in Fig. 7.5 may be favored. This design has the commercial advantages of simplicity and low replacement cost. Some manufacturers favor the sleeve shown in Fig. 7.6, in which the impeller end of the sleeve is threaded and screwed to a matching thread on the shaft. A key cannot be used with this type of sleeve and right- and left-hand threads are substituted so that the frictional grip of the packing on the sleeve will tighten it against the impeller hub. In the sleeve designs shown in Fig. 7.4 and 7.5, right-hand threads are usually used for all shaft nuts because keys prevent the sleeve from rotating. As a safety precaution, the external shaft nuts (Fig. 7.4 and 7.5) and the sleeve itself (Fig. 7.6) use set screws for a locking device.

In pumps with overhung impellers, various types of sleeves are used. Often, stuffing boxes are placed close to the impeller, and the sleeve actually protects the impeller hub

from wear (Fig. 7.7). As a portion of the sleeve in this design fits directly on the shaft, the impeller key can be used to prevent sleeve rotation. Part of the sleeve is clamped between the impeller and a shaft shoulder to maintain its axial position.

Sleeves with leakage seal

In designs with a metal-to-metal joint between the sleeve and impeller hub (see Fig. 7.4), operation under a positive suction head often starts liquid leakage into the clearance between the shaft and sleeve. For a pump operating under negative suction head, the various clearances may cause slight air leakage into the pump. Usually, this leakage is not important; but it occasionally causes trouble, and a sleeve design with a leakage seal may then become desirable. One possible arrangement is shown in Fig. 7.8. The design shown in Fig. 7.9 is used for high-temperature process pumps. The contact surface of the sleeve and shaft is ground at a 45 deg angle. That end of the sleeve is locked, but the other is free to expand with temperature changes.

Material for stuffing box sleeves

Stuffing box shaft sleeves are surrounded in the stuffing box by packing; the sleeve must be smooth so that it can turn without generating too much friction and heat. Thus the sleeve materials must be capable of taking a very fine finish, preferably a

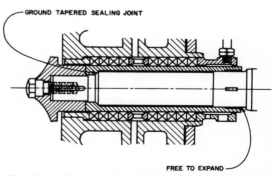

Fig. 7.9 Sleeve with 45-deg bevel contacting surface

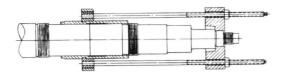

Fig. 7.10 Shaft sleeve puller

polish. Cast iron is therefore not suitable. A hard bronze is generally used for pumps handling clear water, but chrome or other stainless steels are sometimes preferred. For services subject to grit, hardened chrome or other stainless steels give good results. For more severe conditions, stellited sleeves are often used and occasionally sleeves that are chromium plated at the packing area. Sleeves made entirely of a hardened chrome steel are usually the most economical and satisfactory.

Shaft sleeve maintenance

Shaft sleeves are usually the fastest wearing pump part and the one most frequently requiring replacement. Once sleeves are worn appreciably, the packing cannot be adjusted to prevent excessive leakage. As a matter of fact, excessively worn sleeves frequently tear and score any new packing as soon as it is inserted. Thus, sleeves frequently require repair or replacement when no other pump overhaul is necessary.

Sleeves of single-stage and low-head multistage pumps can be removed quite easily. As the long sleeves sometimes used in high-pressure multistage pumps may be harder to remove, they are often fabricated with external grooves so that a sleeve puller can be used (Fig. 7.10). In a design that uses an impeller nut between the sleeve and the impeller (see Fig. 7.5), a tight sleeve can often be loosened by backing off the impeller nut.

Shaft sleeves are occasionally reconditioned by welding or metal spraying and final grinding. This procedure is not recommended for a pump on severe service, or if existing facilities for the final grinding are inadequate. It is necessary to assure both concentricity of grinding and the perpendicularity of the sleeve radial faces to the sleeve bores. Concentricity should be double checked after reassembly on the rotor.

Although it may be easier to pack a pump with brand new shaft sleeves, the sleeves do not have to be replaced each time new packing is installed. The degree of permissible sleeve wear grooving depends on the type of grooving. Usually, the sleeve surface is highly polished under the packing action, and the grooving is undulated rather than composed of sharp separate grooves under each individual packing ring. Sometimes, slight grinding of these worn sleeves is permissible to permit reuse if the pump service is not too severe. The controlling factors are the availability of the necessary tools, shop facilities, and trained shop personnel. Restored sleeves *must* have a good, smooth surface, and the refinished parts should neither be run-out nor distorted. If the facilities are available, it may be advisable to try regrinding and reusing one set of worn sleeves to establish the practicability of this procedure.

The shaft sleeve OD should not be reduced to a point at which excessive clearance at the bottom of the stuffing box permits any packing to be squeezed inside the pump when the glands are tightened. As a rule, sleeves should not be ground down more than 25 to 30 thousandths on the diameter and should be given a 16 micro-inch finish.

Worn sleeves, however, are ordinarily replaced rather than reconditioned. Hammering to expand or crack the material will facilitate their removal, but extra care should be taken to prevent shaft damage.

8 *Stuffing Boxes*

The stuffing box is one of the most important parts of a centrifugal pump. Even slight defects in its arrangement or condition can prevent proper pump operation. Stuffing boxes have the primary function of protecting the pump against leakage at the point where the shaft passes out through the pump casing. However, this function varies both in itself and in the manner in which it is performed. For example, if the pump handles a suction lift and the pressure at the interior stuffing box end is below atmospheric, the stuffing box function is to prevent air leakage into the pump. If this pressure is above atmospheric, the function is to prevent liquid leakage out of the pump.

For general service pumps, a stuffing box usually takes the form of a cylindrical recess that accommodates a number of rings of packing around the shaft or shaft sleeve (Fig. 8.1 and 8.2). If sealing the box is desired, a lantern ring or seal cage (Fig. 8.3) is used that separates the rings of packing into approximately equal sections. The packing is compressed to give the desired fit on the shaft or sleeve by a gland that can be adjusted in an axial direction. The bottom or inside end of the box may be formed by the pump casing itself (see Fig. 7.7), a throat bushing (Fig. 8.1), or a bottoming ring (Fig. 8.2).

For manufacturing reasons, throat bushings are widely used on smaller pumps with axially split casings. In machining casings for these pumps, the diameter of an integrally formed stuffing box throat would limit the boring bar size to an impractically small diameter. Throat bushings are always solid rather than split. The bushing is usually held from rotation by a tongue-and-groove joint locked in the lower half of the casing.

Seal cages

When a pump operates with negative suction head, the inner end of the stuffing box is under vacuum, and air tends to leak into the pump. For this type of service, packing is usually separated into two sections by a lantern ring or seal cage (Fig. 8.1). Water or some other sealing fluid is introduced under pressure into the space, causing flow of sealing fluid in both axial directions. This construction is useful for pumps handling flammable or chemically active and dangerous liquids since it prevents outflow of the pumped liquid. Seal cages are usually axially split for ease of assembly.

75

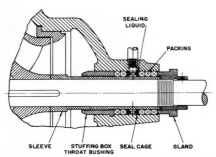

Fig. 8.1 Conventional stuffing box with throat bushing

Fig. 8.3 Lantern gland or seal cage

Some installations involve variable suction conditions, the pump operating part time with head on suction and part time with suction lift. When the operating pressure inside the pump exceeds the atmospheric pressure, the liquid seal cage becomes inoperative (except for lubrication). However, it is maintained in service so that when the pump is primed at starting, all air can be excluded.

Sealing liquid arrangements

When a pump handles clean, cool water, stuffing box seals are usually connected to the pump discharge, or, in multistage pumps, to an intermediate stage. An independent supply of sealing water should be provided if any of the following conditions exist:

1. A suction lift in excess of 15 ft
2. A discharge pressure under 10 lb (or 23-ft head)
3. Hot water (over 250°F) being handled without adequate cooling (except for boiler feed pumps, in which seal cages are not used)
4. Muddy, sandy, or gritty water being handled
5. For all hotwell pumps
6. The liquid being handled is other than water—such as acid, juice, molasses, or sticky liquids—without special provision in the stuffing box design for the nature of the liquid.

When sealing water is taken from the pump discharge, an external connection may be made through small diameter piping (Fig. 8.4) or internal passages. In some

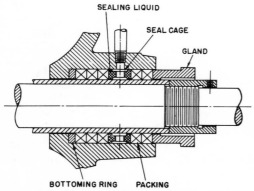

Fig. 8.2 Conventional stuffing box with bottoming ring

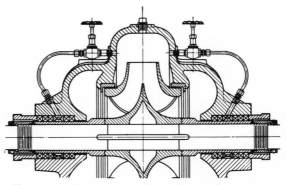

Fig. 8.4 Piping connections from the pump discharge

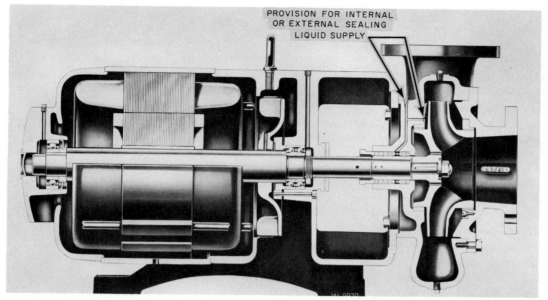

Fig. 8.5 End-suction pump with provision for internal or external sealing liquid supply

pumps, these connections are arranged so that a sealing liquid can be introduced into the packing space through an internal drilled passage either from the pump casing or an external source (Fig. 8.5). Figure 8.6 illustrates the alternate uses of the seal cage connections. In Fig. 8.6A—the general service arrangement in which the liquid pumped is used for sealing—the external connection is plugged. If an external sealing liquid source is required (Fig. 8.6B), it is connected to the external pipe tap with a socket-head pipe plug inserted at the internal pipe tap.

It is sometimes desirable to locate the seal cage with more packing on one side. For example, on gritty-water service, the seal cage location shown in Fig. 8.7 would divert a greater proportion of sealing liquid into the pump, thereby keeping grit from working into the box. Figure 8.8 shows most of the rings between the cage and the inner end of the box. This arrangement would be applied to reduce dilution of the pumped liquid.

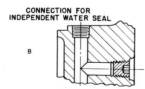

Fig. 8.6 Arrangements of sealing connections for pump in Fig. 8.5

A. Internal seal
B. Independent seal.

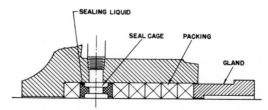

Fig. 8.7 Recommended location of seal cage for gritty or dirty water

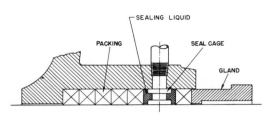

Fig. 8.8 Recommended location of seal cage for pumps when little dilution of liquid is desired

If clean, cool water is not available (as with drainage, irrigation, and sewage pumps), grease or oil seals are often used. Most pumps for sewage service have a single stuffing box subject to discharge pressure and are located with a flooded suction. It is therefore not necessary to seal these pumps against air leakage, but forcing grease into the sealing space and packing helps to exclude grit.

An automatic oil sealer that exerts discharge pressure in a cylinder on one side of a plunger, with oil or light grease on the other side, is available for sewage service. The oil or grease line is connected to the seal connection, which is near discharge pressure. As the inner end of the stuffing box would be at about 80 per cent discharge pressure, there is a slow flow of grease or oil into the pump when the unit is in operation. No flow takes place when the pump is out of service.

Some pumps handle water in which there are small, even microscopic, solids. Using water of this kind as a sealing liquid introduces the solids into the leakage path, shortening the life of the packing and sleeves. It is sometimes possible to remove these solids by installing small pressure filters in the sealing water piping from the volute to the stuffing box.

Water-cooled stuffing boxes

High temperatures or pressures complicate the problem of maintaining stuffing box packing. Pumps in these more difficult services are usually provided with jacketed, water-cooled stuffing boxes. The cooling water removes heat from the liquid leaking through the stuffing box and heat generated by friction in the box, thus improving packing service conditions. In some special cases, oil or gasoline may be used in the cooling jackets instead of water. Two water-cooled stuffing box designs are available. The first (Fig. 8.9) provides cored passages in the casing casting. These passages, which surround the stuffing box, are arranged with in-and-out connections. The second type uses a separate cooling chamber combined with the stuffing box proper, the whole assembly being inserted into and bolted to the pump casing (Fig. 8.10). The choice between the two is based on manufacturing preferences.

Pressure and temperature conditions

In recent years, a more thorough understanding of the interrelation of stuffing box pressures, rubbing speeds, and leakage temperatures has led to an improved water-cooled stuffing box design. Boxes have now been built for temperatures up to 400°F, and stuffing box pressures up to 500 psi, without pressure-reducing breakdowns or labyrinths. This type of stuffing box is il-

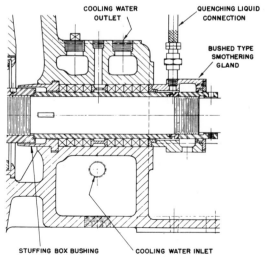

Fig. 8.9 Water-cooled stuffing box with cored water passage cast in casing

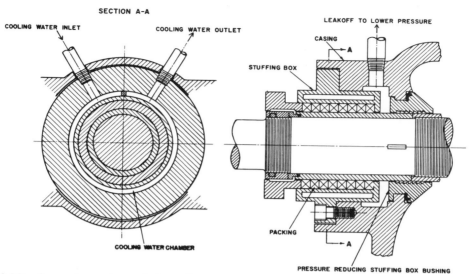

SECTION A-A

COOLING WATER INLET

COOLING WATER OUTLET

LEAKOFF TO LOWER PRESSURE

CASING

STUFFING BOX

COOLING WATER CHAMBER

PACKING

PRESSURE REDUCING STUFFING BOX BUSHING

Fig. 8.10 Separate water-cooled stuffing box assembly with pressure-reducing stuffing box bushing

lustrated in Fig. 8.11. For greatest cooling efficiency, the temperature difference between the cooling liquid and the leakage through the box must be kept to a maximum at all points. In this design (Fig. 8.11), the cooling water is introduced nearer to the outside of the stuffing box. Before moving axially towards the interior of the pump, the cooling water is circulated completely around that portion of the stuffing box which surrounds the packing. A cored passage is provided from this annular chamber towards the interior of the pump. The cooling water then circulates in a secondary annular chamber extending inside the pump beyond the packing. This allows precooling of stuffing box leakage before it reaches the packing. The cooling water then escapes through a second cored passage to the cooling chamber exit. In this design the coldest cooling water is adjacent to the coldest leakage. Having picked up some heat, the cooling water flows into the pump at a higher temperature and cools a higher temperature leakage. One such unit has operated continuously for over a year without renewing the

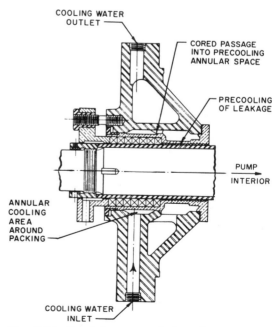

COOLING WATER OUTLET

CORED PASSAGE INTO PRECOOLING ANNULAR SPACE

PRECOOLING OF LEAKAGE

PUMP INTERIOR

ANNULAR COOLING AREA AROUND PACKING

COOLING WATER INLET

Fig. 8.11 Special water-cooled stuffing box for high pressures and high temperatures

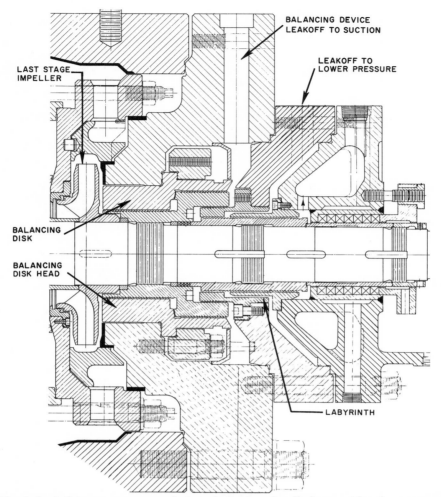

Fig. 8.12 Labyrinth-type pressure reducing device on the discharge side of a pump equipped with a balancing device

packing, under stuffing box operating conditions of 325 psi and 400°F. The shaft diameter at the stuffing box was 4 in., and the operating speed 3,600 rpm.

Stuffing box pressure and temperature limitations vary with the pump type, because it is generally not economical to use expensive stuffing box construction for infrequent high-temperature or high-pressure applications. Therefore, whenever the manufacturer's stuffing box limitations for a given pump are exceeded, the only solution is the application of pressure-reducing devices ahead of the stuffing box.

Pressure-reducing devices

Essentially, pressure-reducing devices consist of a bushing or meshing labyrinth, ending in a relief chamber located between the pump interior and the stuffing box. The relief chamber is connected to some suitable low-pressure point in the installation, and the leakage past the pressure-reducing device is returned to this point. The only drawback to application of these devices is the necessity of bleeding a part of the effective pump capacity back to a lower pressure level, and the resultant reduction in installation efficiency. If the pumped liq-uid must be salvaged, as with treated feed-water, it is returned back into the pump-ing cycle. If the liquid can be wasted, the relief chamber can be connected to a drain.

There are many different pressure-re-ducing device designs. Figure 8.10 illustrates a design for limited pressures. A short, serrated stuffing box bushing is inserted at the bottom of the stuffing box, followed by a relief chamber. The leakage past the ser-rated bushing is bled off to a low-pressure point.

With relatively high-pressure units, in-termeshing labyrinths may be located fol-lowing the balancing device and ahead of the stuffing box (Fig. 8.12). Piping from the chamber following pressure-reducing de-vices should be amply sized so that as wear increases leakage, piping friction will not increase stuffing box pressure. Another very interesting type of pressure-reducing device is shown in Fig. 8.13. Full explanation of its operation is given in the caption.

Stuffing box packing

Basically, stuffing box packing is a pres-sure breakdown device. The packing must be somewhat plastic so that it can be ad-justed for proper operation. It must also absorb energy without failing or damaging the rotating shaft or shaft sleeve. In a breakdown of this nature, friction energy is liberated. This generates heat that must be dissipated in the fluid leaking past the breakdown or by means of cooling water jacketing or both.

There are numerous stuffing box pack-ing materials, each adapted to some par-ticular class of service. Some of the prin-cipal types are the following:

1. *Asbestos packing*—comparatively soft and suitable for cold-water and hot-water applications in the lower temperature range. It is the most common packing ma-terial for general service under normal pressures. For pressures above 200 psi, this packing is only usable at very moderate

rubbing speeds. Asbestos packing is prelu-bricated with either graphite or some inert oil.

2. *Metallic packing*—composed of flexible metallic strands or foil with graphite or oil lubricant impregnation and with either an asbestos or plastic core. The impregna-tion makes this packing self-lubricating for its start-up period. The foils are made of various metals such as babbitt, aluminum, and copper. Babbitt foil is used on water and oil service for low and medium tem-peratures (up to 450°F) and medium to high pressures. Copper is used for medium to high temperatures and pressures with water and low sulphur content oils. Alumi-num is used mainly on oil service and for medium to high temperatures and pres-sures.

Many other types of packing are regu-larly furnished to meet customers' special specifications, for example, hemp, cord, braided type, duck fabric, chevron-type, and others too numerous to mention.

Packing is supplied either in continuous coils of square cross-section or in preformed die-molded rings. When coil-type packing

Fig. 8.13 Pressure-reducing device

Two seals installed in parallel in discharge end of a multistage pump. Seal unit A func-tions as a pressure breakdown and as a hy-draulic balancing device to compensate for the lateral thrust of the rotating element. Seal unit B prevents the liquid being pumped from coming into contact with the packing. This protection is accomplished by injecting a liquid in the chamber between the seal and the stuffing box. (Courtesy Pacific Pumps, Inc.)

(Fig. 8.14 and 8.15) is used, it is cut in lengths that make up individual rings. The ends are cut with a diagonal, or scarf joint, and with a slight clearance to provide for expansion and avoid buckling. The rings have a tendency to swell from the liquid action and the rise in temperature. The scarf joint allows the ends to slide and laterally absorb expansion.

It is preferable, where possible, to use die-molded packing rings (Fig. 8.16), which are available to exact size and in sets. A molded ring insures an exact fit to the

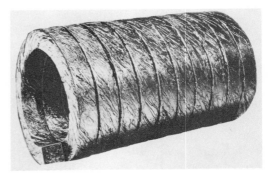

Fig. 8.15   Metallic packing in spiral form
(Courtesy John Crane Co.)

Fig. 8.14   Graphited asbestos packing in continuous coil form

(Courtesy John Crane Co.)

shaft or shaft sleeve and to the stuffing box bore and also establishes equal packing density throughout the stuffing box.

Frequently, more efficient packing life can be gained by a combination of two or more different kinds of packing—for example, alternating hard and soft rings (Fig. 8.17). Such sets are usually available in standard die-formed ring combinations from most reputable packing manufacturers.

For best results, the shaft or shaft sleeves should be in perfect alignment, concentric with the axis of rotation, highly polished, and should operate without vibration. The material of which they are made is also extremely important, as it directly affects the life and maintenance of the packing.

Stuffing box glands

Stuffing box glands may assume several forms, but basically they can be classified into two groups:

1. Solid glands (Fig. 8.18)
2. Split glands (Fig. 8.19).

Split glands are made in halves so that they may be removed from the shaft without dismantling the pump, thus providing more working space when the stuffing boxes

Fig. 8.16   Metallic packing in ring form
(Courtesy John Crane Co.)

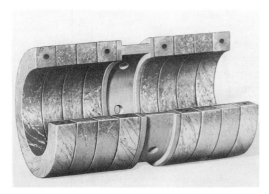

Fig. 8.17 Combination set of hard and soft packing

(Courtesy John Crane Co.)

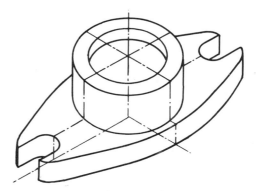

Fig. 8.18 Solid stuffing box gland

are being repacked. Split glands are desirable for pumps that have to be repacked frequently, especially if the space between the box and the bearing is restricted. The two halves are generally held together by bolts (Fig. 8.19), although other methods are also used. Split glands are generally a construction refinement rather than a necessity, and they are rarely used in smaller pumps. They are commonly furnished for large single-stage pumps, for some multistage pumps, and for refinery pumps. Another common refinement is the use of swing bolts in stuffing box glands. Such bolts may be swung to the side, out of the way, when the stuffing box is being repacked.

Stuffing box leakage into the atmosphere might, in some services, seriously inconvenience or even endanger the operating personnel—for example, when such liquids as hydrocarbons are being pumped at vaporizing temperatures or temperatures above their flash point. As this leakage cannot always be cooled sufficiently by a water-cooled stuffing box, smothering glands are used (see Fig. 8.9). Provision is made in the gland itself to introduce a liquid—either water or another hydrocarbon at low temperature—that mixes intimately with the leakage, lowering its temperature, or, if the liquid is volatile, absorbing it.

Stuffing box glands are usually made of bronze, although cast iron or steel may be used for all iron-fitted pumps. Iron or steel glands are generally bushed with a nonsparking material like bronze in refinery service to prevent the ignition of flammable vapors by the glands sparking against a ferrous metal shaft or sleeve.

Stuffing box maintenance

Stuffing box maintenance primarily consists of packing replacement. Although this sounds simple, it must be done correctly, or pump operation will not be satisfactory. The following procedure should be followed in repacking a stuffing box:

1. Never try to add one or two rings to the old packing. This is false economy. Remove the old packing completely, using a packing puller, if available, and thoroughly clean the box. Inspect the sleeve to make

Fig. 8.19 Split stuffing box gland

sure it is in acceptable condition. Putting new packing in a box against a rough or badly worn sleeve will not give satisfactory service.

2. Be sure that the new packing is a proper type for the liquid, operating pressure, and temperature. Unless the packing comes preformed in sets, make sure that each ring is cut square on a mandrel of correct size.

3. Insert each ring of packing separately, pushing it squarely into the box and firmly seating it, using split pusher rings of proper length, fitting the box nicely. Successive rings of packing should be rotated so the joints are 120 or 180 deg apart.

4. When a seal cage is used, make sure to install it between the proper two packing rings so it will correctly handle the sealing liquid supply when the box is fully packed and adjusted.

5. After all the required packing rings have been inserted, install the gland and firmly tighten the gland nuts. Make sure that the gland enters the stuffing box squarely and without cocking, so the full periphery of the packing is under uniform pressure.

6. After the first tightening of the gland, back off the nuts until they are only finger tight. Start the pump with the gland loose, so that there will be excessive initial leakage. Tighten up slightly and evenly on the gland nuts, at fifteen- or twenty-minute intervals, so the leakage is reduced to normal after several hours.

Do not attempt to reduce leakage to the "dripping" state. It must be a steady flow, sufficient to carry away the packing friction heat. Unless enough liquid leaks through the box to remove this heat, the packing will be burnt, and the sleeve scored. (On pumps with quenching glands, stop the supply of quenching water at intervals, and observe actual leakage through the box, otherwise, visual inspection cannot distinguish between leakage past the packing and the quenching liquid supply.)

Repacking and adjusting stuffing boxes should only be done by experienced personnel. Others assigned this work should be cautioned against putting too much pressure on the gland. It should be made clear that excessive leakage is not as harmful as too little.

Packing removed from a stuffing box being repacked, should be examined in order to obtain as much information as possible on the cause of packing wear. Often, correctable operating conditions or inadequate packing procedures are revealed by this examination. Some of the more frequently encountered symptoms are the following:

1. Excessive wear on rings nearest to the gland, while the bottom rings remain in good condition, is caused by overtightening of the packing in one adjustment or by not inserting rings one at a time, and pushing each home before inserting the following ring.

2. Charring or glazing of the inner circumference of the rings is caused by excessive heating, insufficient lubrication, or inadequate packing material for the pressure and temperature conditions.

3. Wear on the outer circumference of the rings occurs when they rotate within the stuffing box.

4. Heavy packing ring wear on one selective portion of the inner circumference may be caused by excessively worn bearings or eccentric rotor operation.

5. If some rings are cut too short or shrink excessively, the adjacent rings will bulge and be extruded into the open space.

9 *Mechanical Seals*

The conventional stuffing box design and composition packing are impractical to use as a method for sealing a rotating shaft for many conditions of service. In the ordinary stuffing box, the sealing between the moving shaft or shaft sleeve and the stationary portion of the box is accomplished by means of rings of packing forced between the two surfaces and held tightly in place by a stuffing box gland. The leakage around the shaft is controlled merely by tightening up or loosening the gland studs. The actual sealing surfaces consist of the axial rotating surfaces of the shaft, or shaft sleeve, and the stationary packing. Attempts to reduce or eliminate all leakage from a conventional stuffing box increase the gland pressure. The packing, being semiplastic in nature, forms more closely to the shaft and tends to cut down the leakage. After a certain point, however, the leakage continues no matter how tightly the gland studs are brought up. The frictional horsepower increases rapidly at this point; the heat generated cannot be properly dissipated; and the stuffing box fails to function. Even before this condition is reached, the shaft sleeves may be severely worn and scored, so that it becomes impossible to pack the stuffing box satisfactorily.

These undesirable characteristics prohibit the use of packing as the sealing medium between rotating surfaces if the leakage is to be held to an absolute minimum under severe pressure. This condition, in turn, automatically eliminates use of the axial surfaces as the sealing surfaces, for a semiplastic packing is the only material that can always be made to form about the shaft and compensate for the wear. Another factor that makes stuffing boxes unsatisfactory for certain applications is the relatively small lubricating value of many liquids frequently handled by centrifugal pumps, such as propane or butane. These liquids actually act as solvents of the lubricants normally used to impregnate the packing. Seal oil must therefore be introduced into the lantern gland of a packed box to lubricate the packing and give it reasonable life. With these facts in mind, mechanical seal designers have had to attempt to produce an entirely different type of seal with wearing surfaces other than the axial surfaces of the shaft and packing.

This form of seal, called the mechanical seal, is a new development compared to regular stuffing boxes but has already found general acceptance in those pumping applications in which the shortcomings of packed stuffing boxes have proved excessive. Fields in which the packed boxes gave good service, however, have shown little tendency to replace them with mechanical seals.

As both packed boxes and mechanical seals are subject to wear, neither are perfect. One or the other proves to be the better according to the application. In some fields both give good service, and choosing between them becomes a matter of personal preference or first cost.

Principles of mechanical seals

Although they may differ in various physical respects, all mechanical seals are fundamentally the same in principle. The sealing surfaces of every kind are located in a plane perpendicular to the shaft and usually consist of two highly polished surfaces running adjacently, one surface being connected to the shaft and the other to the stationary portion of the pump.

Complete sealing is accomplished at the fixed members. The polished or lapped surfaces, which are of dissimilar materials and are held in continual contact by a spring, form a fluid-tight seal between the rotating and stationary members with very small frictional losses. When the seal is new, the leakage is negligible and can actually be considered as nonexistent. (To obtain a pressure breakdown between the internal pressure and the atmospheric pressure outside the pump, a flow of liquid past the seal faces is required. This flow may be only a drop every few minutes or even a haze of escaping vapor—if a liquid such as propane is being handled, for instance. Thus, even though leakage is negligible, technically speaking a rotating mechanical seal cannot *entirely* eliminate it.) Of course, some wear always occurs, and provision must be made for a very small amount of leakage in time.

The wide variation in seal design stems from the many methods used to provide flexibility and to mount the seals. A mechanical seal is similar to a bearing in that

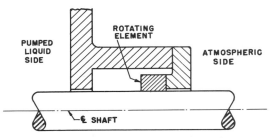

Fig. 9.1   Internal assembly seal

it involves a close running clearance with a liquid film between the faces. The lubrication and cooling provided by this film cuts down the wear, as does a proper choice of seal face materials.

Seals for centrifugal pumps do not operate satisfactorily on air or gas; if run "dry," they will fail rapidly. Seals can be used in pumps handling liquids that contain solids if the solids are prevented from getting between the seal faces or interfering with the flexibility of the mounting.

Comparison of internal and external seals

There are two basic seal arrangements: (1) The *internal assembly* (Fig. 9.1), in which the rotating element is located inside the box and is in contact with the liquid being pumped, and (2) the *external assembly* (Fig. 9.2), in which the rotating element is located outside the box. The pressure of the liquid in the pump tends to force the rotating and stationary faces together in the inside assembly and to force them apart in the external assembly. But both internal and external types always have three primary points (Fig. 9.3) at which sealing must be accomplished:

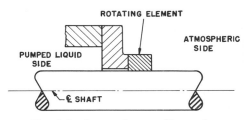

Fig. 9.2   External assembly seal

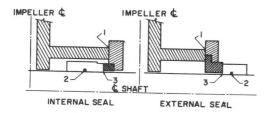

Fig. 9.3 The three sealing points in a mechanical seal

1. Between the stationary element and the casing
2. Between the rotating element and the shaft (or the shaft sleeve, if one is used)
3. Between the mating surfaces of the rotating and stationary seal elements.

To accomplish the first seal, conventional gaskets or some form of a synthetic O-ring are used. Leakage between the rotating element and the shaft is stopped by means of O-rings, bellows, or some form of flexible wedges. Leakage between the mating surfaces cannot be entirely stopped but can be held to an insignificant amount by maintaining a very close contact between these faces.

Comparison of balanced and unbalanced seals

The pressure within the pump just ahead of the mechanical seal tends to keep the mating faces of the internal seal together. In the simplest design (as in Fig. 9.1), the entire internal pressure acts to close the faces. If the liquid handled is a good lubricant and the pressures are not excessive,

this loading will not be harmful. The design is known as the *unbalanced seal*. A graphic description of the forces and area relationships in this seal are given in Fig. 9.4. If P = pressure of liquid in the box and P' = average pressure across the seal faces, then:

$$\text{Closing force} = (P)(\text{area } A) + \text{Spring loading}$$

$$\text{Opening force} = (P')(\text{area } B)$$

Generally, the application of unbalanced seals is limited to pressures lower than 100 to 150 psig and to liquids with lubricating properties equal to or better than gasoline.

When these criteria are not met, it has been found preferable to so proportion the areas subject to pressure as to reduce the loading on the mating faces, providing what is known as a *balanced seal* (Fig. 9.5). Although such seals have been applied very successfully for quite high internal pressures, they are not particularly suitable for low pressures (under 50 psig) as the sealing force is reduced to a point at which contact between the mating faces may not be sufficient to provide adequate sealing. If P and P' are the quantities described above, then:

$$\text{Closing force} = (P)(\text{area } A - \text{area } C) + \text{Spring loading}$$

$$\text{Opening force} = (P')(\text{area } B)$$

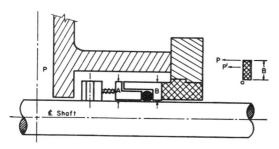

Fig. 9.4 Unbalanced seal construction

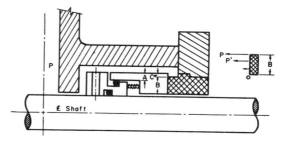

Fig. 9.5 Balanced seal construction

Double seals

Two mechanical seals may be mounted inside a stuffing box to make a *double seal* assembly, shown diagrammatically in Fig. 9.6. Such an arrangement is used for pumps handling toxic or highly inflammable liquids that cannot be permitted to escape into the atmosphere. It is also applicable to pumps handling corrosive or abrasive liquids at very high or very low temperatures. A clear, filtered, and generally inert sealing liquid is injected between the two seals at a pressure slightly in excess of the pressure in the pump ahead of the seal. This liquid prevents the pumped liquid from coming into contact with the seal parts or from escaping into the atmosphere.

Seal designs

Some manufacturers of seals have found it advisable to provide the prospective user with a very complete engineering service in order to insure that every application is given the most through analysis. Such an analysis is an essential key to the success of a mechanical seal application, and users should take full advantage of this service.

The operation of a mechanical seal can best be understood by reference to a few standard commercial units. The following discussion treats of the general characteristics of several typical mechanical seals. It should be borne in mind, however, that every sealing problem, like every pumping problem, is different. When positive and dependable results are desired, the me-

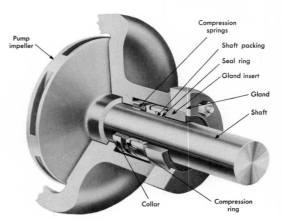

Fig. 9.7   Dura-seal mechanical seal
(*Courtesy Durametallic Co.*)

chanical seal should be specifically designed and fitted for the unit in question. It should be mentioned at the same time that the art of mechanical seal design is anything but static. The severity of the problems of new applications on the one hand and the rewards to be achieved for their satisfactory solution on the other have fostered a vital process of continuous development among a large number of seal manufacturers. Because of this continuous growth and improvement, seal descriptions run the risk of obsolescence in the short time between the preparation of a manuscript and the publication of a book. The following descriptions, therefore, are primarily intended to illustrate the basic principles of mechanical seal design and operation.

One typical seal construction is illustrated in Fig. 9.7. The gland with its gland insert is fitted into the casing and constitutes the stationary member of the seal assembly. It provides a seal at two points: (1) Between the gland and the face of the stuffing box, by means of the gaskets, and (2) between the gland insert and the seal ring face, by contact. The mating seal ring with a hardened steel surface rotates with the shaft and is held against the stationary member by the compression ring. The latter supports a nest of springs that are connected at the opposite end to the collar,

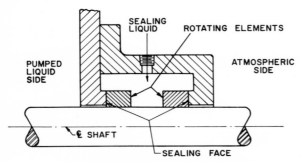

Fig. 9.6   Double mechanical seal

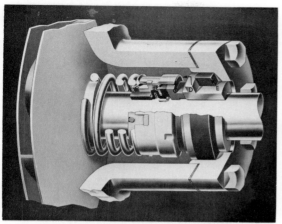

Fig. 9.8 Mechanical seal (rubber bellows type)

(Courtesy John Crane Co.)

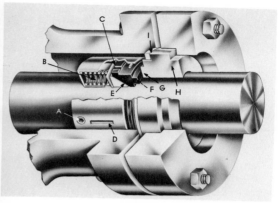

Fig. 9.9 Mechanical seal (Teflon type)

(Courtesy John Crane Co.)

which, in turn, is fixed to the shaft. The seal ring is fitted with packing in the shape of an O-ring, which prevents all leakage between the seal ring and the shaft. It is essential in such a seal that one face be flexibly mounted so as to keep the surfaces in full contact with reasonable shaft deflection. The springs and an ample clearance between the shaft and the seal ring proper accomplish this end in the unit illustrated. At the same time, the collar keeps all the other rotating members of the seal in proper position.

The gland insert in contact with the seal ring is made of antifrictional material. When necessary, it can be designed for lubrication by a liquid other than the liquid pumped.

A typical spring-loaded, synthetic rubber bellows mechanical seal is illustrated in Fig. 9.8. The tail of the synthetic rubber bellows (*A*) seals against leakage between the rotating element and the shaft; the head is flexible and adjusts automatically for washer wear and shaft end play. The protective ferrule (*B*) prevents the flexing area of the bellows from sticking to the shaft. The sealing washer (*C*) has a positive drive through metal parts and seals against the stationary floating seat (*E*). The two sealing faces (*D*) are lapped at the factory and provide a seal against leakage.

For extremely high- or extremely low-temperature applications (for instance, the circulation of a very low-temperature refrigerant), the bellows seal is not satisfactory. A Teflon seal is manufactured for this type of service (Fig. 9.9). The metal retainer locked to the shaft by set screws (*A*) provides a positive drive from the shaft to the carbon sealing washer (*F*) through dents (*D*), which fit into corresponding washer grooves. The seal between the shaft and the washer is insured by the Teflon wedge ring (*E*), which is preloaded by the action of multiple springs (*B*). The spring pressure is uniformly distributed by a metal disk (*C*). The lapped raised face of the rotating sealing washer mates against the precision lapped face (*G*) of the stationary seat (*H*) to provide a positive leakproof seal with minimum running friction between the faces. The spring pressure keeps the faces in constant contact, providing automatic adjustment for wear and shaft end play. A Teflon ring (*I*) acts as a static seal between the stationary seat and the end play. When this seal is applied to pumps on vacuum service or operating with a high suction lift, a supply of lubricating liquid is provided through a connection (Fig. 9.10). If the liquid pumped is clear, this tap can be connected to the discharge of the pump.

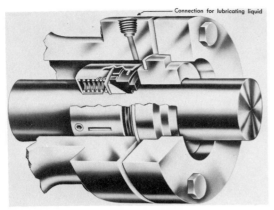

Fig. 9.10 Mechanical seal (Teflon type)

A connection for liquid lubrication of the seal faces is shown at the top of the figure. (Courtesy John Crane Co.)

A number of pump manufacturers have lines of single-stage end-suction pumps that can be equipped with mechanical seals instead of conventional packing. In many cases, such seals are offered as a standard alternative construction and can be applied

to the pumps without any change in machining. A few manufacturers build some sizes of standard pumps with mechanical seals only, but such practice makes these pumps less flexible as a line.

Cooling the seal

Cooling of the seal faces is important for satisfactory seal life, and a seal installed inside a pump without a suitably directed flow of liquid for cooling and flushing (to prevent solid material from settling on the springs, for instance) may have a high failure incidence.

A number of different methods are used to provide cooling and flushing. Sometimes all that may be necessary is to direct some of the pumped liquid at the sealing faces. When the pumped liquid is not suitable for this purpose, or when it must first be filtered, external circulation must be provided. Standard lines of pumps arranged

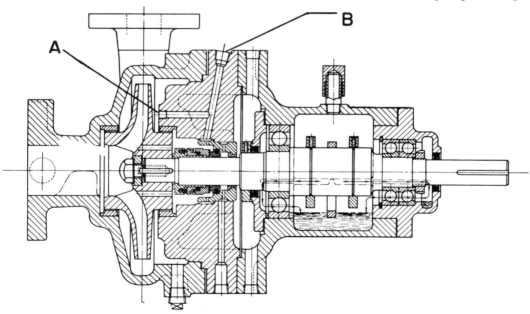

Fig. 9.11 Cooling circulation provisions for mechanical seal

KEY: (A) Use this circulation for cooling seal faces when clean liquid is pumped. For this installation, ½-in. pipe must be plugged; (B) provide ½-in. piping from discharge for cooling seal faces when liquid contains gritty particles. A strainer should be installed for keeping these particles out of seal area. For this installation, ¼-in. pipe must be plugged.

for mechanical seals are therefore generally built in such a manner that either type of cooling circulation can be provided (Fig. 9.11).

In both methods, the circulation is taken from a pressure higher than that in the stuffing box. This increased pressure provides positive circulation and prevents flashing at the seal faces caused by the heat generated by the seal when the pump handles liquids near their boiling point.

When the pumping temperatures reach 350°F, it is advisable to provide some means for cooling the chamber surrounding the seal as well. A typical arrangement for accomplishing this is shown in Fig. 9.12. The seal is equipped with a water-circulator ring and a heat-exchanger hookup to the seal chamber. The water-circulator ring acts as a miniature pump, causing the water to flow through the outlet piping located at the top of the seal chamber. The water then passes through the heat exchanger, from which it returns directly to the seal faces through the bottom inlet at the end-plate. As the water circulates back through the circulation ring, it picks up heat from the pump housing and the shaft. Because this is a closed circulating system, none of the hot pumped water enters the seal chamber.

Limitations on seal applications

The substitution of mechanical seals for packed stuffing boxes is not always an unmixed blessing, and for some services the mechanical seal is not as desirable as packing, including those with conditions tending to cause the pumped liquid to form crystals after temperature changes or on settling. If mechanical seals are used for such services, it is very important to provide adequate flushing. Another condition unfavorable to mechanical seals is a pump service with long idle periods, when the pump may even be drained. The flexible materials used in the seal may harden or slight rusting may occur, thereby possibly

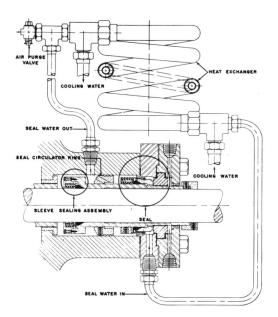

Fig. 9.12 Mechanical seal with external cooling arrangement

(*Courtesy John Crane Co.*)

causing the seal to stick and become damaged on restarting. Finally, mechanical seals are still subject to failure on occasion, and their failure may be more rapid than that of conventional packing. If packing fails, the pump can usually be kept running by temporary adjustments until it is convenient to shut it down. If a mechanical seal fails, the pump must be shut down at once in nearly every case.

The mechanical seal has its place in centrifugal pumps as it will effectively prevent stuffing box leakage when properly applied under favorable conditions.

Maintenance of mechanical seals

Those responsible for maintenance of pumps employing mechanical seals should carefully read the seal manufacturer's instructions for the operation and maintenance of the seals. An adequate supply of spare parts (according to the service and equipment availability) should be on hand so that the seal can be repaired if it fails.

10 *Condensate Injection Sealing*

The possibility of applying mechanical seals to boiler feed service has long attracted interest. As a matter of fact, a certain number of such installations were made some years ago. But results had been inconclusive until recently because of the need for long-time operation to prove their acceptability. Another serious difficulty was the lack of experience in the operating speeds of 6,000 to 9,000 rpm that have become commonplace for boiler feed pumps.

Before these high-speed pumps could be built and operated, on the other hand, the problem of stuffing boxes had to be solved, because it was impossible to operate conventional packed boxes successfully at such speeds. It was this requirement that led to the development of the condensate injection sealing design, or "packless" stuffing boxes as they are commonly called.

The construction of a pump with condensate injection sealing is illustrated in Fig. 10.1. A labyrinth breakdown bushing is substituted for the conventional packing, and the pump shaft sleeve runs within this bushing with a reasonably small radial clearance. Cold condensate, available at a pressure in excess of the boiler feed pump suction pressure, is introduced centrally in this breakdown bushing. A small portion of

the injection water flows inwardly into the pump proper; the remainder flows out into a collecting chamber that is vented to the atmosphere. From this chamber, the leakage is piped back to the condenser.

Source of supply

Cold condensate (at temperatures from 80° to 100°F) is available at pressures in excess of the boiler feed pump suction pressure in closed cycles (Fig. 10.2) as well as open cycles (Fig. 10.3), in which the condensate pump discharges into a deaerating heater from which the pump takes its suction. The water for the injection in both should be taken immediately from the condensate pump discharge before it has gone through any closed heaters. It is preferable to use injection water at temperatures below 120°F so as to avoid the slight steaming at the seal covers (steaming takes place if injection temperatures in excess of that figure are used). This steaming is undesirable partly because of the concern it may arouse in the operators and partly because of the possibility of its condensation near the pump bearings.

Pumps arranged for condensate injection sealing, moreover, are usually not provided

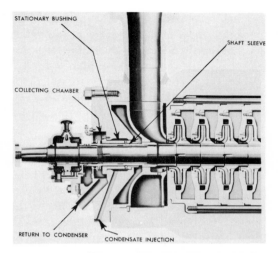

Fig. 10.1 Packless stuffing box construction for high-pressure boiler feed pump

with any other cooling means in that area. If the shaft sleeves rotating within the condensate seal breakdown bushing are not adequately cooled, the heat from inside the pump will travel through the shaft to the pump bearings and may be injurious to the bearing life.

In the closed feedwater cycle, all deaeration takes place in the condenser, and therefore the injection water is fully deaerated. Although this is not quite true in the open cycle, an appreciable amount of deaeration takes place in the condenser even though a deaerating heater is provided in the feed cycle. Thus the injection condensate in a modern steam power plant will contain almost no oxygen (0.01 cc per liter or less). As the saturation level at a temperature of 100°F and atmospheric pressure is about 4.7 cc per liter, the oxygen in the injection supply itself has no significance. Moreover, the amount of injection water that enters the pump proper and is not returned to the condenser is very small. Thus no appreciable contamination of the feedwater will take place through the condensate injection sealing.

The amount of the injection water will depend upon (1) the diameter of the running joint between the shaft sleeve and the pressure breakdown bushing, (2) the clearance at that running joint, and (3) the injection pressure. To give some general idea of the values in question, if the sleeve diameter is 5 in. and the diametral clearance 0.009 in., the amounts measured in a 3,600-rpm pump will be approximately as follows:

1. Total injection, per box—8 to 10 gpm
2. Leakage into the pump interior, per box—2 to 4 gpm

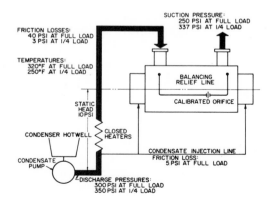

Fig. 10.2 Application of injection packless boxes to a closed feedwater cycle

Pressure distribution indicated for full and one-quarter load.

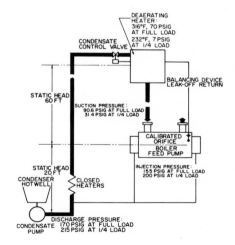

Fig. 10.3 Application of injection packless boxes to an open feedwater cycle

Pressure distribution indicated for full and one-quarter load.

3. Return to condenser, per box—6 to 8 gpm.

It is essential that the injection supply be absolutely clear and free of foreign matter. It is, therefore, necessary to install filters or strainers in the injection line to avoid the entrance of fine mill scale or oxide particles into the close clearances between the stationary bushings and the sleeves. Pressure gages should be installed upstream and downstream of these filters to permit the operator to follow the rate at which foreign matter clogs up the filters and to clean these when the pressure drop across them becomes excessive.

Drains from condensate injection sealing

Two different systems are used to dispose of the drains coming from the collecting chambers. The first utilizes traps that drain directly to the condenser. The second collects the drains in a condensate storage tank into which various other drains are also returned. As this tank is under atmospheric pressure, it must be set at a reasonable elevation below the pump centerline so that the static elevation difference will overcome friction losses in the drain piping. A pump then transfers the condensate drains from the storage tank into the condenser.

To our knowledge, no specific difficulties have ever taken place in installations in which the injection sealing condensate is evacuated through traps, except for an isolated case of trap malfunctioning. Proper maintenance of this equipment should hold such occurrences to an absolute minimum. A minor problem may arise if boiler feed pumps are operated during the start-up before condenser vacuum is established. This operation causes a rise in the back pressure on the seal drains and, unless provision is made to relieve this back pressure, some overflow of injection condensate may take place at the collecting chamber covers.

Neither system of evacuation has major advantages over the other, and the choice

between the two is dictated primarily by personal preferences. However, if the boiler feed pumps are located at the lowest plant elevation (in some outdoor plants, for instance), it becomes necessary to use traps because there is insufficient elevation difference to drain from the collecting chambers into an open tank.

The clearances between the sleeves and the breakdown bushings will double in a time approximately equal to the life of the internal wearing parts. With doubled clearances, the leakage will double. This factor should be considered when sizing the return drain piping back to the condenser or to the collecting tank if friction losses are to be kept to a minimum in this piping. The collecting chamber at the pump stuffing box is vented to the atmosphere; the only head available to evacuate it is the static head between the pump and the point of return. This head must always be well in excess of the frictional losses (even after the leakage doubles); otherwise the drains will back up and run off at the collecting chamber.

The water thrown off the shaft into the ventilated collection chamber will probably reach 75 per cent saturation or more before it reaches the drain pipes. Assuming that there is a considerable length of partially full piping between the collecting chamber and the trap in the drain line to the condenser, the oxygen content of the returned condensate can well be assumed to have 100 per cent saturation. Although this figure may seem very high, the deaerating capacity of a modern condenser is greatly in excess of average requirements and should be amply capable of handling the oxygen in a saturated return, for it makes up 2 per cent or less of the normal flow. Its presence, nevertheless, has made it necessary in a few installations to provide breather pipes in the drain lines to avoid the accumulation of air pockets caused by the separation of the entrained air. Until these breather pipes were installed, the evacuation of the drains was erratic, and some "cyclic" spillover took place at the seal covers each time that a slug of air formed in the drain piping. A small "geyser" would rise from the collecting chamber vents; building up of the static head in the collecting chambers would then force the slug of air out through the piping, and the geyser would subside. The installation of a breather, or "burp" pipe, at a suitable location eliminated the difficulty entirely.

Control of the injection

For all the advantages to be gained from packless stuffing boxes in maintenance and availability, it may still sometimes be desirable to limit the amount of injection water to as little as possible. Throttling control valves are used in the injection lines in some installations for this purpose. Their function is to maintain a constant minimum differential between the injection pressure and the pressure at the pump interior. A separate throttling valve is sometimes installed at each injection point, that is, two valves per pump. Other times the desire for simplifying the installation has led to the use of two valves for each group of pumps serving a common turbo-generator unit. Two valves are used because the pressure on the suction and discharge sides will not necessarily be the same. An examination of Fig. 10.1 will show that the

inward flow on the suction side has to overcome a pressure exactly equal to the pump suction pressure. On the discharge side, the flow will proceed into the balancing-device relief chamber. The pressure there will exceed the suction pressure by the amount of loss through the calibrated orifice in the balancing relief line, which is used to measure the leakage past the balancing device. This loss may be quite appreciable after the pump has become worn.

The injection control arrangement with two valves per set of pumps is illustrated in Fig. 10.4. On the suction side, a pilot line is connected to the suction header. On the discharge side, a header is provided with lines from each balancing-device relief chamber and a small check valve in each line so that the highest of the three back pressures is measured. There is always the possibility that one of the pumps will have experienced greater wear in the balancing device, and therefore a greater pressure drop will take place through the leakoff orifice. Then the pressure in the balancing relief chamber of that pump will be greater than in the other two pumps. If all three pilot lines were connected, there would be some crossflow from the relief chamber under the highest pressure to the other two chambers. The individual check valves prevent this. The highest pressure of the three

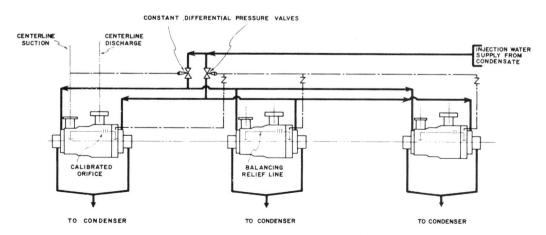

Fig. 10.4 Typical method of controlling injection flow in packless stuffing boxes

relief chambers will be registered, but no flow will take place through these lines into the remaining pumps.

It is of prime importance to hold the pressure drop through the control valve to a minimum if little pressure difference exists between the condensate pump discharge and the feed pump suction when the control valve is wide open. This consideration dictates the use of a relatively large valve. It is also important to avoid control valve complications or possibility of malfunctioning, because cutting off the injection water to a pump handling high-temperature water is not to be tolerated. Obviously, if no cold water is available to seal off the leakage from inside the pump, water at the temperature handled by the pump (sometimes as high as 400°F) would flow through the breakdown, flash, and envelop the pumps in a cloud of steam. As already stated, the interruption of injection flow and the consequent heating of the shaft sleeves, shaft, and bearings can have serious effects on the pump.

We must now examine the relationship between the pressure available at the injection of the condensate and the internal pressure over the complete range of operating station loads. For reasons that will become obvious, this relationship is vastly different for open and for closed feedwater cycles. The application of packless boxes to the two cycles should thus be studied separately.

Open cycle

The relationship of the various pressures under consideration is illustrated in Fig. 10.3, which represents a typical installation with injection labyrinths in an open cycle. The condensate pump discharges into a deaerating heater through a series of closed heaters. The discharge pressure at the condensate pump rises from 170 psig at full load to 215 psig at one-quarter load. The static head between the condensate pump and the deaerating heater is 80 ft, or ap-

proximately 35 psi. The friction losses in the piping and through the closed heaters are 45 psi at full load and only 3 psi at one-quarter load. Thus the pressure immediately ahead of the heater will be 90 psig at full load and 177 psig at one-quarter load. The condensate control valve located at the entrance to the deaerating heater will vary the admission of condensate in accordance with load requirements, throttling off approximately 20-psi pressure at full load and as much as 170-psi pressure at one-quarter load.

The boiler feed pump centerline is located 60 ft below the waterlevel in the deaerating heater. The friction losses in the suction piping are 3 psi at full load and 0.2 psi at one-quarter load. Thus, the suction pressure at the boiler feed pump varies from 90.6 psig at full load down to 31.4 psig at one-quarter load. The internal pressure on the suction side of the pump varies the same.

On the discharge side, the internal pressure will be somewhat higher. In an open feedwater cycle, the balancing device leak-off is returned to the deaerating heater. The calibrated orifice in this return line can be assumed to have a loss of 5 psi and the return piping itself another 5-psi friction loss. Thus the internal pressure on the discharge side will exceed that on the suction side by approximately 10 psi and will therefore range from 100.6 psig at full load to 41.4 psig at one-quarter load.

As the condensate pump is located 20 ft below the boiler feed pump, the injection pressure will range from 155 psig at full load to 200 psig at one-quarter load if we assume friction losses of approximately 6 psi in the injection line. Thus the injection pressure will exceed the internal pressures at all loads. If it is desirable to minimize both the amount of inward flow and of the condensate being returned to the condenser and being repumped, the injection lines may well be provided with control throttling valves. If it is desirable to maintain,

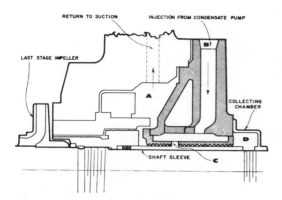

Fig. 10.5   Section through packless stuffing box on the discharge side

KEY: (A) Suction pressure plus pressure drop through calibrated orifice in balancing device leak-off line; (B) discharge pressure of condensate pump, less friction loss through supply piping; (C) essentially same as (B); (D) essentially atmospheric pressure.

let us say, a 5-psi differential between the injection pressure and the internal pressure, the amounts of pressure to be throttled will be the following:

1. At the suction side—64.4 psi at full load and 168.6 psi at one-quarter load
2. At the discharge side—54.4 psi at full load and 158.6 psi at one-quarter load.

There should be no difficulty in selecting control valves that will maintain the desired pressure differential, and the rather high value of the pressure to be throttled will permit selection of a reasonably small valve for the purpose.

Closed cycle

When the packless stuffing box is applied to a closed feedwater cycle, the conditions prevailing at the discharge (balancing-device) side of the stuffing box are actually more severe than at the suction side (Fig. 10.5). At full load and at some reduced load conditions, the injection pressure, P_B, is greater than the pressure in the balancing-device relief chamber, P_A, because of the friction losses between the discharge of the condensate pump and the suction of the

boiler feed pump. As the pressure, P_C, is essentially the same as the pressure at the injection point, it also exceeds the pressure in the balancing-device relief chamber, and flow takes place inwardly from point C to point A as well as outwardly from point C into the collecting chamber, D.

As the load is reduced, the friction losses between the condensate pump and the main feed pump decrease approximately with the square of the capacity. Thus at some extremely low pump loads, the boiler feed pump suction pressure may be only 1 or 2 psi lower than the condensate pump discharge pressure (neglecting the static elevation difference, which is the same for both suction and injection piping).

A typical example of what happens is shown in Fig. 10.2. At full load, there is ample excess pressure in the injection line to produce flow into the pump. When the load is reduced to one-quarter flow, the following situation prevails:

1. At the suction side, the suction pressure becomes 337 psi whereas the discharge pressure of the condensate pump (less the static head) is 340 psi. If the injection flow remains essentially unchanged and the friction losses in the injection piping are still assumed to be 5 psi, the available injection pressure is reduced to 335 psi, or 2 psi less than the suction pressure.

2. At the discharge side, the pressure in the balancing-device relief chamber is equal to the suction pressure plus the loss through that calibrated orifice in the balancing relief line which is used to measure the leakage past the balancing device. When the pump is new, this loss is approximately 5 psi, and the pressure in this chamber at full load is 255 psi. The injection-line pressure is 285 psi, and a 30-psi differential is available to cause inward flow of injection water. At one-quarter load, the pressure in the relief chamber is 342 psi, whereas the injection-line pressure is only 335 psi, or 7 psi less than the relief chamber pressure.

If the pump is worn and the pressure drop through the calibrated orifice is permitted to go up to 35 psi, the relief chamber pressure will become 285 psi, balancing exactly the injection line pressure. At one-quarter load, the relief chamber pressure becomes 372 psi whereas the injection pressure is only 335 psi. Thus no excess pressure exists in a worn pump between the injection pressure and the relief chamber pressure under any load conditions.

What actually does take place, then, in the packless stuffing box under these conditions? As the difference between pressures at C and A in Fig. 10.5 diminishes, less and less flow takes place inwardly from point C to point A. Finally, a condition prevails in which the pressure at A slightly exceeds the pressure at C. At that time, a small amount of feedwater flows from point A to point C and mixes with the injection water. The mixture proceeds as before towards the collecting chamber, from which it is returned to the condenser.

Because of the breakdown between points A and C and because the pressure differential between them is only a fraction of the pressure differential between points C and D, the amount of this "reverse" flow is relatively small and should not raise the temperature of the mixture appreciably. Thus, even though the reduction of the pressure drop between the condensate pump discharge and the boiler feed pump suction

Fig. 10.7   Floating seal ring design

results in a change of flow direction between points C and A, the operation of the packless stuffing box remains acceptable.

This description of the flow process eliminates the effects of the balancing device on the pressure at point A for the sake of simplicity. Actually, this device develops a pumping action that leaves the pressure at the shaft sleeve near point A some 20 to 30 psi below the pressure at the periphery of the relief chamber. Figure 10.6, which shows the inward leakage (from points C to A) under varying pressure differences between injection line and the relief chamber, graphically illustrates this condition. Paradoxically, some flow still takes place inwardly, even though the pressure difference is negative, as long as the pump is running. When the pump is idle, of course, the condition disappears.

Of course, if a continuous flow of injection water from point C to point A is de-

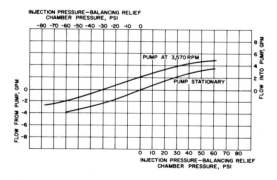

Fig. 10.6   Effect of pump rotation on injection flow

sired regardless of load, a small booster pump should be installed in the injection line, taking its suction from the condensate pump discharge and raising that pressure by some 50 psi. It should be noted that no flow takes place through the balancing device when the pump is idle and that the inward pressures at the suction and discharge sides are essentially equal. Thus there should be no problem in maintaining inward sealing flow to a pump kept idle on standby service.

The condensate injection sealing arrangement, which has been so successfully applied to boiler feed pump service, is of course applicable to a number of other services; in fact, it is rapidly becoming a worthy competitor to the mechanical seal. For instance, it is very suitable for cold-water high-pressure pumps applied to hydraulic descaling or hydraulic press work. In such services, of course, there is no need to bring in injection supply water to the breakdowns (unless this water is not clear and free of gritty material), for the water handled by the pump is already cold.

Mechanical modifications

A number of mechanical modifications have been developed by various pump manufacturers. One of these, illustrated in Fig. 10.7, consists of substituting a stack of individual solid rings for the serrated breakdown bushing shown in Fig. 10.1. Each ring is mounted in a holder, spring-loaded to produce a stationary seal face in an axial direction, and locked against rotation by a pin-and-slot arrangement. A small radial clearance is provided between the rings and the shaft sleeve. The length of each ring varies with the diameter of the condensate injection seal but is generally about $\frac{1}{2}$ in. The advantage of this construction is that the individual seal rings are "floating" to a certain degree and can find their own position relative to the shaft. Their short length reduces the effect of angular displacement between the stationary and rotating components, whether this displacement arises from errors in original assembly or from distortions caused by temperature changes.

11   *Bearings*

The function of bearings in centrifugal pumps is to keep the shaft or rotor in correct alignment with the stationary parts under the action of radial and transverse loads. Those that give radial positioning to the rotor are known as *line bearings,* whereas those that locate the rotor axially are called *thrust bearings.* In most applications, the thrust bearings actually serve both as thrust and radial bearings.

Types of bearings used

All types of bearings have been used in centrifugal pumps. Even the same basic design of pump is often made with two or more different bearings, required either by varying service conditions or the preference of the purchaser.

Two external bearings are usually used for the double-suction, single-stage general-service pump, one on either side of the casing. These were formerly of the babbitted type using oil lubrication, but in recent years most manufacturers have changed to antifriction bearings using either grease or oil lubrication.

Some of the small, inexpensive centrifugal pumps used for pumping clear liquids are provided with an internal sleeve bearing (Fig. 11.1). The liquid itself is used as a lubricant, although separate grease lubrication through an alemite fitting is used in some designs.

In horizontal pumps with bearings on each end, the bearings are usually designated by their location as *inboard* and *outboard bearings,* the former being those between the casing and the coupling. Pumps with overhung impellers have both bearings on the same side of the casing so that the bearing nearest the impeller is called inboard and the one farthest away outboard. In a pump provided with bearings at both ends, the thrust bearing is usually placed at the outboard end and the line bearing at the inboard end.

The bearings are mounted in a housing that is usually supported by brackets attached to or integral with the pump casing. The housing also serves the function of containing the lubricant necessary for proper operation of the bearing. Occasionally, the bearings of very large pumps are supported in housings that form the top of pedestals mounted on soleplates or on the pump bedplate. These are called pedestal bearings.

Because of the heat generated by the bearing itself or the heat in the liquid being

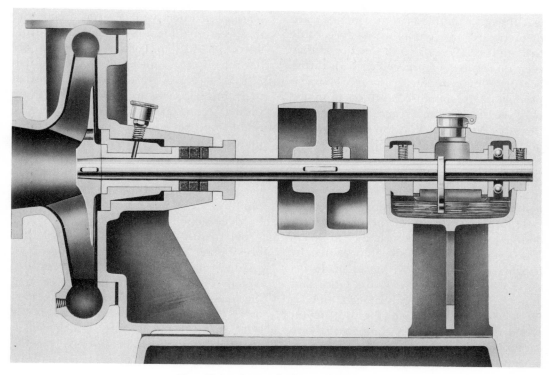

Fig. 11.1   Internal sleeve bearing

pumped, some means other than radiation to the surrounding air must occasionally be used to keep the bearing temperature within proper limits. If the bearings have a forced-feed lubrication system, cooling is usually accomplished by circulating the oil through a separate water-to-oil cooler. Otherwise, a jacket through which a cooling liquid is circulated is usually incorporated as part of the housing.

Pump bearings may be rigid or self-aligning. A self-aligning bearing will automatically adjust itself to a change in the angular position of the shaft. In babbitted or sleeve bearings, the name "self-aligning" is applied to bearings that have a spherical fit of the sleeve in the housing. In antifriction bearings, it is applied to bearings the outer race of which is spherically ground or the housing of which provides a spherical fit.

The most common antifriction bearings used on centrifugal pumps are the various types of ball bearings. Roller bearings are used less often, although the spherical roller bearing (Fig. 11.2) is used frequently for large shaft sizes for which there is a limited choice of ball bearings. As most roller bearings are suitable only for radial loads, their use on centrifugal pumps tends to be limited to applications in which they are not required to carry a combined radial and thrust load.

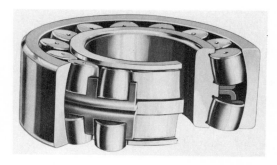

Fig. 11.2   Self-aligning spherical roller bearing
(Courtesy SKF Industries, Inc.)

Although double-suction pumps are theoretically in hydraulic balance, this balance is rarely realized in practice so that even these pumps are provided with thrust bearings. A centrifugal pump, being a product of the foundry, is subject to minor irregularities that may cause differences in the eddy currents set up on the two sides of the impeller. As this disturbance can create an axial hydraulic thrust, some form of thrust bearing that is capable of taking thrust in either direction is necessary to maintain the rotor in its proper position.

The thrust capacity of the bearing of a double-suction pump is usually far in excess of the probable unbalance caused by irregularities. This provision is made because (1) unequal wear of the rings and other parts may cause unbalance, and (2) the flow of the liquid into the two suction eyes may be unequal and cause unbalance because of an improper suction piping arrangement.

BALL BEARINGS

Basic principles

As the coefficient of rolling friction is less than that of sliding friction, one must not consider a ball bearing in the same light as a sleeve bearing. In the former, the load is carried on a point contact of the ball with the race, but the point of contact does not rub or slide over the race and no appreciable heat is generated. Furthermore, the point of contact is constantly changing as the ball rolls in the race, and the operation is practically frictionless. In the sleeve bearing there is a constant rubbing of one surface over another, and the friction must be reduced by the use of a lubricant.

Ball bearings operated at an absolutely constant speed theoretically would require no lubricant. However, no speed can be called absolutely constant, for the conditions affecting the speed always vary slightly. For instance, a motor with a full-load speed rated at 3,510 rpm might vary in speed in the course of a minute from 3,505 to 3,515 rpm. Each variation in speed has the effect of causing the balls in a ball bearing to lag or lead the race because of their inertia. Consequently, a very slight sliding action takes place, almost immeasurable but nevertheless existent. Another limiting condition is that the hardest of metals suffer minute deformations on carrying load, thus upsetting perfect point contact and adding another slight sliding action. For these reasons ball bearings must be given some lubrication.

Ball thrust bearings are built to carry heavy loads by pure rolling motion on an angular contact. As thrust load is axial, it is equally distributed to all the balls around the race, and the individual load on each ball is only a very small fraction of the total thrust load. In such bearings it is essential that the balls be very equally spaced, and for this purpose a retaining cage is used between the balls and between the inner and outer race. This cage carries no load, but the contact between it and the ball produces sliding friction that generates a small amount of heat. It is for this reason that ball thrust bearings are generally water-jacketed.

Types and applications

A pump designer has a wide variety of antifriction bearings to choose from. Each type has characteristics that could make it a good or bad selection for a specific application. Although several types might sometimes be acceptable, it is best for purchasers to leave the choice to the manufacturer. For example, some purchasers specify double-row bearings whatever the size or type of pump, even though single-row bearings are often equally suitable or better.

The most common ball bearings used on centrifugal pumps are (1) the single-row, deep-groove, (2) the double-row, deep-groove, (3) the double-row, self-aligning, and (4) the angular-contact, either single-

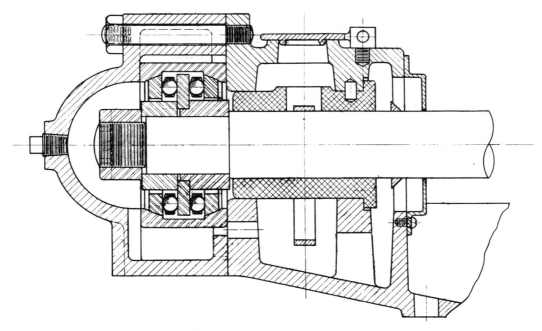

Fig. 11.3 Ball thrust bearing used with sleeve bearing

or double-row. All except the double-row, self-aligning bearings are capable of carrying thrust loads as well as radial loads. The ball thrust bearing of the type shown in Fig. 11.3 was formerly used for carrying thrust only (in conjunction with babbitt sleeve bearings) but is now rarely used on centrifugal pumps.

Sealed ball bearings, adapter ball bearings, and other modifications have also found special applications. It is important to remember that sealed prelubricated bearings require special attention if the unit in which they are installed is not operated for a long period of time (for in-

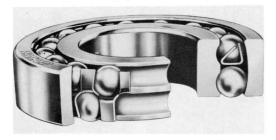

Fig. 11.4 Self-aligning double-row ball bearing

(Courtesy SKF Industries, Inc.)

stance, one kept in stock or storage). The shaft should be turned over occasionally, say once every three months, to agitate the lubricant and maintain a film coating of the balls of such units.

The self-aligning ball bearing (Fig. 11.4) is the most serviceable bearing for heavy loads, high speeds, long-bearing spans, and no end thrust. For this reason, it is ideally adapted for service as a line bearing on a centrifugal pump. Its double row of balls runs in fixed grooves in the inner or shaft race; its outer race is ground to a spherical seat. Any slight vibration or shaft deflection is therefore taken care of by this bearing, which operates as a pivot. In lightly constructed pumps, it will also compensate for the slight misalignment caused by the "breathing" that takes place in the casing when pressure is developed.

The self-aligning ball bearing has proved very satisfactory for high speeds and has long life, even with long-bearing spans. It has very little thrust capacity, however, and is not used for combined radial and thrust loads in centrifugal pumps. For large

Fig. 11.5 Single-row deep-groove ball bearing

(Courtesy SKF Industries, Inc.)

shafts, the self-aligning spherical roller bearing (Fig. 11.2) is used instead, for it can carry such loads with a considerable thrust component.

The single-row, deep-groove ball bearing (Fig. 11.5) is the most commonly used bearing on centrifugal pumps except for the larger sizes. It is good for both radial, thrust, and combined loads but requires careful alignment between the shaft and the housing in which the bearing is mounted. It is sometimes used with seals built into the bearing in order to exclude dirt, retain lubricant, or both.

The double-row, deep-groove ball bearing—in effect, two single-row bearings placed side by side—has greater capacity both for radial and thrust loads (Fig. 11.6). It is used quite commonly if the loading is more than that permitted by the single-row bearing.

The angular-contact ball bearing operates on a principle that makes it good for heavy thrust loads. The single-row type

Fig. 11.6 Double-row deep-groove ball bearing

(Courtesy SKF Industries, Inc.)

(Fig. 11.7) is good for thrust in only one direction, whereas the double-row type (Fig. 11.8), which is basically two single-row bearings placed back-to-back, can carry thrust in either direction (Fig. 11.9). Two single-row angular-contact bearings are frequently matched and the faces of the races ground by the manufacturer so they can be used in tandem for large, one-directional thrust loads or back-to-back (Fig. 11.10) for two-directional thrust loads. The two bearings are sometimes locked together by recessing the inner races and pressing them onto a short ring (Fig. 11.11). If two separate angular-contact bearings are used,

Fig. 11.7 Single-row angular-contact ball bearing

(Courtesy New Departure.)

care must be taken to mount them correctly on the shaft.

The single-row angular-contact ball bearing can be used singly on centrifugal pumps only if the thrust is always in one direction. Its field of application is thus limited primarily to vertical pumps. Another very interesting application is the use of two such bearings in an end-suction pump to take care of axial thrust in either direction. This arrangement permits a certain amount of axial adjustment of the impeller in its volute, accomplished by loosening up one bearing nut and tightening the other. Unfortunately, such adjustment is extremely delicate and requires a first-class mechanic; its commercial practicability is therefore somewhat limited.

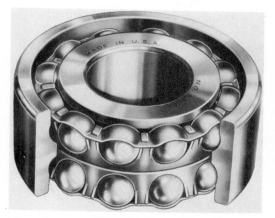

Fig. 11.8 Double-row angular-contact ball bearing

(Courtesy New Departure.)

Fig. 11.10 Two single-row angular-contact bearings mounted back-to-back

(Courtesy New Departure.)

The double-row angular-contact bearing, or its equivalent of a matched pair mounted back-to-back, has been found very satisfactory for pumps capable of a high thrust load in either direction. Some pump manufacturers standardize on this bearing for many applications.

Lubrication of antifriction bearings

In the layout of a line of centrifugal pumps, the choice of the lubricant for the pump bearings is dictated by application requirements, by cost considerations, and sometimes by the preferences of a group of purchasers committed to the major portion of the output of that line.

For example, the application of vertical wet-pit condenser circulating pumps dictates the choice of water in preference to grease or oil lubrication. If oil or grease were used in such pumps and the lubri-

cant leaked into the pumping system, the condenser operation might be seriously affected because the tubes would become coated with the lubricant.

Most centrifugal pumps for refinery service are presently supplied with oil-lubricated bearings because of the insistence of refinery engineers on this feature. In the marine field, on the other hand, the preference lies with grease-lubricated bearings. For very high pump operating speeds (5,000

Fig. 11.9 Method of operation of angular-contact ball bearing

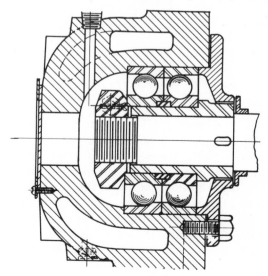

Fig. 11.11 Double row angular-contact ball thrust bearing (grease lubricated, water cooled)

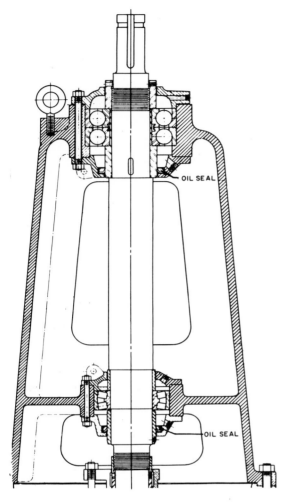

Fig. 11.12   Ball-bearing construction with seal in vertical pump

Seal guards against escape of grease.

inner race. (Even if the grade of grease is relatively light, it is still a semisolid and flows slowly. As heat generates in the bearing, however, the flow of the grease is accelerated until the grease is thrown out at the outer race by the rotation.) As the expelled grease is cooled by contact with the housing and thus attracted to the inner race, there is a continuous circulation of grease to lubricate and cool the bearing. This method of lubrication requires a minimum amount of attention and has proved itself very satisfactory.

As housings of bearings in vertical pumps require seals to prevent the escape of the lubricant, grease is usually preferred, for it lessens the chance of leakage (Fig. 11.12).

A bearing fully packed with grease prevents proper grease circulation in itself and its housing. As a rough rule, therefore, it is recommended that only one-third of the void spaces in the housing be filled. An excess amount of grease will cause the bearing to heat up, and grease will flow out of the seals to relieve the situation. Unless

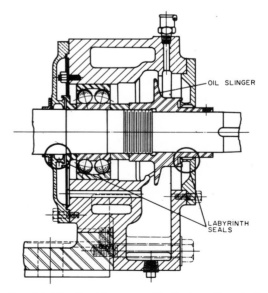

Fig. 11.13   Double-row self-aligning line ball bearing (oil lubricated, water cooled)

An oil slinger circulates the oil; labyrinth seals prevent loss of oil.

rpm and above), oil lubrication is found to be the most satisfactory.

For highly competitive lines of small pumps, the main consideration is cost, and the most economical lubricant is chosen, depending upon the type of bearing used.

Ball bearings used in centrifugal pumps are usually grease-lubricated, although some services use oil lubrication. In grease-lubricated bearings, the grease packed into the bearing is thrown out by the rotation of the balls, creating a slight suction at the

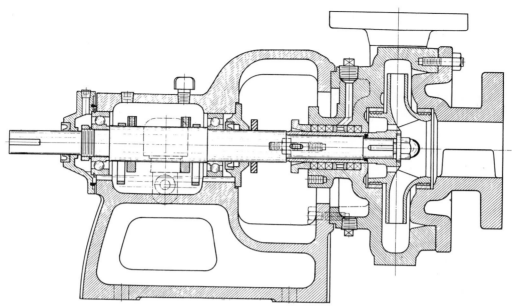

Fig. 11.14 Ball bearing pump with oil rings

the excess grease can escape through the seal or through the relief cock that is used on many large units, the bearing will probably fail early. More trouble is usually caused in grease-lubricated bearings by an excess of lubrication than by a lack of it— an exceptional situation in which too much attention is worse than too little.

Oil-lubricated ball bearings require an adequate method for maintaining a suitable oil level in the housing. This level should be at about the center of the lowermost ball of a stationary bearing. It may be achieved by a dam and an oil slinger to maintain the level behind the dam and thereby increase the leeway in the amount of oil the operator must keep in the housing (Fig. 11.13). Oil rings are sometimes used to supply oil to the bearings from the bearing housing reservoir (Fig. 11.14). In other designs, a constant-level oiler is used (Fig. 11.15).

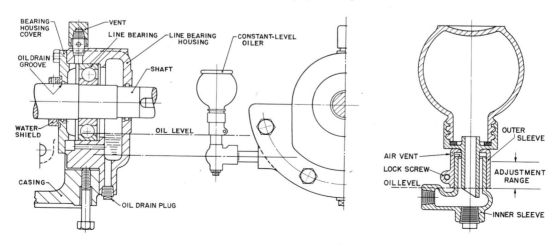

Fig. 11.15 Constant-level oiler

As ball bearings (unlike sleeve bearings) do not require forced-feed lubrication, their lubrication systems do not need to be discussed.

Maintenance of antifriction bearings

If properly applied and lubricated, antifriction bearings in centrifugal pumps have long life and are unusually troublefree. Failure can result, however, from the following: (1) Use of the wrong type or size for a particular application, (2) faulty mounting because of improper workmanship in manufacture or during maintenance, (3) improper design of the mounting, (4) improper lubricant or lubricating practice, (5) entry of water, dirt, or grit into the bearing, and (6) mechanical damage to the balls, rollers, or races.

Pump designers base their selection of bearing type, size, and lubrication to suit the field or fields of service for which the lines of pumps will be used. Occasionally through a misunderstanding, a pump will be used for conditions or in surroundings not suitable for its bearing design and consequently suffers from short bearing life.

The inner race of antifriction bearings must not turn on the shaft; the outer race must not turn in its housing; and the bearing must be in correct alignment. Antifriction bearings are usually pressed or shrunk on their shafts; if thrust loads are involved, they are further held in axial position on their shafts by shoulders and shaft nuts. If the shaft is undersize, the fit will be too loose, allowing rotation of the inner race on the shaft with resulting damage to the bearing, the shaft, or both. On the other hand, too large a shaft diameter can result in expansion of the inner race, causing insufficient clearances between the balls or rollers and their inner and outer races. Likewise, the mounting must provide sufficient holding force through proper gripping of the outer race in the housing to prevent the outer race from turning in the housing. This force is generally more of a

problem with radial bearings than with combined radial and thrust bearings or straight-thrust bearings because the outer race is clamped between two shoulders in the housing assembly if thrust is involved. In radial bearings, however, the outer race must be able to move axially in its housing if temperature changes cause unequal expansion of the shaft and casing. The fit of the outer race in its housing is therefore in the nature of a push fit. It is also very important for antifriction bearings to be squarely mounted on their shafts and in their housing and to not be cocked. A pump designer has to make sure that the casing will not distort unduly when pressure is applied, as distortion would throw the bearing out of line. Antifriction bearings have close tolerances; pump design and workmanship must meet them.

Some bearing housing designs incorporate means to assure initial centering of the bearings (Fig. 11.16). In such cases, proper alignment must be reached after the pump is reassembled with new bearings.

Many failures of antifriction bearings (like other bearings) can be traced to the use of improper lubricants. The following extracts from SKF Catalog No. 350 outline the basic requirements for suitable greases and oils for antifriction bearings:

> Only good quality, neutral mineral greases and mineral oils should be used. Generally speaking, lubricants of animal or vegetable oils should not be used because of possible damage by deterioration or the forming of acid.
>
> Greases satisfactory for antifriction bearings are, broadly speaking, divided into two classes, those with a lime soap base and those with a soda soap base. At temperatures between 32° and 115°F, a properly made lime soap grease is usually satisfactory. Above a temperature of 115°F, lime soap grease used for any length of time is unsuitable, because the oil will separate from the base and the residue is detrimental to the bearing.
>
> For higher temperatures, a soda soap base is necessary. A suitable grease of this type can be used at temperatures from minus 5° to plus 160°F and, for short periods, up to

210°F. If this grease melts, it will separate but will return to its original state when the temperature is lowered. Under ordinary temperatures, this grease is in most cases comparatively stiff; it therefore facilitates sealing of the housing, particularly if the labyrinth form of design is used. It also has emulsifying properties so that a certain amount of water can be absorbed, thus forming an excellent protection for the bearing against rust.

The many greases available on the market are of different qualities and compositions. It is therefore necessary to select the kind of grease carefully in order not to jeopardize dependable bearing service.

A good bearing grease must have the following properties:

1. Freedom from chemically or mechanically active ingredients such as free lime, iron oxide, and similar mineral or solid substances of any character.

2. The slightest possible tendency of change in consistency, whether by thickening, separation of oil, acid formation, hardening, or the like.

3. A melting point considerably higher than the operating temperature.

The choice of lubricating oils is easier. They are more uniform in their character and if they are resistant to oxidation, gumming and evaporation can be selected primarily on the basis of a suitable viscosity.

To maintain lubrication safely, use larger quantities of oil, usually of somewhat higher viscosity in order to reduce losses from evaporation or leakage.

It is a good rule to select an oil that will have a satisfactory viscosity at the operating temperature, keeping in mind that the temperature of the oil is usually 5° to 10°F higher than that of the bearing housing. If possible, this viscosity at operating temperature should be at least 70 Saybolt Seconds for ball bearings, 100 Saybolt Seconds for radial roller bearings, and 150 Saybolt Seconds for spherical roller thrust bearings.

Care should be exercised to prevent water from entering the bearing. If water gets into the housing—except for small amounts with soda-soap base grease lubricant—the bearing parts are sure to become rusted and hence fail. Too much cooling of the housing has been known to cause

condensation of atmospheric moisture inside the housing. In liquid-jacketed bearings, the flow of cooling liquid should be regulated so that the bearing is reasonably warm and the supply cut off when the pump is idle.

Dirt or grit allowed into the bearing will naturally cause damage. As grease makes a good seal against dust and dirt, grease lubrication is generally preferred if the pump is to be installed in a dusty location.

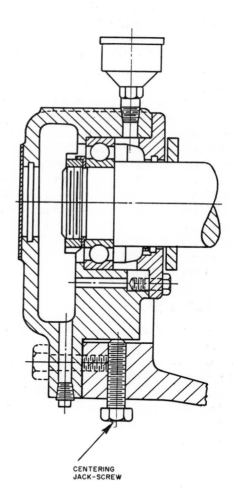

CENTERING
JACK-SCREW

Fig. 11.16 Typical mounting of ball bearings in double-suction pump

The vertical jack-screw is for initial centering of the bearing in the casing bore; pump uses grease lubrication.

Mechanical damage to the balls, rollers, or races causes early bearing failure. For that reason, proper mounting and dismounting procedures should be followed.

Relubrication periods

It is not advisable to schedule any fixed time period for adding or renewing the lubricant in an antifriction bearing but to follow instead the specified period set by the machine manufacturer. The time interval for grease lubrication may be determined by the combination of the bearing size and the rotative speed. A large grease-lubricated bearing operating at high speed may require additional grease every two months or less if operated continuously; a more normal combination might last without relubrication for four to six months; whereas a small bearing at low rotative speed or one seeing little service might need relubrication only once every year or two, merely to offset possible deterioration of the grease itself. Oil-lubricated bearings are generally supplied with oil gages, and the proper oil level should be restored if these indicate a loss of oil for any reason. Yearly replacement of oil might be considered normal.

Relubricating procedure

In relubricating grease-lubricated bearings having housings with drain plugs, the usual practice is to remove the drain plug and force grease through the bearing until new grease starts to come through the drain opening. The machine should then be allowed to run at least 20 min before replacing the drain plug so that the excess lubricant in the housing can escape.

If it is desired to clean ball bearings without removing them from the pumps, the following method may be used:

1. Wipe the bearing housing and remove the grease fitting and the drain plug
2. Clean the openings of hardened grease with a screw driver or similar tool
3. Inject some solvent, such as carbon tetrachloride, into the bearing housing while the pump is running; as the grease thins out, it will run out through the drain opening; add solvent until it runs clear
4. Flush out the solvent with a light oil
5. Add new grease; replace drain plug and grease fitting.

Relubrication of oil-lubricated bearings is usually merely a matter of draining the old oil and adding the proper amount of fresh oil. If the old oil is dirty, the bearing and housing should be cleaned by some approved method before new oil is added.

Mounting and dismounting antifriction bearings

As the fit between the outer race of an antifriction bearing and its housing classifies as a push fit, the mounting or dismounting of a bearing in its housing offers little problem. Some housing designs make it impossible however, particularly when dismantling, to apply the force that is necessary to pull the bearing out of its housing anywhere except through the balls or rollers. Such a force can easily damage the bearing.

It is desirable to mount a bearing on its shaft with the equivalent of a press fit. Actually, the bearing may be pressed on the shaft or shrunk on. Bearings to be shrunk on are first heated in an oil bath to about 200°F and then slipped into place on the shaft, the inner race being tapped lightly with a tube over the shaft if necessary. If the bearing is pressed on the shaft, the use of an arbor press is desirable (Fig. 11.17). The force should be applied to the inner race through a tubular sleeve or pipe, a ring, or small blocks of equal thickness. If an arbor press is not available, the bearing can be driven onto the shaft by hammering alternately on opposite points on the circumference of a tubular sleeve held against the inner race. Care must be taken to keep the bearing from being cocked, and feeler gages should be used to make sure

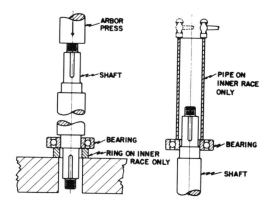

Fig. 11.17   **Two methods of mounting ball bearing on a shaft**

it is pressed firmly against the shaft shoulder.

Bearings to be dismounted from a shaft must usually be forced off, as the use of heat is seldom feasible. The technique followed will depend upon the design and the equipment available, but a split washer is usually employed to bear against the inner race or against a shaft sleeve on which the bearing has been pressed. A firm, steady pressure is applied through the split washer by an arbor press or a form of wheel puller (Fig. 11.18). Care must be taken to keep the shaft straight, to avoid damage from cocking. With proper tools the mounting or dismounting of antifriction bearings is no problem. Improper tools usually cause damage.

SLEEVE BEARINGS

Although the plain cylindrical journal or sleeve bearing has been replaced by antifriction bearings in most designs, it still has a large field of application. At one extreme, it is used for reasons of economy of construction, for instance, in certain small pumps used strictly for pumping clear liquids. The internal sleeve bearing in these pumps (see Fig. 11.1) depends primarily on the liquid pumped for its lubrication (sometimes a water seal is centered

in the bearing and connected to the pump discharge through which the liquid is introduced under pressure).

At the other extreme, sleeve bearings are used for large heavy-duty pumps with shaft diameters of such proportions that the necessary antifriction bearings are not commonly available. Another typical application is in high-pressure multistage pumps, like boiler feed pumps for pressures of 1,500 psi and higher, which require sleeve bearings because of a combination of fairly large shaft diameters and high speeds (3,600 to 9,000 rpm). Still another application is in vertical submerged pumps, like vertical turbine pumps, in which the bearings are subject to a water contact, a condition that precludes the use of antifriction bearings. Moreover, a personal preference for sleeve bearings is sometimes strong enough to make a pump purchaser even pay a premium for special design modifications of a standard pump so that those bearings may be substituted for the usual ball bearing construction.

Most sleeve bearings are oil lubricated. As a proper means of oil feed must be provided, vertical and horizontal shaft bearings vary considerably.

In especially heavy-duty applications, the line sleeve bearings are usually self-aligning. A self-aligning bearing adjusts itself

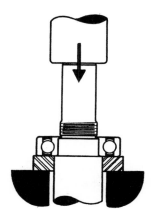

Fig. 11.18   **Removing ball bearing with an arbor press**

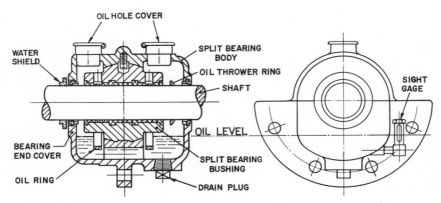

Fig. 11.19 Self-aligning sleeve bearing, spherically seated

automatically to small changes in the angular position of the shaft. This adjustment may be accomplished by providing the bearing shell with a spherical fit in the bearing housing (Fig. 11.19) or—an equally satisfactory solution and one that results in a shorter bearing and shorter bearing span—by reducing the length of the bearing bushing engaged in the housing (Fig. 11.20). The second method permits the

bushing to rock slightly in its seat and thus compensate for changes in the angular position of the shaft.

Babbitted bearings

Various materials are used for the bearing bushings, but babbitted bearings are generally preferred for heavy-duty service. The bearing bushing may consist of a babbitt lining ($\frac{1}{8}$ in. thick or more) that is anchored in the cast-iron bearing shell by means of dovetailed grooves. To insure a perfect bond, the shells are first tinned and the babbitt poured at the melting point of tin.

The "precision" automotive bearing has been widely applied of late in high speed pumps (Fig. 11.21). This bearing consists of a split thin steel shell with a similarly thin deposit of babbitt. It is available in a wide range of sizes.

To build up bearing load, the bearings of high-speed pumps are made shorter than those in a conventional unit. In addition, they may incorporate a so-called "anti-oil whip" construction. (High-speed construction results in lighter rotor weights that tend to induce oil whip if ignored.) A pocket is provided in the upper half of the bearing in which a pressure pad of oil builds up during operation. This pressure creates a downward force on the journal holding it down in the bearing and keeping the vibration to a minimum.

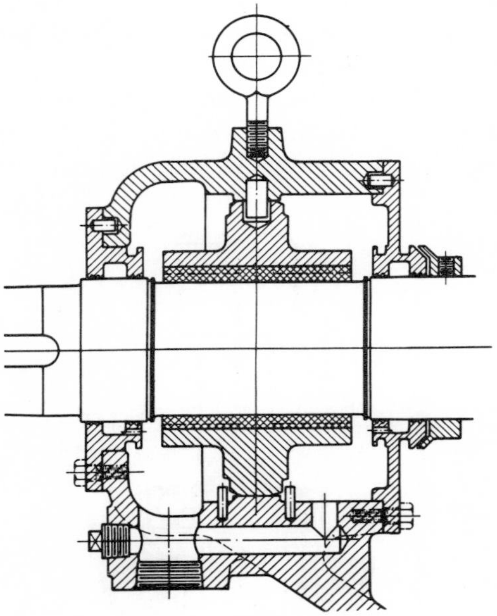

Fig. 11.20   Self-aligning sleeve bearing, with short seating of bearing bushing

Sleeve bearings are seldom applied to horizontal pumps with an overhung impeller arrangement, but rather for those designed for bearings on both ends and for drive through a flexible coupling. For such pumps, provision must be made for thrust in one of the following ways:

1. A sleeve radial bearing at one end and a babbitted, combined radial and thrust bearing at the other end

2. Two sleeve bearings with a separate thrust-bearing element of either the sleeve or antifriction type (see Fig. 11.3)

3. Sleeve radial bearings with babbitted shoulders or faces on the ends of the bushings, these acting in conjunction with collars on the shaft (Fig. 11.22).

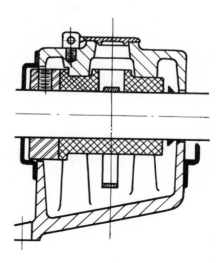

Fig. 11.22   Simple babbitted thrust bearing

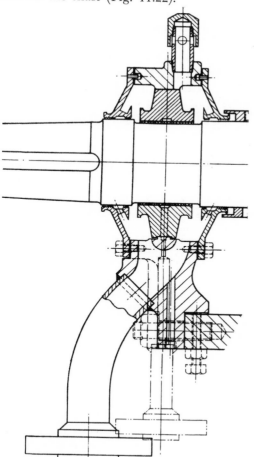

Fig. 11.21   "Precision" automotive sleeve bearing

This third type of combined thrust and radial bearing is, of course, the simplest design. It was quite popular when centrifugal pumps operated at much slower speeds and carried lower thrust loads than they do today. Lubrication was usually supplied under a slight pressure between the shaft collar and the babbitted shoulder in the bearing shell. A more refined construction, used very widely in the past under the name of "marine type thrust bearing," incorporated as many as three to five collars on the pump shaft; these were located with axial clearance in corresponding babbitted shoulders in the bearing shell. (This construction failed, however, to compensate for the angular displacement of the shaft under deflection.)

For thrust loads heavier than those usually encountered, it is now the practice to apply a Kingsbury or a Kingsbury type thrust bearing, which is also suitable for high speeds.

Kingsbury thrust bearings

This bearing was first developed to meet the need for a suitable pivot bearing for vertical-shaft turbines and has gradually been applied to other rotary apparatus,

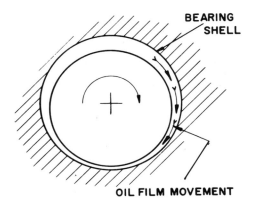

Fig. 11.23 Ordinary cylindrical bearing with oil film formed by pumping action

such as the centrifugal pump. The operating principle is simple. An ordinary cylindrical or sleeve bearing has a running clearance between the bearing shell and the journal. Because of the relation of the curved surfaces and the capillary attraction of the oil particles, a "pumping" action takes place that draws a lubricating oil film into this clearance (Fig. 11.23). If the oil is of correct viscosity, it will resist the breakdown of the film except at excessive loads. To provide a positive and ample supply of cool oil to the bearing, a simple gravity device is ordinarily used, although

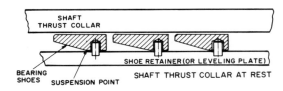

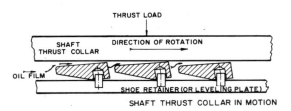

Fig. 11.24 Principle of Kingsbury thrust bearing

operation at higher speeds resulting in maximum tendency to heat requires some form of forced feed lubrication. In an ordinary thrust collar subjected to high pressures and high speeds, the parallel surfaces tend to squeeze out the oil film. The metal-to-metal contact that results makes this type of bearing unsuitable for heavy loads.

The principle of the Kingsbury bearing can be described as follows: Suppose that a circular collar is cut into little segments and that each block is suitably supported on its underside so that it may rock slightly on the point indicated as the suspension point and yet stay in place. When the shaft begins to rotate, the film of oil tends to be dragged in under the slightly rounded edges of the blocks. As the speed of the shaft increases, this tendency increases, the block adjusting itself slightly by tipping at a greater angle, riding up on the oil film as a sled runner rides up upon meeting the surface resistance of snow underneath (Fig. 11.24). The higher the speed, the greater this tendency for the block to rock forward, permitting an increased "sledding" action, and the greater the tendency to adjust itself to the increasing oil film dragged underneath it. Construction details of a typical Kingsbury bearing can be examined more closely in the sectional assembly shown in Fig. 11.25.

The thrust mounting of Kingsbury bearings used in horizontal pumps is arranged to take thrust in both directions. Sometimes both loads are approximately equal; other times there may be a major thrust in one direction and an occasional minor thrust in the opposite direction. In any event, the Kingsbury bearing is provided with thrust shoes on each side to limit the axial motion of the rotor. The number of shoes on each side may or may not be equal, depending on the application.

Kingsbury bearings are capable of taking care of unit thrust loads and linear speeds so far in excess of those suitable for the ordinary straight collar thrust that there can be no comparison between the

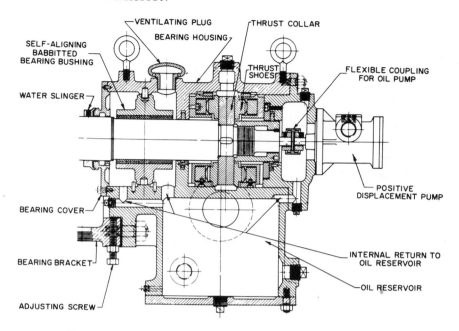

VENTILATING PLUG
THRUST COLLAR
BEARING HOUSING
SELF-ALIGNING
BABBITTED
BEARING BUSHING
THRUST
SHOES
FLEXIBLE COUPLING
FOR OIL PUMP
WATER SLINGER
POSITIVE
DISPLACEMENT PUMP
BEARING COVER
INTERNAL RETURN TO
OIL RESERVOIR
BEARING BRACKET
OIL RESERVOIR
ADJUSTING SCREW

Fig. 11.25 Sectional assembly of Kingsbury thrust bearing

two. As their cost is relatively high, however, their use may be warranted only for extreme thrust conditions.

Sleeve bearing lubrication

A ring oiled bearing is furnished with a soft steel oil ring that rides on the pump shaft through a slot cut in the middle of the top half of the bearing shell. This ring rotates as the shaft turns and picks up oil from the reservoir in the bearing housing. The oil is wiped off on the top of the pump shaft, flows between the bearing bore and the shaft, and is discharged at the ends of the bearing (see Fig. 11.19 and Fig. 11.22). Lubrication by means of oil rings is fully satisfactory only at relatively low operating speeds. A provision for automatic circulation of the oil—and if necessary, for cooling it—is an essential feature of all higher speed sleeve bearings, especially thrust bearings.

In some bearings, the oil circulation is effected by a rotary positive-displacement gear pump directly connected to the out-

board end of the pump shaft by means of a flexible coupling (see Fig. 11.25). The oil pump takes the oil from a reservoir, located either in the bearing housing itself or separately on the pump baseplate, and delivers it under pressure through the oil cooler. From the cooler, the oil flows in part to the outboard thrust bearing, from which it flows into the reservoir located in the lower half of the bearing housing. It then overflows by gravity from this reservoir into the main reservoir on the baseplate. This lubricating system is illustrated in Fig. 11.26.

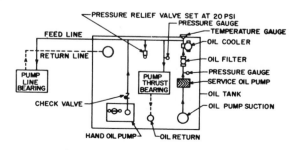

PRESSURE RELIEF VALVE SET AT 20 PSI
PRESSURE GAUGE
FEED LINE
TEMPERATURE GAUGE
OIL COOLER
RETURN LINE
OIL FILTER
PRESSURE GAUGE
PUMP
LINE
BEARING
PUMP
THRUST
BEARING
SERVICE OIL PUMP
OIL TANK
CHECK VALVE
OIL PUMP SUCTION
HAND OIL PUMP
OIL RETURN

Fig. 11.26 Typical forced-feed oil piping diagram

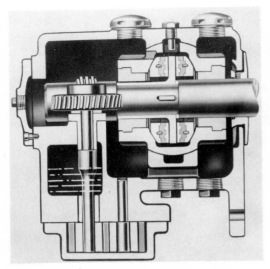

Fig. 11.27 Vertical oil pump driven from main pump shaft by worm gear

(Courtesy Ingersoll-Rand.)

General practice supplies the inboard line bearing of this system with oil under pressure through a branch line in the discharge from the oil cooler. The oil from the inboard bearing is returned by gravity through large return lines into the main reservoir. It is essential to provide an adequate pressure drop from all bearings so that the oil will not overflow because of unsatisfactory evacuation.

Numerous alternative methods exist for supplying the bearing with forced-feed lubrication. For example, some arrangements use a vertical oil pump driven from the main pump shaft by means of a worm gear (Fig. 11.27). Other bearings employ the Kingsbury "adhesive lubrication" system (Fig. 11.28). In this system, oil from the reservoir beneath the thrust bearing is drawn into a bronze ring (Fig. 11.28a), called the "circulator" or "oil pumping ring," which is around the collar. The adhesion of oil to the collar carries the oil around in the groove in the ring (Fig. 11.28b). The oil travels with the collar for almost a complete revolution. It then meets a dam in the groove and is pushed by the stream behind it into a port leading to spaces between the two lowest shoes on both sides of the thrust collar. Shaft rotation carries it to the other shoes, and it finally escapes, above the collar, into a passage leading down to a cooler. From the cooler it returns to the reservoir. The oil will circulate equally well with the collar running the other way. When the collar changes direction, the adhesiveness of the oil carries the circulator with it through a short angle, until the lug at the top of the circulator meets a stop. In either of the two "stop" positions, oil enters the groove in the circulator by the proper port for the direction of rotation and is discharged through the middle port.

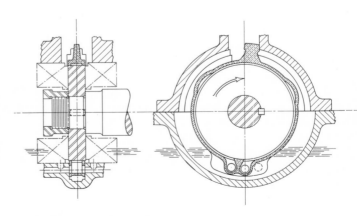

Fig. 11.28 Pumping ring of Kingsbury bearing

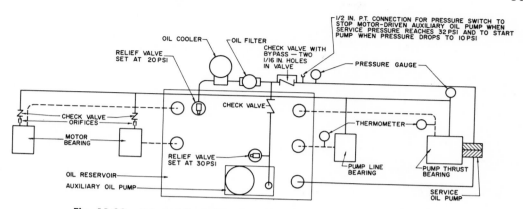

Fig. 11.29   Integral forced-feed system for pump and motor bearings

Sometimes the forced-feed lubrication system supplies oil to the driver bearings as well, although this arrangement is usually restricted to electric motor drivers. A typical system combining pump and driver lubrication is shown in Fig. 11.29 and 11.30. If pumps are driven by steam turbines or through gears, it is customary to have the turbine or the gear supply oil to the pump bearings. Such arrangements require reconcilement of the lubricating oil characteristics and of the operating temperatures established by the manufacturers of the individual pieces of equipment.

The use of oil rings for line sleeve bearings normally supplied with oil under pressure is optional and not always justified. Their function is basically that of supplying oil to the bearing at the start of the pump operation, supposedly before the forced-feed system has had the time to do so. It should be remembered that sufficient oil is generally retained in the bearings to take care of their needs before forced-feed delivery takes place.

If the normal retention of oil in the bearing or the use of oil rings will not afford adequate protection, auxiliary oil pumps are called upon. These may be manually operated gear pumps (Fig. 11.31) intended for use at scheduled intervals when the pump is standing idle. Operation of this auxiliary pump at weekly or bi-weekly intervals is usually sufficient to keep the oil

from draining out completely from the bearings or the oil piping.

More elaborate lubricating systems incorporate a motor-driven auxiliary oil pump, which is started before the main pump begins operating. The motor starter controls are interlocked in such a manner that the main motor cannot be started until the oil pressure in the system reaches a predetermined value. As soon as the oil pump driven from the main pump shaft develops sufficient pressure, the auxiliary pump is shut down by means of a pressure switch. A second pressure switch setting automatically restarts the auxiliary pump on failure of the regular pump to maintain the desired pressure. This arrangement was illustrated in Fig. 11.29 and 11.30. The settings for the pressure switch are indicated in the former, and both drawings show the arrangements of the oil cooler, oil filter, oil flow indicators, relief valves, and the like.

Sleeve bearing maintenance

It is generally recommended that the clearance between the shaft and the bushings should ordinarily not be allowed to exceed 150 per cent of the original clearance before it is renewed. If the diametral clearance is not given in the instruction book, it can be approximated on the basis of allowing 0.001 in. per inch of diameter.

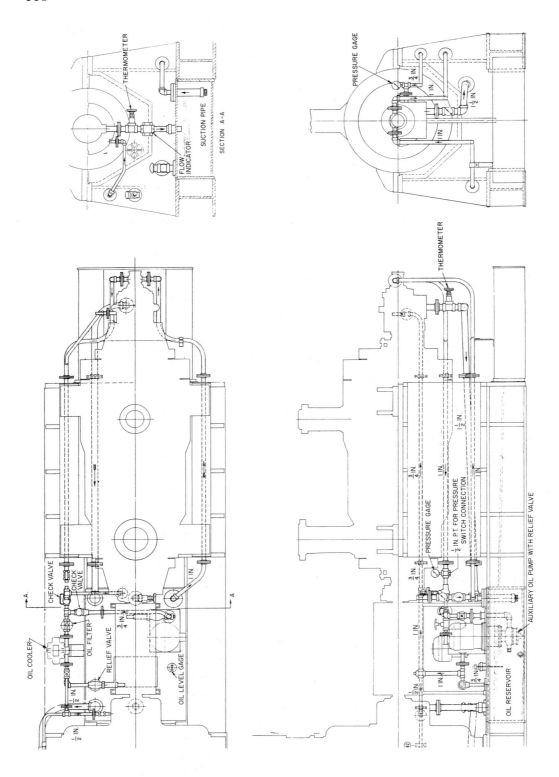

Fig. 11.30 Oil piping layout of lubrication system in Fig. 11.29

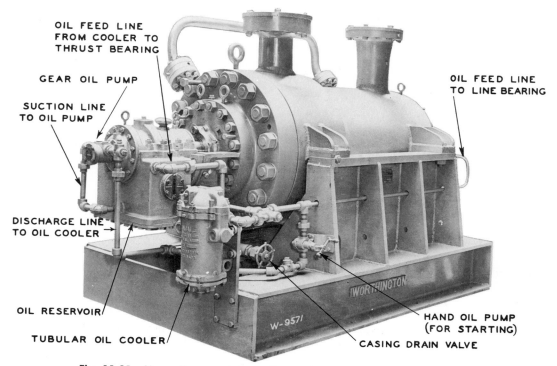

OIL FEED LINE
FROM COOLER TO
THRUST BEARING

GEAR OIL PUMP

SUCTION LINE
TO OIL PUMP

OIL FEED LINE
TO LINE BEARING

DISCHARGE LINE
TO OIL COOLER

OIL RESERVOIR

TUBULAR OIL COOLER

HAND OIL PUMP
(FOR STARTING)

CASING DRAIN VALVE

W-9571

Fig. 11.31 Manually operated auxiliary oil pump for forced-feed system

When bearings are inspected, it is very important to examine the condition of the shaft at the journals. Likewise, bushings should be examined for pitting, which is a definite sign of stray electric currents. If pitting marks are discovered, means for eliminating the stray currents must be devised or the bearing housing or pedestal must be insulated.

To renew clearances, the bearings may either be replaced or rebabbitted. Although spare bushings are frequently carried in stock, the maintenance crew may occasionally be faced with the problem of a scored shaft. Such a shaft must be turned down

and polished, and spare bushings will not then be suitable. Bearings can be rebabbitted by pouring after the original babbitt has been bored out or melted out of the shell. An undersize shaft should be used for the core when pouring and the new bearing then bored to size to minimize scraping. Care should be taken to restore the oil grooving in its proper location so as to permit proper lubrication.

Kingsbury thrust bearings should be reconditioned with great care and the manufacturer's instructions followed exactly. It is usually the safest practice to replace worn parts from the stock of spares.

12 *Couplings*

Centrifugal pumps are connected to their drivers through couplings of one sort or another, except for close-coupled units, in which the impeller is mounted on an extension of the shaft of the driver. Couplings can be either rigid or flexible. A coupling that permits neither axial nor radial relative motion between the driving and the driven shafts is called a rigid coupling. It connects the two shafts solidly and, in effect, makes them a single shaft. The use of rigid couplings is primarily restricted to vertical pumps.

A flexible coupling, on the other hand, is a device that connects two shafts but is capable of transmitting torque from the driving to the driven shaft while allowing for the minor misalignment (angular, parallel, or a combination) of both. Contrary to some popular conceptions—and despite the fact that flexible couplings take care of slight errors in emergencies—misalignment is still undesirable and should not be tolerated permanently. It causes whipping of the shaft, adds thrust to the pump and driver bearings, and usually leads to excessive maintenance and potential failure of the equipment.

A flexible coupling must also permit some lateral float of the shafts so that the two shaft ends may move closer together or far-ther apart under the influence of thermal expansion, hydraulic float, or shifting of the magnetic centers of electric motors, and so move without introducing excessive thrusts on the bearings. This aspect of flexible coupling design will be discussed in greater detail subsequently.

Rigid couplings

Clamp couplings

The clamp coupling (Fig. 12.1) is a typical rigid coupling. It consists basically of a split sleeve provided with bolts so that it can be clamped on the adjoining ends of the two shafts and form a solid connection. Both axial and circular keys are commonly incorporated in the clamp coupling so that the transmission of torque and thrust is not made solely dependent upon the frictional grip.

Compression couplings

A compression coupling (Fig. 12.2) is likewise essentially a rigid coupling. The central portion of the coupling is made up of a slotted bushing, bored to fit the two shafts and taper-machined on its outside diameter from the center out to both ends. The two coupling halves themselves are finish-bored to suit this taper. When drawn together by bolting, the bushing is compressed to the two shafts, and the frictional grip transmits the torque without the use of keys.

Flexible couplings

Pin and buffer couplings

A pin and buffer coupling is a flexible coupling with pins attached to one half of the coupling; these project into the buffers,

which are mounted in the half of the coupling on the other shaft (Fig. 12.3). The buffers are made of rubber or other compressible material to provide the necessary flexibility. The driving bolts have an easy sliding fit in the bushings; slight longitudinal variations are therefore taken care of whereas slight errors in angularity are compensated for by the flexibility of the rubber.

The Lovejoy coupling shown in Fig. 12.4 is actually a modification of the pin and buffer coupling principle. It consists of two flanged hubs mounted on the driving and driven shafts respectively, with lug projections or jaws in the flanges. These jaws mesh with a central flexible element in the form of a spider (generally made of rubber), which absorbs minor misalignments and vibration.

All-metal flexible couplings

An all-metal flexible coupling is one whose parts are all made of metal. Some of these couplings depend on the flexibility of metal plates or springs whereas others depend upon the angular displacement made possible by two splines with a connecting splined sleeve.

The gear type of "Fast" flexible coupling is illustrated in Fig. 12.5. In the outer shell of the coupling, inside each end, a ring of internally cut gear teeth meshes with the gears on the driving and driven halves of the coupling. The torque is transmitted

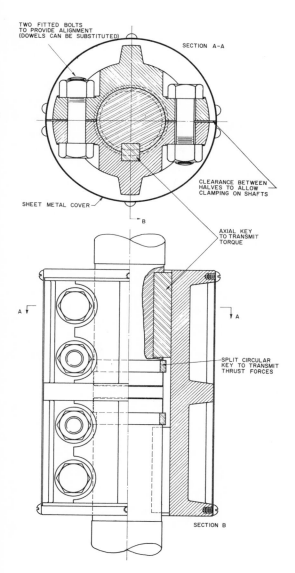

Fig. 12.1 Clamp coupling

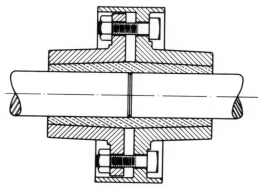

Fig. 12.2 Compression coupling

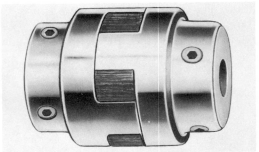

Fig. 12.4 Lovejoy coupling

Fig. 12.3 Pin and buffer coupling

through the gear teeth, whereas the necessary sliding action and ability for slight adjustments in position comes from a certain freedom of action provided between the two sets of teeth. To prevent any tendency to bind because of friction, the gears work in a constant bath of oil that is re-

tained within the outer shell. Some high-speed applications use light grease.

Another type of all-metal coupling is the Falk flexible coupling (Fig. 12.6). It consists of two flanged steel hubs, a special tempered-steel spring forming a complete cylindrical grid, and a steel-shell cover. The peripheries of the hubs are slotted to receive the spring member. The slots widen inwardly towards each other in the form of an arc that bears a definite relation to the thickness of the grid spring bars. The curvature is such that the points of support approach each other as the load increases (Fig. 12.7). In fact, the slots are so formed that the stress on the spring members remains almost constant throughout the entire elastic range of the coupling. During light loading (Fig. 12.7A), the springs fit in the grooves closely only at their outer ends. Thus there is a long free

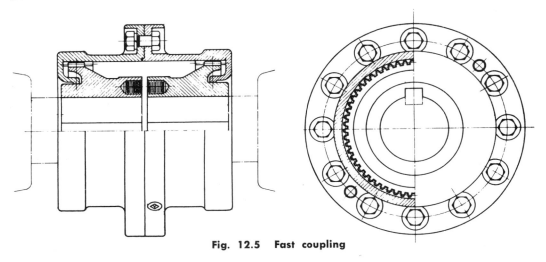

Fig. 12.5 Fast coupling

Fig. 12.6 Falk coupling

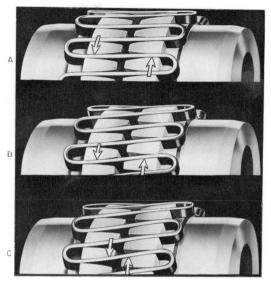

Fig. 12.7 Falk coupling in operation

span of spring between the point of support, and the power is transmitted through almost the entire length of the flexible rungs of the springs. During normal loading (B), the distance between supports on the grooves is automatically shortened as the load increases, thus stiffening or strengthening the spring against bending. During extreme shock loading (C), the load becomes so great that the springs bear on the entire length of the grooves, which

makes possible the transmission of severe overloads. The Falk flexible coupling is grease-lubricated.

If interruption of operation for the purpose of relubricating couplings with oil cannot be tolerated, constantly lubricated couplings are used, for example, in boiler feed pump installations without spares or

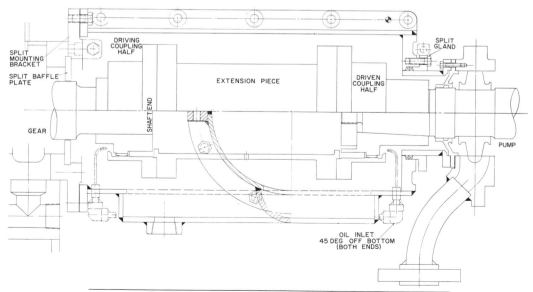

Fig. 12.8 Continuously lubricated coupling.

Fig. 12.9   Thomas couplings

boiler feed pumps driven directly from the main generator shaft. A typical arrangement is illustrated in Fig. 12.8. It consists of an oil-tight enclosure bolted at one end to the stationary portion of either the driving or driven piece of equipment. The other end of the enclosure has a slip fit inside a cover that is bolted to the other piece of equipment. Some form of packing is used to prevent loss of lubricant at the slip joint. Oil under pressure is brought through the enclosure and impinges upon the meshing gear teeth of the coupling, the excess being collected at the bottom of the enclosure and returned to the oil reservoir.

The Thomas coupling shown in Fig. 12.9 is a nonlubricated all-metal flexible coupling made up of two hubs, mounted respectively on the driving and driven shafts, and of a central flange connected to the flanges of the two hubs through a series of flexible disks. Power is transmitted in tension by these disks, which are bolted alternately to the end flanges and to the center member. Flexing of the disks compensates for the misalignment.

Numerous other types of all-metal flexible couplings exist, of course; the three illustrated here are given only as general examples.

Limited end-float travel

Horizontal sleeve-bearing electric motors are usually not equipped with thrust bearings but rather with babbitted faces or shoulders on the line bearings. The motor rotor, which is allowed to float, will seek

the magnetic center, but a rather small force can cause it to move off this center. This movement may sometimes be sufficient to cause the shaft collar to contact the bearing shoulders, causing heat and bearing difficulties.

This effect is particularly noticeable in large electric motors of 200 hp and more. As all horizontal centrifugal pumps are equipped with thrust bearings, it has become the practice to use "limited end-float" couplings between pumps and motors in this horsepower range to keep the motor rotor within a restricted location. The motors are built so that the total clearance between shaft collars and bearing shoulders is not less than $\frac{1}{2}$ in. In turn, the flexible couplings are arranged to restrict the end float of the motor rotor to less than $\frac{3}{16}$ in. To keep the gap open between the shaft collar and the shoulders, one of the following methods is used:

1. *For gear or grid couplings*—by a "button" at the end of the pump shaft or by a predimensioned plate between the two shaft ends (Fig. 12.10).

2. *For flexible-disk couplings, such as the Thomas coupling*—by the stiffness of the flexible disks themselves, which have inherent float-restricting characteristics.

Contact between the hubs and the coupling covers prevents excessive movement in the opposite direction in gear or grid

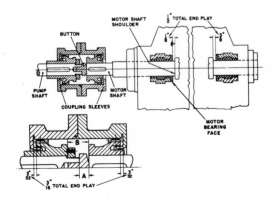

Fig. 12.10   Limited end-float travel

Fig. 12.11   Application of extension couplings—pump mounted

Fig. 12.12   Application of extension couplings—pump dismounted

*An extension coupling enables a pump to be dismantled without moving either
the driver or the pump casing.*

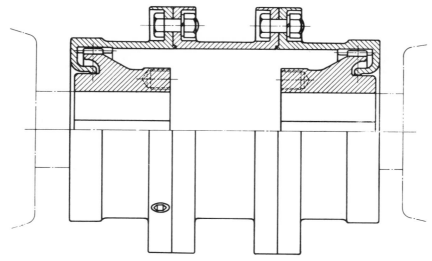

Fig. 12.13   Extension coupling

couplings. The stiffness of the flexible disks is the restraining force in both directions in Thomas couplings.

Floating-shaft couplings

Regular flexible couplings are made to connect driving and driven shafts with relatively close ends and are suitable for limited misalignment. Provision sometimes has to be made for greater misalignment, however, or the ends of the driver and pump shafts have to be separated by a considerable distance for special reasons. Such is the case, for example, with end-suction pump designs in which the rotor and bearing assembly is removed by withdrawing it axially toward the driver. If neither the pump nor the driver can be readily removed, it becomes desirable to separate the driver and pump shaft ends sufficiently to permit withdrawal of the pump rotor (Fig. 12.11 and 12.12). For this

purpose, a readily removed flexible driver of sufficient length is necessary.

The extension or spacer sleeve coupling is commonly used in pumping units handling hot liquids and therefore subject to expansion and possible misalignment. Their purpose is to prevent harmful misalignment with a minimum separation of the driving and driven shaft ends. Usually they consist of two single-engagement elements connected by a sleeve (Fig. 12.13).

The floating-shaft coupling consists of two flexible elements connected by a shaft that must be supported on each end by the flexible elements themselves. Different manufacturers use different approaches as required by their basic coupling designs. For instance, each of the two couplings may be of the single-engagement type, may consist of a flexible half-coupling and a rigid half-coupling at each end, or may be completely flexible couplings with some piloting or guiding construction.

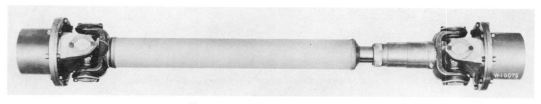

Fig. 12.14   Flexible drive shaft

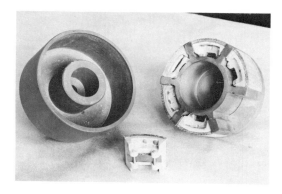

Fig. 12.17   Clutch coupling
(Courtesy Centric Clutch Co.)

Fig. 12.15   Chain coupling (quick disconnect)

The coupling chain is connected by a single pin. The revolving coupling casing contains the grease lubricant. (Courtesy Diamond Chain Co.)

In the smaller horsepower field (below 25 hp per 100 rpm), "flexible drive shafts" are commercially available. These use universal joints at each end with a tubular floating shaft and a splined portion to provide for length variation (Fig. 12.14).

The floating shaft and flexible drive shaft are frequently used in vertical dry-pit pumps, an application that is discussed with that type of pump.

Clutch couplings

Regular disk clutches are rarely used to connect a centrifugal pump to its driver for two major reasons. The first is that most clutch designs impose a high additional thrust load on the pump thrust bearing; the second is that very accurate alignment between the clutch parts is necessary, and this is difficult to maintain. The overrunning clutch design has been used to connect drivers to pumps, particularly in dual-

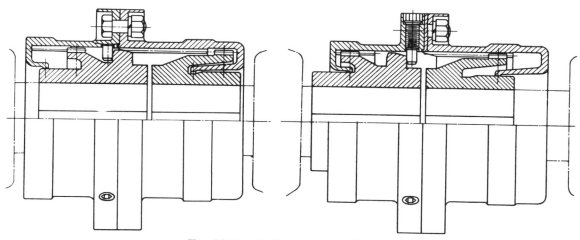

Fig. 12.16   Fast's cut-out coupling

driven units, and the most successful of these designs have a flexible coupling incorporated into the clutch unit. Also used is a clutch coupling with faced weights in the driver half that are pressed against a drum surface on the driven half by centrifugal force.

Couplings for dual-drive

In dual-driven pump installations, it is generally desirable to have one driver idle either to save power or to save wear. Internal combustion engines, however, cannot be allowed to turn over idle and must be disconnected. The ideal type of couplings for such units are those that can be readily disengaged and re-engaged. An example of the chain quick-disconnecting coupling is shown in Fig. 12.15.

Another cut-out coupling is illustrated in Fig. 12.16. The left view shows the coupling in the connected position; the right view, in the disconnected position. It is a quick and simple operation to release the location pins, slide the sleeve into or out of engagement, and thereby connect or disconnect the driving unit and the pump.

If time is of extreme importance or if the starting of the standby driver is automatically controlled, a device like the freewheeling clutch is necessary. A clutch coupling with spring-retained shoes in the driven half is very suitable for this service (Fig. 12.17). The centrifugal action on these shoes can be controlled to any predetermined speed, and no shoe engagement takes place until this speed is reached. Above this speed, the coupling automatically picks up the load. The tension of the springs can be made just sufficient to overcome the weight of the shoe or to allow a pump to be driven by a motor on one side while an internal combustion engine is idling without load on the other. Bringing the engine up to speed engages the coupling.

Coupling guards

Coupling guards are stationary enclosures surrounding a coupling with the basic function of protecting the operator from the possible danger of getting caught by its revolutions. Usually made of steel mesh or steel plate, they are fastened to the bedplate or to the pump foundation in the absence of a bedplate. They are often desirable and even mandatory in some localities.

Magnetic clutches, magnetic drives, and hydraulic couplings

Magnetic clutches, magnetic drives, and hydraulic couplings are not couplings in the strict sense of the word, as their function is to vary the speed of the driven unit rather than to provide merely a connecting device between pump and driver.

Magnetic clutches are rarely used to connect a centrifugal pump to its driver because they require accurate alignment and the few installations that have been made have not been very successful. Their maintenance costs are also high. The only advantageous application of this device is in accumulator-tank pumping or similar services for which the demand varies over a wide range. It is now the practice either to start and stop the entire pumping unit or, if the cycle is too frequent for that, to allow the pump to operate at reduced capacity during the period of small demand, incorporating a bypass so the capacity will never fall below a safe value if the demand drops too low for proper operation.

Both hydraulic couplings and magnetic drives are used in centrifugal pumps if variations in operating conditions warrant the use of variable output speed devices. Although they have approximately the same over-all efficiency as slip ring motors with speed control, they have the advantage of easily producing any desired output speed whereas the regular control for slip ring motors permits adjustment of speed only by steps. A more complete discussion of these devices appears in Chap. 20.

13 *Bedplates and Other Pump Supports*

For very obvious reasons, it is desirable that pumps and their drivers be removable from their mountings. Consequently, they are usually bolted and doweled to machined surfaces that in turn are firmly connected to the foundations. To simplify the installation of horizontal-shaft units, these machined surfaces are usually part of a common bedplate on which either the pump or the pump and its driver have been pre-aligned.

Bedplates

The primary function of a pump bedplate is to furnish mounting surfaces for the pump feet that are capable of being rigidly attached to the foundation. Mounting surfaces are also necessary for the feet of the pump driver or drivers or of any independently mounted power transmission device. Although such surfaces could be provided by separate bedplates or by individually planned surfaces, it would be necessary to align these separate surfaces and fasten them to the foundation with the utmost care. Usually this method requires in-place mounting in the field as well as drilling and tapping for the holding-down bolts after all parts have been aligned. To minimize such "field work," coupled horizontal-shaft pumps are usually purchased with a continuous base extending under the pump and its driver; ordinarily, both these units are mounted and aligned at the place of manufacture.

Although such bases are designed to be quite rigid, they deflect if improperly supported. It is therefore necessary to support them on foundations that can supply the required rigidity. Furthermore, as the base can be sprung out of shape by improper handling during transit from the place of manufacture to point of installation, it is imperative that the alignment be carefully rechecked during erection and prior to starting the unit.

As the unit size increases so does the size, weight, and cost of the base required. The cost of a prealigned base for most large units would exceed the cost of the field work necessary to align individual bedplates or soleplates and to mount the component parts. Such bases are therefore used only if appearances require or if their function as a drip collector justifies the additional cost. Even in fairly small units, the height at which the feet of the pump and the other elements are located may differ considerably. A more rigid and pleasant

129

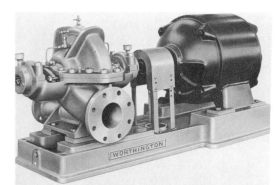

Fig. 13.1 Horizontal-shaft centrifugal pump and driver on cast-iron bedplate

looking installation can frequently be obtained by using individual bases or soleplates and building up the foundation to various heights under the separate portions of equipment.

Cast-iron baseplates are usually provided with a raised edge or "raised lip" around the base to prevent dripping or draining onto the floor (Fig. 13.1). The base itself is suitably sloped towards one end so as to collect the drainage for further disposal. A drain pocket is provided near the bottom of the slope, usually with a mesh screen. A tapped connection in the pocket permits piping the drainage to a convenient point.

Bedplates fabricated of steel plate and structural steel shapes, now used very extensively, do not easily permit incorporation of a raised lip or drip pocket (Fig. 13.2 and 13.3). From a utility viewpoint, however, the customary use of bearing brackets as drip pockets (to collect leakage from the stuffing boxes) now makes the use of a raised-lip bedplate unnecessary in horizontal pumps except if the pump handles a cold liquid in a moist atmosphere and thus must contend with considerable condensation on its surface.

Fig. 13.2 Pump and internal combustion engine mounted on portable steel skid base

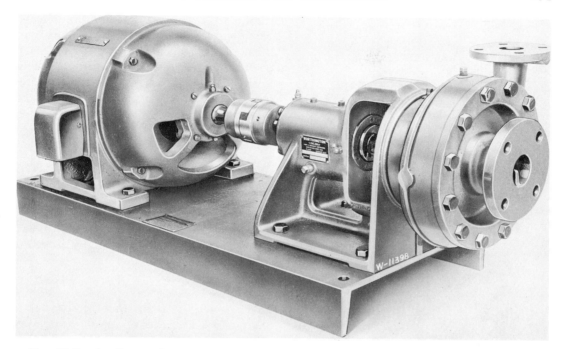

Fig. 13.3 Small centrifugal pump on structural steel bedplate made of a simple channel

Soleplates

Soleplates are cast-iron or steel pads located under the feet of the pump or its driver and embedded into the foundation. The pump or its driver are doweled and bolted to them. Soleplates are customarily used for vertical dry-pit pumps and also for some of the larger horizontal units to save the cost of the large bedplates otherwise required.

Centerline support

For operation at high temperatures, the pump casing must be supported as near to its horizontal centerline as possible in order to prevent excessive strains caused by temperature differences. These might seriously disturb the alignment of the unit and eventually damage it. Centerline construction is usually employed in boiler feed pumps or hot-water circulating pumps operating at temperatures around 300°F (Fig. 13.4).

Horizontal units using flexible pipe connections

The foregoing discussion of bedplates and supports for horizontal shaft units assumed their application to pumps with piping set-ups that do not impose hydraulic thrusts on the pumps themselves. If flexible pipe connections or expansion joints are desirable in the suction or discharge piping of a pump (or in both), however, the pump manufacturer should be so advised for several reasons. First, the pump casing will be required to withstand various stresses

Fig. 13.4 Single-stage hot-water circulating pump with centerline support

Fig. 13.5 Vertical-shaft installation of double-suction single-stage pump

Casing is provided with mounting support flange.

caused by the resultant hydraulic thrust load. Although this is rarely a limiting or dangerous factor, it is best that the manufacturer have the opportunity to check the strength of the pump casing. Second, the resulting hydraulic thrust has to be transmitted from the pump casing through the casing feet to the bedplate or soleplate and then to the foundation. Usually, horizontal-shaft pumps are merely bolted to their bases or soleplates so that any tendency to displacement is resisted only by the frictional grip of the casing feet on the base and by relatively small dowels. If flexible pipe joints are used, this attachment may not be sufficient to withstand the hydraulic thrust. If high hydraulic thrust loads are to be en-

countered, therefore, the pump feet must be keyed to the base or supports. Similarly, the bedplate or supporting soleplates must be of a design that will permit transmission of the load to the foundation. (For a more complete discussion of flexible expansion joints, see Chap. 24.)

Bases and supports for vertical pumping equipment

Vertical-shaft pumps, like horizontal-shaft units, must be firmly supported. Depending upon the installation, the unit may be supported at one or several elevations. Vertical units are seldom supported from walls, but even that type of support is sometimes encountered.

Occasionally, a nominally horizontal-shaft pump design is arranged with a vertical shaft and a wall used as the supporting foundation. The regular horizontal-shaft units shown in Fig. 13.1 and 13.4 could be used for this purpose without modification, except that the bedplate is attached to a wall. For such installations, it is advisable to lock the pump feet to the bedplate by keys or dowels rather than to rely strictly on the friction between the pump feet and the pads of the bedplate. Of course, it is assumed that careful attention will have been given to the arrangement of the pump bearings to prevent the escape of the lubricant.

Installations of double-suction single-stage pumps with the shaft in the vertical position are relatively rare, except in some marine or navy applications. Hence manufacturers have very few standard pumps of this kind arranged so that a portion of the casing itself forms the support (to be mounted on soleplates). Figure 13.5 shows such a pump, which also has a casing extension to support the driving motor.

A complete discussion of the methods of supporting pumps that are specifically designed for vertical mounting will be found in Chap. 14.

14 *Special Designs: Vertical Pumps*

Preceding chapters on centrifugal pumps with horizontal-shaft construction should not obscure the fact that many centrifugal pumps utilize vertical-shafting. Vertical-shaft pumps fall into two separate classifications: (1) The dry pit, and (2) the wet pit. The former operate surrounded by air, whereas the latter are either fully or partially submerged in the liquid handled.

VERTICAL DRY-PIT PUMPS

Dry-pit pumps with external bearings include most small, medium, and large vertical sewage pumps; most medium and large drainage and irrigation pumps for medium and high head; many large condenser circulating and water supply pumps; and many marine pumps. Sometimes the vertical design is preferred (especially for marine pumps) because it saves floor space. Other times it is desirable to mount a pump at a low elevation because of suction conditions and it is then also preferable or necessary to have its driver at a high elevation. The vertical pump is normally used for very large capacity applications because it is more economical than the horizontal type, all factors considered.

Many vertical dry-pit pumps are basically horizontal designs with minor modifications (usually in the bearings) to adapt them for vertical-shaft drive (see Chap. 13, Fig. 13.5). The reverse is true of small- and medium-sized sewage pumps; a purely vertical design is the most popular for that service. Most of these sewage pumps have elbow suction nozzles (Fig. 14.1–14.3) because their suction supply is usually taken from a wet well adjacent to the pit in which the pump is installed. The suction elbow usually contains a handhole with a removable cover to provide easy access to the impeller.

To dismantle one of these pumps, the stuffing box head must be unbolted from the casing after the intermediate shaft or the motor and motor stand have been removed. The rotor assembly is drawn out upwards, complete with the stuffing box head, the bearing housing, and the like. This rotor assembly can then be completely dismantled at a convenient location.

Vertical-shaft installations of single-suction pumps with a suction elbow are commonly furnished with either a pedestal or a base elbow (see Fig. 14.1). These may be bolted to soleplates or even grouted in. The grouting arrangement is not too desirable

Fig. 14.1   Small vertical sewage pump with intermediate shafting

unless there is full assurance that the pedestal or elbow will never be disturbed or that the grouted space is reasonably regular and the grout will separate from the pump without excessive difficulty.

Vertical single-suction pumps with bottom suction are commonly used for larger sewage, water supply, or condenser circulating applications. Such pumps are provided with wing feet that are bolted to soleplates grouted in concrete pedestals or piers (Fig. 14.4). Sometimes the wing feet may be grouted right in the pedestals. These must be suitably arranged to provide proper access to any handholes in the pump and to allow clearance for the elbow suction nozzles if these are used.

If a vertical pump is applied to condensate service or some other service for which the eye of the impeller must be vented to prevent vapor binding, a pump with a bottom single-inlet impeller is not desirable because it does not permit effective venting. Neither does a vertical pump employing a double-suction impeller (Fig. 14.5). The most suitable design for such applications incorporates a top single-inlet impeller (Fig. 14.6).

If the driver of a vertical dry-pit pump can be located immediately above the pump, it is often supported on the pump itself (see Fig. 14.3). The shafts of the pump and driver may be connected by a flexible coupling, which requires that each have its own thrust bearing. If the pump shaft is rigidly coupled to the driver shaft or is an extension of the driver shaft, a common thrust bearing is used, normally in the driver.

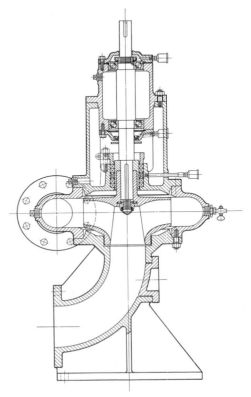

Fig. 14.2   Section of pump in Fig. 14.1

Although the driving motors are frequently mounted right on top of the pump casing, one important reason for the use of the vertical-shaft design is the possibility of locating the motors at an elevation sufficiently above the pumps to prevent their accidental flooding. The pump and its driver may be separated by an appreciable length of shafting, which may require steady bearings between the two units. It is extremely important that these steadying bearings be rigidly supported and maintained in strict alignment. The support is generally provided by horizontal structural steel beams tied into the wall structure, although occasionally a similar vertical support is used. For proper operation of the vertical shafting, the deflection of the vertical guide bearings under any operating conditions must be kept within the limits set by the design of the shafting and the operating speed. In small units, a channel located between the walls of the station usually gives adequate support in all directions. Larger units with larger reaction loads on the guide bearings may require two channels or beams with lattice bars. Some installations incorporate reinforced concrete beams in the structure. Naturally, if the design of the building requires the construction of an intermediate floor, this floor can be used to support the guide bearings.

The most common shafting connecting a small- or medium-size centrifugal pump with its driver makes use of the universal joint with hollow tubing (Fig. 14.7). The lower section has a universal joint at both ends whereas the upper sections (if more than one is used) have a guide bearing supporting the lower end and a universal joint at the upper end. Such shafting compensates for angular misalignment and, as the lower section incorporates a splined joint, also compensates for any minor discrepancy in length. If speed permits, shaft sections as long as 120 in. or more can be obtained. Sections longer than 10 ft are easily sprung and must be handled carefully. As this

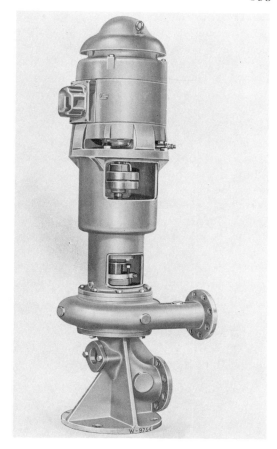

Fig. 14.3 Vertical sewage pump with direct mounted motor

shafting does not transmit thrust, both pump and driver must have a thrust bearing.

Although a vertical motor may be mounted directly on soleplates grouted into the floor, a separate stand is sometimes necessary so that the motor may be raised to provide access to the coupling. Occasionally, removable beams are placed directly across a large opening in the floor to serve as the motor mounting. This method permits easy access to the pumps for servicing and simplifies lowering them into place during the initial installation.

A driver supported on a stand above the floor provides access to the flange connection and upper universal joint for bolting purposes and for relubrication. If the driver uses hollow-shaft rather than solid-shaft construction, it must be provided with a head shaft guided by a lower bearing to act in the same capacity. The weight of this shafting (excluding that of the lowest universal joint) is carried on the motor; provided it is not extremely long, the total weight involved is relatively small and a normal thrust motor can be used. Actually, hollow shafting is more expensive than solid shafting. But the basic universal joint is so widely used on automobiles and trucks that it is somewhat of a production item, and the increase in cost it entails over solid shafting is very reasonable.

Units requiring more torque in their intermediate shafting than can be carried by the available sizes of universal-joint shafting use solid shafting, either with solid or with flexible couplings. If solid or rigid

couplings are used, only one thrust bearing is needed (usually in the driver), and all other bearings are merely guide bearings. This shafting has the disadvantage of requiring very accurate alignment of all bearings, a difficult feat for open shafting employing more than three bearings.

Solid vertical shafting using flexible couplings usually consists of several shaft sections (including pump and driver), each having two or possibly three bearings connected by floating shaft sections and a piloted or guided flexible coupling at each end, thus acting in effect like a universal joint (see Fig. 14.4). Naturally each section has to have a thrust bearing to carry the weight of the shaft section.

The intermediate shafting for large pumps requiring large shafts is usually of solid construction with solid flanged cou-

Fig. 14.4 Vertical bottom-suction volute pumps with piloted couplings

if a shaft is to run at twice the speed of another, it must be twice as large in diameter
for the same bearing span, or its permissible
bearing span will be reduced to 70 per cent
of that permissible with the lower speed.

Bearings for vertical dry-pit pumps and
for intermediate guide purposes are usually
antifriction bearings that are grease lubricated to simplify the problem of retaining
a lubricant in a housing with a shaft projecting vertically through it. Typical ball
steady bearings used as intermediate shaft
steady bearings are shown in Fig. 14.9.
Larger units, for which antifriction bearings are not available or desirable, use self-
oiling babbitt bearings or forced-feed-oiled
babbitt bearings with a separate oiling system (Fig. 14.10 and 14.11). Figure 14.11 illustrates a vertical dry-pit pump design
with a single-sleeve type line bearing. The
pump is connected by a rigid coupling to

**Fig. 14.5   Vertical double-suction volute pump
with direct mounted motor**

plings that are often forged onto the shaft
sections (Fig. 14.8).

The size of the shafting used for an installation may finally be determined by
the torque to be transmitted. However, if
a certain span between bearings is desirable
because of existing supports (floors or
beams), a shaft larger than that required
by the torque may be necessary so that the
operating speed will be sufficiently below
the critical speed. It is thus general practice to have the first critical speed higher
than any operating or runaway speed of
the pump. The critical speed of a vertical
solid shaft is a direct function of the diameter and an inverse function of the
square of the span between bearings. Thus

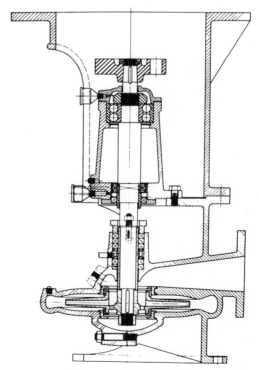

**Fig. 14.6   Section of vertical pump with top-
suction inlet impeller**

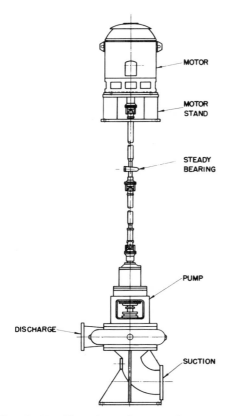

Fig. 14.7   Elevation of vertical pump with tubular shafting

its motor (not shown in the illustration), which is provided with a line and a thrust bearing.

The supports for the guide bearings of vertical shafting connecting a centrifugal pump and its driver must be sufficiently rigid. The radial load is usually assumed to be the same as if the unit were in a horizontal position. With this loading, the deflection of the supports in any direction should not exceed Δ in the following equation:

$$\Delta = (187.7/n_c)^2$$

in which:

Δ = the deflection, in inches

n_c = critical speed of the shafting, in rpm.

This critical speed is usually 125 per cent of the pump rotative speed or some value

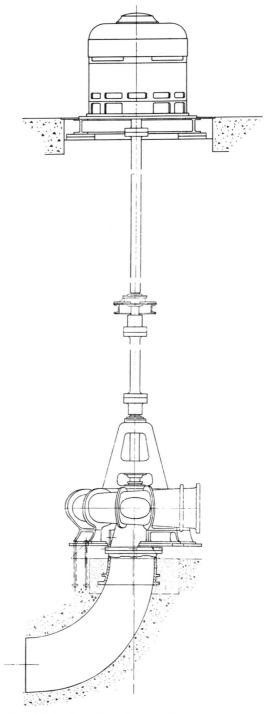

Fig. 14.8   Elevation of vertical pump with solid shafting

Motor supports rotating parts.

above the possible runaway speed to allow for back flow through the pump. If beams or channels support the bearings, the design of the latter naturally depends on the span between them, the radial force, and the permissible deflection. Small pump installations with short spans usually require a single channel (to which vertically mounted bearings are most easily attached). Larger units with long spans often require fairly widely spaced channels or beams with strengthening lattice work (for which a horizontally mounted bearing resting directly on the beams or on a bridging plate is more convenient). These considerations tend to make vertically mounted bearings preferable for small units and horizontally mounted bearings preferable for large units.

Vertical dry-pit centrifugal pumps are structurally similar to horizontal-shaft pumps. It is to be noted, however, that many of the very large vertical single-stage single-suction (usually bottom) volute

Fig. 14.9 Ball bearings used for intermediate shaft steady bearings

(*Courtesy Seal Master.*)

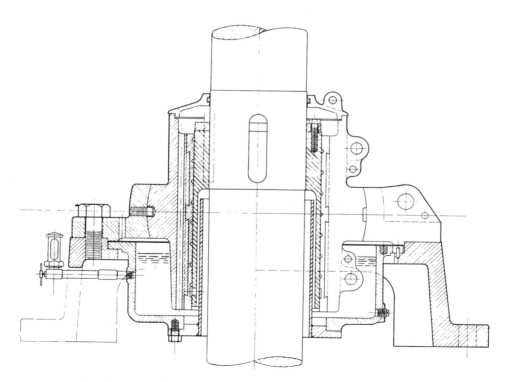

Fig. 14.10 Self-oiling steady bearing for large vertical shafting

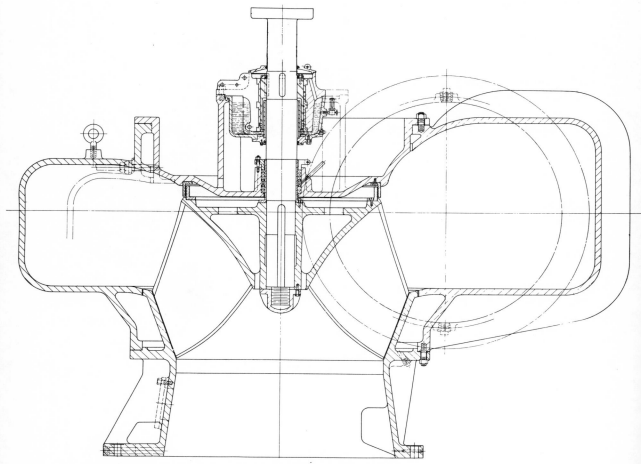

Fig. 14.11 Section of large vertical bottom-suction volute pump with single guide bearing

pumps that are preferred for large storm
water pumpage, drainage, irrigation, sewage, and water supply projects have no comparable counterpart among horizontal-shaft
units. The basic U-section casing of these
pumps, which is structurally weak, often
requires the use of heavy ribbing to provide sufficient rigidity. In comparable water turbine practice, a set of vanes (called
a speed-ring) is employed between the casing and runner to act as a strut. Although
the speed-ring does not affect the operation
of a water turbine adversely, it would function basically as a diffuser in a pump because of the inherent hydraulic limitations
of that construction. Some high-head

pumps of this type have been made in the
twin-volute design. The wall separating
the two volutes acts as a strengthening rib
for the casing, thus making it easier to design a casing strong enough for the pressure
involved (see Fig. 2.7).

Vertical pumps equipped with bottom
single-inlet impellers (see Fig. 14.2) have a
leakage joint between the wearing ring hub
of the impeller and the suction head. When
pumps of this type handle gritty water, the
grit separates out during periods of shut-
down and concentrates at or near this joint.
As soon as the pump is started again, this
concentration of grit is washed through the
leakage joint, causing wear. Large pumps

may resort to a ring construction like that shown in Fig. 14.12, in which the stationary ring is extended above the suction head to form a pocket for the grit to be deposited in and from which it can be periodically flushed. These and other refinements are feasible in large but not in small pumps.

VERTICAL WET-PIT PUMPS

Vertical pumps intended for submerged operation are manufactured in a great number of designs, depending mainly upon the service for which they are intended. Thus wet-pit centrifugal pumps can be classified in the following manner:

1. Vertical turbine pumps
2. Propeller or modified propeller pumps
3. Sewage pumps
4. Volute pumps
5. Sump pumps.

Vertical turbine pumps

Vertical turbine pumps were originally developed for pumping water from wells and have been called "deep-well pumps," "turbine-well pumps," and "borehole pumps." As their application to other fields has increased, the name "vertical turbine pumps" has been generally adopted by the manufacturers. (This is not too specific a designation because the term "turbine pump" has been applied in the past to any pump employing a diffuser. There is now a tendency to designate pumps using diffusion vanes as "diffuser pumps" to distinguish them from "volute pumps." As that designation becomes more universal, applying the term "vertical turbine pumps" to the construction formerly called "turbine-well pumps" will become more specific.)

The largest fields of application for the vertical turbine pump are pumping from wells for irrigation and other agricultural purposes, for municipal water supply, and for industrial water supplies, processing,

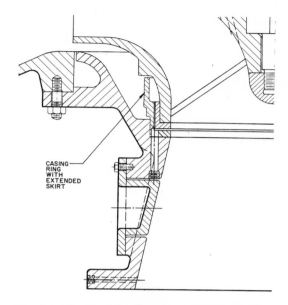

Fig. 14.12 Wearing ring construction with extended skirt for large vertical pumps

circulating, refrigerating, and air conditioning. This type of pump has also been used for brine pumping, mine dewatering, oil field repressuring, and other purposes.

These pumps have been made for capacities as low as 10 or 15 gpm and as high as 25,000 gpm or more, and for heads up to 1,000 ft. Most applications naturally involve the smaller capacities. The capacity of the pumps used for bored wells is naturally limited by the physical size of the well as well as by the rate at which water can be drawn without lowering its level to a point of insufficient pump submergence.

Vertical turbine pumps should be designed with a shaft that can be readily raised or lowered from the top to permit proper adjustment of the position of the impeller in the bowl. An adequate thrust bearing is also necessary to support the vertical shafting, the impeller, and the hydraulic thrust developed when the pump is in service. As the driving mechanism must also have a thrust bearing to support its vertical shaft, it is usually provided with one of adequate size to carry the pump parts as well. For these two reasons, the hollow-shaft motor or gear is most commonly used for vertical turbine pump drive. In addition, these pumps are sometimes made with their own thrust bearings to allow for belt drive or for drive through a flexible coupling by a solid-shaft motor, gear, or turbine. Dual-driven pumps usually employ an angle gear with a vertical motor mounted on its top.

The design of vertical pumps illustrates how a centrifugal pump can be specialized to meet a specific application. Figure 14.13 illustrates a turbine design with closed im-

pellers and enclosed line shafting; Fig. 14.14 illustrates another turbine design with closed impellers and open line shafting.

The bowl assembly or section consists of the suction case (also called suction head or inlet vane), the impeller or impellers, the discharge bowl, the intermediate bowl or bowls (if more than one stage is involved), the discharge case, the various bearings, the shaft, and miscellaneous parts such as keys, impeller locking devices, and the

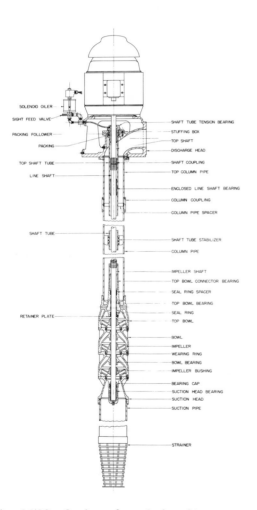

Fig. 14.13 Section of vertical turbine pump with closed impellers and enclosed line shafting (oil lubrication)

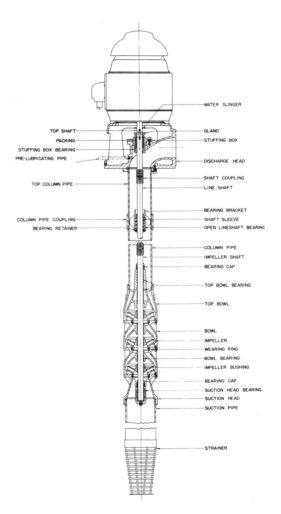

Fig. 14.14 Section of vertical turbine pump with closed impellers and open line shafting (water lubrication)

like. The column pipe assembly consists
of the column pipe itself, the shafting above
the bowl assembly, the shaft bearings, and
the cover pipe or bearing retainers. The
pump is suspended from the driving head,
which consists of the discharge elbow (for
above-ground discharge), the motor or
driver support, and either the stuffing box
(in open-shaft construction) or the assembly for providing tension on and the introduction of lubricant to the cover pipe.
Below-ground discharge is taken from a tee
in the column pipe, and the driving head
functions principally as a stand for the
driver and support for the column pipe.

Liquid in a vertical turbine pump is
guided into the impeller by the suction case
or head. This may be a tapered section
(Fig. 14.15 and 14.16) for attachment of
a conical strainer or suction pipe, or it may
be a bellmouth.

Semiopen and enclosed impellers are both
commonly used. For proper clearances in
the various stages, the semiopen impeller
requires more care in assembly on the impeller shaft and more accurate field adjustment of the vertical shaft position in order
to obtain the best efficiency. Enclosed impellers are favored over semiopen ones,
moreover, because wear on the latter reduces capacity, which cannot be restored
unless new impellers are installed. Normal
wear on enclosed impellers does not affect
impeller vanes, and worn clearances may be
restored by replacing wearing rings. The
thrust produced by semiopen impellers may
be as much as 150 per cent greater than
that by enclosed impellers.

Occasionally in power plants, the maximum water level that can be carried in the
condenser hotwell will not give adequate
NPSH (net positive suction head) for a conventional horizontal condensate pump
mounted on the basement floor, especially
if the unit is one that has been installed in
an existing plant in a space originally al-
lotted for a smaller pump. To build a pit
for a conventional horizontal condensate

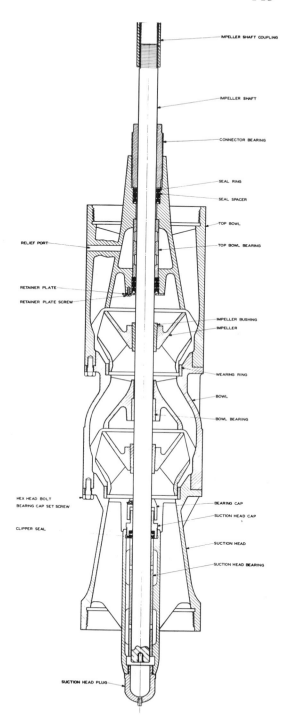

**Fig. 14.15 Section of bowl of vertical turbine
pump (closed impellers) for connection to enclosed shafting**

pump or a vertical dry-pit pump that will
provide sufficient submergence involves
considerable expense. Pumps of the design
shown in Fig. 14.17 have become quite
popular in such applications. This is basically a vertical turbine pump mounted in a
tank (often called a can) that is sunk into
the floor. The length of the pump has to be
such that sufficient NPSH will be available
for the first-stage impeller design, and the
diameter and length of the tank has to al-

low for proper flow through the space between the pump and tank and then for a
turn and flow into the bellmouth. Installing
this design in an existing plant is naturally
much less expensive than making a pit, because the size of the hole necessary to install the tank is much smaller. The same
basic design has also been applied to pumps
handling volatile liquids that are mounted
on the operating floor and not provided
with sufficient NPSH.

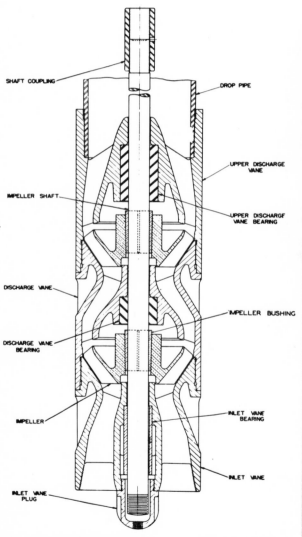

Fig. 14.16 Section of bowl of vertical turbine pump (open impeller) for connection to open line shafting

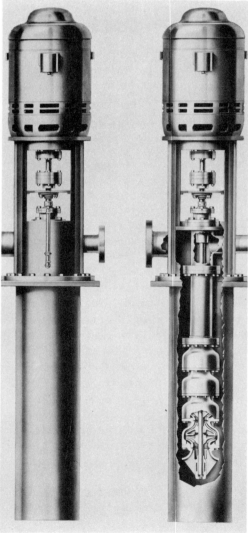

Fig. 14.17 Vertical turbine "can" pump for condensate service

Propeller pumps

Originally the term "vertical propeller pump" was applied to vertical wet-pit diffuser or turbine pumps with a propeller or axial-flow impellers, usually for installation in an open sump with a relatively short setting (Fig. 14.18 and 14.19). Operating heads exceeding the capacity of a single-stage axial-flow impeller might call for a pump of two or more stages or a single-stage pump with a lower specific speed and a mixed-flow impeller. High enough operating heads might demand a pump with mixed-flow impellers and two or more stages. For lack of a more suitable name, such high-head designs have usually been classified as propeller pumps also.

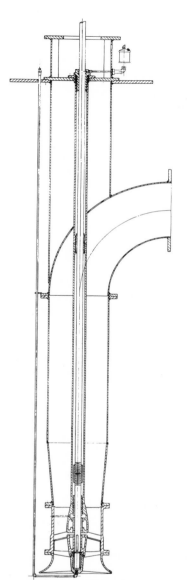

Fig. 14.18 Section of vertical propeller pump with below-ground discharge

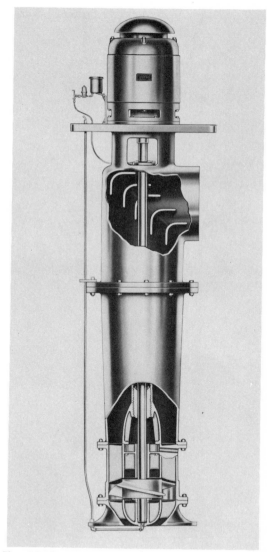

Fig. 14.19 Vertical propeller pump with below-ground discharge

(Courtesy Peerless Pump Co.)

Although vertical turbine pumps and vertical modified propeller pumps are basically the same mechanically and even could be of the same specific speed hydraulically, a basic turbine pump design is one that is suitable for a large number of stages, whereas a modified propeller pump is a mechanical design basically intended for a maximum of two or three stages.

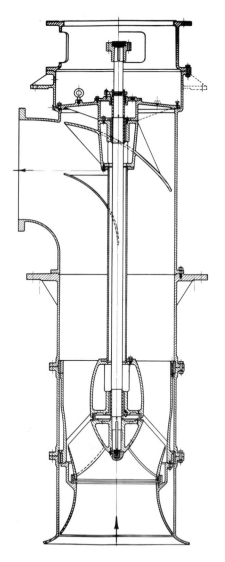

Fig. 14.20 Section of vertical modified-propeller pump with removable bowl and shafting assembly

Most wet-pit drainage, low-head irrigation, and storm-water installations employ conventional propeller or modified propeller pumps. These pumps have also been used for condenser circulating service, but a specialized design dominates this field. As large power plants are usually located in heavily populated areas, they frequently have to use badly contaminated water (both fresh and salt) as a cooling medium. Such water quickly shortens the life of fabricated steel. Cast iron, bronze, or an even more corrosion resistant cast metal must therefore be used for the column pipe assembly. This requirement means a very heavy pump if large capacities are involved. To avoid the necessity of lifting this large mass for maintenance of the rotating parts, some designs (one of which is illustrated in Fig. 14.20) are built so that the impeller, diffuser, and shaft assembly can be removed from the top without disturbing the column pipe assembly. These designs are commonly designated as "pull-out" designs.

Like vertical turbine pumps, propeller and modified propeller pumps have been made with both open and enclosed line shafting. Except for condenser circulating service, enclosed shafting—using oil as a lubricant but with a grease-lubricated tail bearing below the impeller—seems to be favored. Some pumps handling condenser circulating water use enclosed shafting but with water (often from another source) as the lubricant, thus eliminating any possibility of oil getting into the circulating water and coating the condenser tubes.

Propeller pumps have open propellers. Modified propeller pumps with mixed-flow impellers are made with both open and closed impellers.

Sewage pumps

Except for some large vertical propeller pumps that handle dilute sewage (basically storm water contaminated by domestic sewage), vertical wet-pit sewage pumps have a bottom-suction volute design with impellers capable of handling solids and stringy materials with minimum clogging. Usually suspended from a higher floor by means of a drop pipe, these pumps often employ covered or enclosed shafting like that used in vertical turbine pumps. Except for a bellmouth suction inlet and certain differences in bearing and stuffing box construction, they usually are hydraulically and mechanically similar to their dry-pit counterparts.

Three basic constructions have been used for such pumps. The first employs impellers without back rings and a water- or grease-lubricated bearing with a seal at its lower end immediately above the impeller (Fig. 14.21). The upper end is vented to the suction pit to prevent any appreciable hydraulic pressure on the seal at the lower end of the shaft cover pipe; otherwise water would work into the cover pipe. The seal at the lower end of the impeller bearing must be especially effective with high pump heads; otherwise a considerable amount of water will leak through the bearing, with some cutting if grit is present.

The second construction is similar but employs pump-out vanes or wearing-ring joints on the back side of the impeller (the latter necessitates balance holes through the impeller hub) so that the bearing is subjected only to suction pressure. The third construction, used primarily with impellers having no back rings or pump-out vanes, retains a stuffing box in some form, with bearings above and separate from the box.

Although shaft seals and packing used to seal the lower end of the cover pipe or bottom bearing are intended to exclude as much water as possible, some leakage is to be expected at high suction-water levels even when the seal is new. As some of the shaft bearings may have to operate in water or a mixture of oil and water, the bearing may wear relatively faster than one lubricated positively with oil or grease. Wet-pit sewage pumps should usually be limited to services requiring operation for a very limited period of the day.

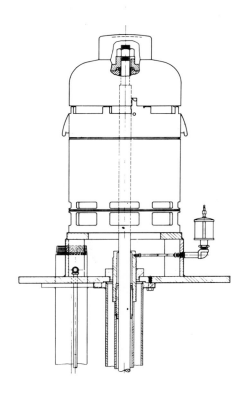

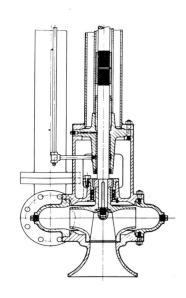

Fig. 14.21 Section of vertical wet-pit sewage (non-clogging) pump

Volute pumps

Besides the more common wet-pit sewage pumps, some single-suction designs with either bottom or top suction and double-suction designs that are supported on the pit floor are sometimes used on this service. Except for some floating dry-dock installations, such pumps have little modern application.

Sump pumps

The term "sump pump" ordinarily conveys the idea of a vertical wet-pit pump that is suspended from a floor plate or sump cover or supported by a foot on the bottom of a well, that is motor-driven and automatically controlled by a float switch, and that is used to remove drains collected in a sump. The term does not indicate a specific construction, for both diffuser and volute designs are used; these may be single-stage or multistage and have open or closed impellers of a wide range of specific speeds.

For very small capacities serviced by fractional hp motors, "cellar drainers" can be obtained. These are small and usually single-stage volute pumps with single-suc-

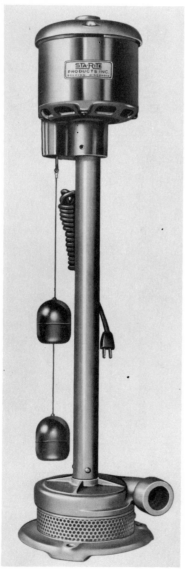

Fig. 14.22 Typical cellar-drainer sump pump
(*Courtesy Sta-Rite Products.*)

Fig. 14.23 Typical duplex sump pump
(*Courtesy Economy Photo.*)

TOP VIEW

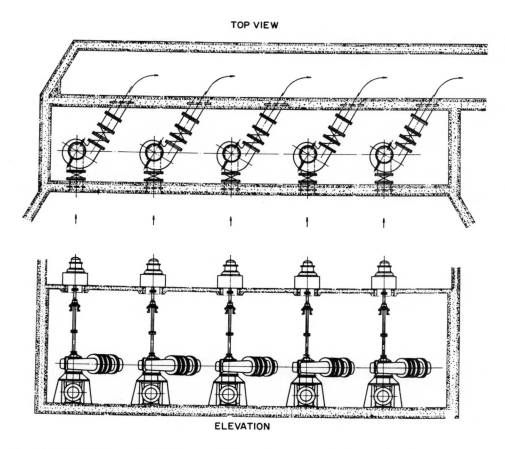

ELEVATION

Fig. 14.24 Multiple-unit station with vertical dry-pit volute pumps lengthwise along long suction bay

tion impellers (either top or bottom suction) supported by a foot on the casing; the motor is supported well above the impeller by some form of a column enclosing the shaft. These drainers are made as complete units, including float, float switch, motor, and strainers (Fig. 14.22).

Sump pumps of larger capacity may be vertical propeller or turbine pumps (single stage or multistage) or vertical wet-pit sewage or volute pumps. If solids or other waste materials may be washed into the sump, the vertical wet-pit sewage pump with a nonclogging impeller is preferred. The larger sump pumps are usually standardized but obtainable in any length, with covers of various sizes (on which a float

switch may be mounted), and the like. Duplex units, that is, two pumps on a common sump cover (sometimes with a manhole for access to the sump) are often used (Fig. 14.23). Such units may operate their pumps in a fixed order, or a mechanical or electrical alternator may be used to equalize their operation.

Application of vertical wet-pit pumps

Like all pumps, the vertical wet-pit pump has advantages and disadvantages, the former mostly hydraulic and the latter primarily mechanical. If the impeller (first-stage impeller in multistage pumps) is submerged, there is no priming problem, and

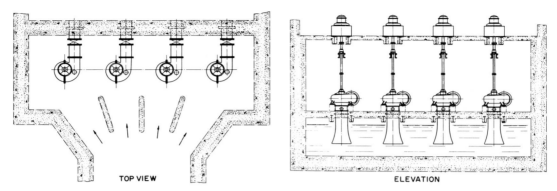

Fig. 14.25   Multiple-unit station with vertical wet-pit volute pumps at end of conduit

the pump can be automatically controlled without fear of its ever running dry. Moreover, the available NPSH is greater (except in closed tanks) and often permits a higher rotative speed for the same service conditions. The only mechanical advantage is that the motor or driver can be located at

any desired height above any flood level. The mechanical disadvantages are the following: (1) Possibility of freezing when idle, (2) possibility of damage by floating objects if unit is installed in an open ditch or similar installation, (3) inconvenience of lifting out and dismantling for inspection and re-

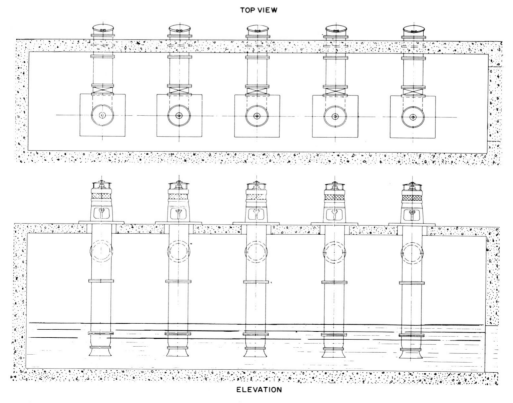

Fig. 14.26   Multiple-unit station of vertical propeller pumps with suction supply from one end of well

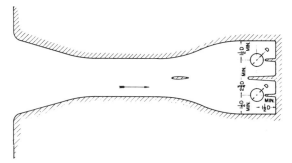

Fig. 14.27 Pump installation with good intake design

D = suction bellmouth diameter.

pairs, no matter how small, and (4) the relative short life of the pump bearings unless the water and bearing design are ideal. The vertical wet-pit pump is the best pump available for some applications, not ideal but the most economical for other installations, a poor choice for some, and the least desirable for still others.

Typical arrangements of vertical pumps

A pump is only part of a pumping system. The hydraulic design of the system external to the pump will affect the over-all economy of the installation and can easily have an adverse effect upon the performance of the pump itself. Vertical pumps are particularly susceptible because the small floor space occupied by each unit offers the temptation to reduce the size of the station by placing the units closer together. If the size is reduced, the suction arrangement may not permit the proper flow of water to the pump suction intake. As many factors are involved in the design of a suction well and the location of a bellmouth and no simple rules or relations can be reliably applied, none are included in this discussion. The physical size of the pumps (whether propeller or volute) rarely affects the design of the suction well, the location of the bellmouth, or the spacing of the units. These are usually controlled by factors governing the proper flow of the water to the bellmouth.

Figure 14.24 illustrates an ideal arrangement for a multiple-unit station with dry-pit pumps. It provides an unrestricted flow on the suction side to all the units. Stations using this arrangement for a group of vertical volute pumps often have the suction bellmouths and elbows formed right in the concrete substructure. If wet-pit pumps are installed with vertical bellmouths, adequate clearance must be provided at the back wall and between the units (Fig. 14.25). This arrangement illustrates a common situation in which the suction is located at the end of a conduit the width of which is less than the length of the suction well. Without a flared section with division walls to guide the distribution of the incoming water to the various units, the flow would

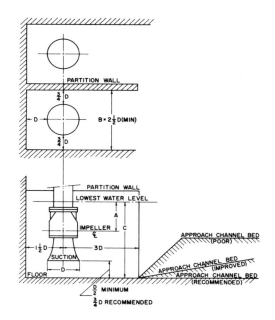

Fig. 14.28 Recommended channel and pit design

KEY: A = Minimum submergence above impeller centerline, approximately 1½ to 2D depending on pump cavitation characteristics; B = minimum width of sump or pit; C = minimum depth of sump or pit; D = suction bell diameter (normally same as bowl diameter). Cross-sectional area of sump (B × C) shall not be less than ten times the suction bell area ($\pi D^2/4$).

be badly disturbed and the operation of the pumps adversely affected.

A propeller-pump arrangement that is often troublesome (vertical volute-pump arrangements with suction bellmouths like those in Fig. 14.25 have the same problem) is shown in Fig. 14.26. Unless the width of the suction well provides sufficient area and unless the locations of the bellmouths permit good flow, the demand of the units first in line will disturb the flow in more removed units. Very often installations of this general arrangement require extensive baffling to correct the distribution. Some stations are made with walls that form individual wells for each pump, a channel to supply these wells running lengthwise of the station.

Various recommendations have been developed over the years for the dimensioning of intake channels and approaches. If feasible, an intake like that illustrated in

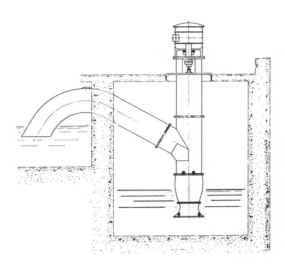

Fig. 14.30 Vertical wet-pit propeller pump with siphon discharge

Fig. 14.27 will give excellent results. The dimensions for the channel width and spacing are given in terms of the suction bellmouth diameter. Another example of good channel and pit design for vertical turbine pumps is given in Fig. 14.28, which also indicates recommended clearances between the suction bellmouth and the bottom of the pit and between the pump, the pit back wall, and the partition walls.

If long discharge lines are involved, valves are required in the piping. Normally, both a gate valve and a check valve are used (Fig. 14.29). The check valve acts to prevent reverse flow, whereas the gate valve functions when the unit is shut down for an extended period. In some installations, the gate valve is omitted, and stop planks or a sluice gate are used. A cone valve that acts both as a check and a stop valve appears in other installations. The high cost of this valve, however, usually restricts its use to installations requiring a flow that is started and stopped gradually to prevent water hammer. A few installations with long discharge lines for single pumps have no valve other than a flap valve at the discharge end. If the unit is

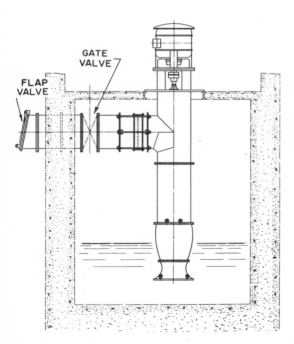

GATE
VALVE

FLAP
VALVE

Fig. 14.29 Vertical wet-pit propeller pump with gate valve and flap valve

stopped, the water in the discharge line flows back through the pump until the pipe is emptied.

If the design of an installation or the failure of a check or flap valve to close permits a reverse flow of water through a pump, the pump acts as a water turbine. The torque developed by the pump as a turbine will cause reverse rotation in freely rotating drivers like electric motors. Usually it is not sufficient to cause reverse rotation in internal combustion engines. In motors, the reverse speed that will be attained will depend both upon the net head and the runaway speed of the pump acting as a water turbine. The net head is then less than the static head because of friction losses. The runaway speed is dependent upon the specific speed of the pump. Higher specific speeds have higher runaway speeds (measured as a percentage of normal speeds). The reverse speed obtainable in an actual installation is usually below the safe operating speed of its component parts, and it is not necessary to use a special design.

The use of a siphon discharge eliminates the necessity for valves in the discharge line (Fig. 14.30). The high point of the siphon must be above high-water level on the discharge in order to break the siphon and prevent backflow of the water when the pump is shut down. When a pump operating on a siphon discharge is started, the usual procedure is to exhaust air from the system by a priming device until the pump is primed. The pump may then be started to help fill the siphon. The connection to the high point of the siphon is also provided with a valved opening so that air can be admitted and the siphon broken when it is desired to stop the unit. It is possible to control the admission of air automatically so that the valve functions if the unit stops for any reason.

Although siphons with short legs are relatively simple and troublefree both in design and operation, more care must be taken if they have long legs. Some siphons operate successfully, with legs exceeding 25 ft, but these are primarily limited to circulating systems in power plant installations. The use of a siphon discharge is desirable in drainage installations for pumping over a levee because it provides a lower head than would be obtained if the water were discharged at the top of the levee.

15 *Special Designs: Self-Priming Pumps*

The standard centrifugal pump cannot handle air or vapors. Unless it is located beneath its source of supply, some means must be found of filling both the pump and its suction piping with liquid, that is, to prime it. A demand naturally developed, therefore, for a centrifugal pump able to handle appreciable quantities of air and to reprime itself automatically when located above the water supply. This requirement is especially important in the construction field because pumps may be used to dewater areas into which seepage is slower than the pump can handle. A standard pump will operate until it uncovers the entrance to the suction pipe, get air-bound, and then be unable to reprime itself even after sufficient seepage has accumulated to prevent further air infiltration.

A true "self-priming pump" is one that will clear its passages of air if it becomes air-bound and resume delivery of the pumped liquid without outside attention. Therefore, its basic requirement is that the pumped liquid entrain air (in the form of bubbles) so that the air will be removed from its suction side. The air must be allowed to separate from the liquid once the mixture of the two has been discharged by the impeller, and the separated air must

be allowed to escape or to be swept out through the pump discharge. A self-priming pump therefore requires an air-separator, which is a large stilling chamber or reservoir provided on its discharge side to effect this separation.

Several ways exist of making a centrifugal pump self-priming, the most important being the following:

1. Recirculation from discharge back into suction
2. Recirculation within the discharge and impeller itself.

These two basic methods have many variations; only one example of each will be discussed here.

Recirculation to suction

A pump made self-priming by this method contains a liquid reservoir either attached to or built in the casing. The first time the pump is to be started, this reservoir is filled. A recirculating port is provided in the reservoir, communicating with the suction side of the impeller. As the pump is started, the impeller handles whatever liquid comes to it through the recirculating port plus a certain amount of air

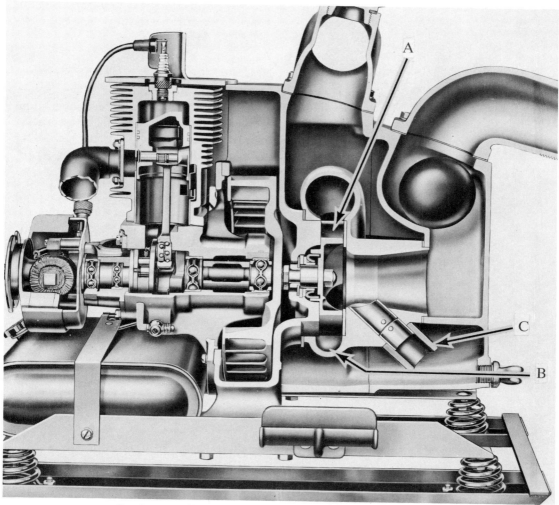

Fig. 15.1   Self-priming pump with valved recirculation to suction
(*Courtesy Homelite Corp.*)

from the suction line. This mixture of air and liquid is discharged into the water reservoir where the two elements separate, the air passing out of the pump discharge and the liquid returning to the suction of the impeller through the recirculating port. This operation continues until all the air has been exhausted from the suction line. The vacuum thus produced draws the liquid from the suction supply right up to the impeller. It is essential that the reservoir remain filled with liquid when the pump is brought to a stop. This is accom-

plished by incorporating either a valve or some form of trap between the suction line and the impeller.

A typical self-priming pump operating on this principle is illustrated in Fig. 15.1. The pump housing, *B*, consists of a conventional volute and an inlet passage. The inlet has a priming passage with a priming valve, *C*, attached to it. This priming valve is a cylindrical rubber tube. The impeller, *A*, is of a conventional semienclosed design. During priming, the pump body is filled with water. This water is drawn into the

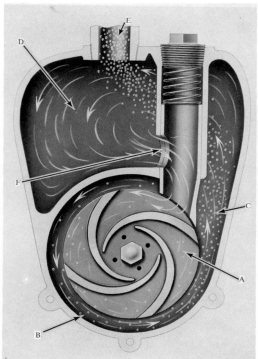

Fig. 15.2 Self-priming pump with volute recirculation

valve is built into the suction line to maintain the vacuum in the line between operations.

Recirculation at discharge

This form of priming is called "volute priming" or "diffuser priming," depending on the design of the discharge casing. It may be distinguished from the preceding method by the fact that the priming liquid is not returned to the suction of the pump but mixes with the air either within the impeller itself or at its periphery. Its principal advantage, therefore, is that it eliminates the complexity of internal valve mechanisms.

A typical "volute priming" self-priming pump is illustrated in Fig. 15.2. An open impeller, *A*, rotates within a volute casing,

pump housing through the priming valve and discharged from the volute back into the pump body. The suction created by the impeller draws air from the inlet passage at the same time that it is drawing water through the priming valve. The air is mixed with the water and discharged into the pump body along with the water. In the pump body, the air bubbles separate from the water, rise to the surface, and pass out through the pump discharge while the priming valve picks up water that is relatively free of air.

After the air has been exhausted from the suction piping or hose and water is drawn into the pump, sufficient pressure difference exists between the pump body and the inlet passage to cause the rubber priming valve to collapse. The recirculation thereby being stopped, all the water that goes through the impeller is discharged from the pump body. A ball check

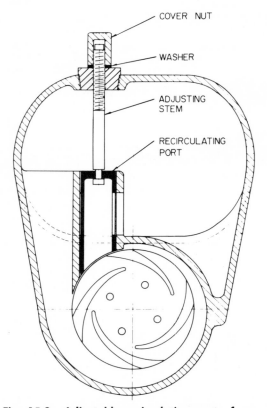

Fig. 15.3 Adjustable recirculating port of volute recirculation pump

Fig. 15.4 Self-priming pump with separate motor drive

B, discharging the pumped liquid through passage C into the sealing reservoir, D. When the pump starts, the trapped liquid carries entrained air bubbles from the suction to the discharge chamber. There, the air separates from the liquid and escapes into the discharge chamber, E. The liquid in the reservoir returns to the impeller through the recirculation port, F, reenters the impeller, and mixing once more with air bubbles is discharged through C. This operation is repeated continuously until all the air has been expelled through E. Once the pump is primed, the uniform pressure distribution established around the impeller prevents further recirculation, and the liquid is discharged into reservoir D both at C and at F.

Some sizes of this pump incorporate an externally adjustable recirculation port (Fig. 15.3). The original clearance between the impeller and the casing can be restored, when these parts become worn, by the following steps.

1. Remove cover nut

2. Turn adjusting stem until recirculating port touches impeller
3. Back off adjusting stem 1½ turns
4. Replace cover nut.

This adjustment appreciably extends the usable life of the pump casing. An added advantage of this design is the ability to use impellers of different diameters in the same casing without losing priming capabilities.

Such pumps are built with as many combinations of drives as ordinary nonpriming

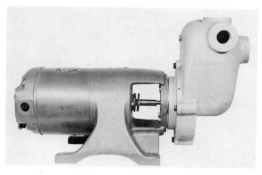

Fig. 15.5 Close-coupled self-priming pump

**Fig. 15.6 Portable engine-driven self-priming
pump**

pumps. They are commonly available
either with separate drive (Fig. 15.4), close
coupled (Fig. 15.5), or engine drive (Fig.
15.6).

REGENERATIVE PUMPS

The name "regenerative pump" describes
a unit with a multiblade impeller that develops head or pressure by a principle considerably different from that of a centrifugal pump. These pumps have had a number of other names given to them, for example, "turbulence pumps," "peripheral
pumps," "vortex pumps," and "turbine
pumps." The term "regenerative," however,
best describes the actual pumping principle
involved.

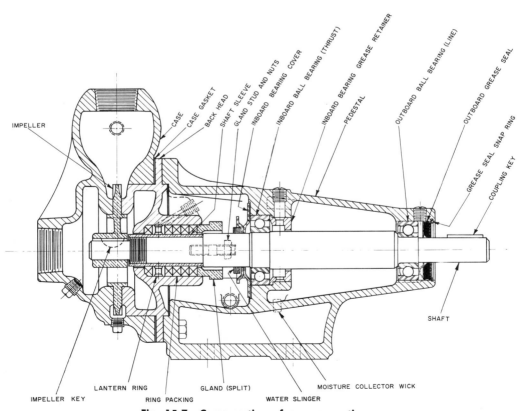

Fig. 15.7 Cross section of a regenerative pump

Fig. 15.8   "Exploded" regenerative pump

Principle of operation

Figure 15.7 shows a cross section of a regenerative pump; Fig. 15.8 is an "exploded" photograph of the same unit. The impeller has a multiplicity of radial vanes cut into its rim that rotate within an annular chamber. The liquid enters the pump casing and flows to both sides of the impeller either through a cored passage in the casing or through ports or openings provided for this purpose in the web of the impeller. This design, in effect, makes the pump a double-suction unit and balances the axial hydraulic thrust.

At one point of the periphery, there is a separating wall or "stripper" that the impeller passes, in its rotation, with a very narrow clearance. Passages are provided from the suction into the annular chamber surrounding the impeller rim, immediately beyond this dividing wall. The liquid is picked up in the spaces between the impeller vanes and then thrown out again into the annular chamber because of the kinetic energy it gains from the centrifugal force action in the impeller. The kinetic energy is transformed into pressure energy as the liquid slows down in the casing.

The manner in which a regenerative pump develops head is illustrated in Fig. 15.9. The liquid enters the casing and flows to both sides of the impeller, twin passages leading the liquid to the impeller blades (Fig. 15.9a). Each casing is equipped with a dividing wall (or stripper) through which the impeller passes with close clearance, as shown in Fig. 15.9b (at A). Just beyond this wall in the direction of rotation, the twin suction passages, which have passed around the sealing wall, come into the impeller chamber (at B). The impeller blade engages the liquid as it comes out of the suction passage, and centrifugal force throws the liquid out to the periphery of the impeller (Fig. 15.9c). The liquid leaving the impeller blade has had velocity energy added and leaves the impeller as shown in the vector diagram (Fig. 15.9d). The casing passage causes a gradual reduction of velocity with the accompanying increase in potential energy (pressure). The pump has thus generated head. The shape of the space between the impeller vanes imparts a rotating motion to the liquid as it leaves the impeller cavities (Fig. 15.9e). As the rotating motion continues in the annular chamber, the liquid is guided back into the "root" of the cavities, proceeding circumferentially around the chamber (Fig. 15.9f). The cycle is then repeated, adding energy to the liquid every time it leaves and reenters the

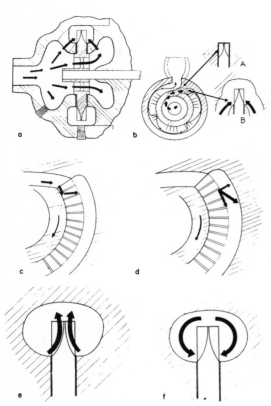

Fig. 15.9 Development of pressure in regenerative pumps

speed variation: (1) The capacity varies directly with the speed, (2) the head varies as the square of the speed, and (3) the power consumption varies as the cube of the speed.

The efficiency of regenerative pumps is considerably lower than that of centrifugal pumps. On the other hand, they develop much higher heads and are therefore useful for small capacities that would otherwise require multistage centrifugals. They are usually applied to capacities less than 100 gpm and for heads up to 500 or 600 ft. A few sizes are available for capacities up to 200 gpm and for 1,200- to 1,500-ft heads, but these are very special applications. The regenerative pump can handle viscous liquids up to about 400 ssu; if viscosities exceed this value, performance falls off very rapidly, and the pump ceases to be practical.

Self-priming features

As long as sufficient liquid remains within the pump to seal the clearance between the impeller and the separating wall

impeller. The number of times the process repeats itself may vary from 2 to 50 depending upon the head to be developed by the pump. The more times the liquid reenters and is discharged from the impeller, the higher the head. When the liquid finally reaches the discharge side of the separating wall, it flows into the discharge passage and out the discharge nozzle.

Performance characteristics

The performance of a regenerative pump resembles that of a high specific-speed centrifugal pump in that the head rises very rapidly with a reduction in capacity, as does power consumption. Typical performance characteristics are illustrated in Fig. 15.10. Both pumps follow the same laws of

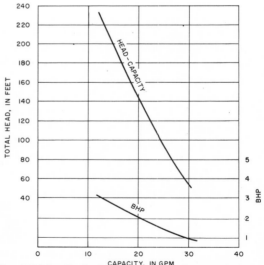

Fig. 15.10 Typical performance characteristics of a regenerative pump

Shut-off head of 355 ft, 1,750 rpm, and 15 ft suction lift.

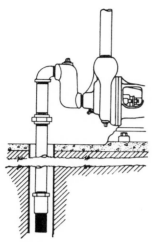

Fig. 15.11 Arrangement of trap in the suction of a regenerative pump

in the casing, the impeller cavities will take up all fluid present, whether simple liquid or a mixture of liquid and vapor or air. The regenerative pump will therefore always prime itself—by evacuating the air out of its suction line—provided the installation is arranged to trap sufficient liquid on shut-down. This condition is usually met by building in a trap in the pump suction (Fig. 15.11). In addition, an enlargement is provided in the discharge of the pump to slow down the velocity of the delivered liquid and to permit its separation from any vapor or air.

General application

Because the satisfactory operation of a regenerative pump depends on the close clearance between the impeller and the separating wall, or stripper, this pump is not too suitable for handling corrosive liquids or liquids containing abrasive foreign particles. The first may attack the metal at the running clearance joint to a point that the pump will lose a major part of its capacity through internal recirculation. Solid particles of the products of corrosion may also build up and wear upon the pump surfaces at the running clearances just as grit or other abrasive particles in the liquid would. The regenerative pump should ordinarily be used to handle clean, clear liquids. To prevent the entry of foreign material, a 40-mesh strainer is desirable.

The clearance at the dividing wall, moreover, has a greater effect on the effective capacity of this pump than the clearances at the wearing ring of a centrifugal pump. Regenerative pumps thus require more frequent maintenance and renewal of internal clearances. An ample margin over the maximum requirement for pump capacity is recommended. Depending upon the pump's construction, clearances can be renewed either by replacement of parts (side-plates, casing heads, and the like) or by changing the thickness of the gaskets that determine the relative location of the casing walls and the impeller itself.

Because of the steepness of its head-capacity curve, a regenerative pump operated at excessively low capacities may develop excessive pressures. Consequently, a relief valve is usually arranged in the discharge line to bypass some of the capacity back to the suction line whenever the discharge pressure reaches a predetermined maximum.

16 *Materials of Construction*

Centrifugal pumps can be fabricated of almost all the known common metals or metal alloys, as well as of porcelain, glass, and even synthetics. The conditions of service and the nature of the liquid to be handled finally determine which materials will be the most satisfactory. A listing of materials commonly recommended for various liquids can be readily found in the Standards published by the Hydraulic Institute, as well as in the catalogs and bulletins of pump manufacturers, particularly those who specialize in centrifugal pumping equipment for chemical service, the field that presents the greatest variety of highly specialized problems.

Some of the service conditions that affect the selection of materials are the following:

1. Corrosion resistance
2. Electrochemical action
3. Abrasiveness of suspended solids
4. Pumping temperature
5. Head per stage (affects both the peripheral velocity of the impeller and the liquid velocities in the waterways)
6. Operating pressure
7. Suitability of the material for particular structural features
8. Load factor and expected life.

Our analysis will first concentrate on the materials most commonly used for individual parts.

Casing materials

Centrifugal pump casings are usually made of cast iron. As cast iron at normal temperatures has definite strength limitations, however, a cast-iron casing of any given design will be suitable only for a definite pressure limit. If higher pressures are to be encountered (because of operation at higher than design speeds or of high suction pressures), either the design must be modified to obtain greater strength or another metal that can be stressed to a higher value (cast steel, for example) must be substituted. A stronger metal than cast iron will also be necessary if pumping systems are subjected to water hammer or shock pressures far in excess of normal operating pressures.

Although it is theoretically possible to use a cast-iron casing under high pressure, the thickness of metal required would violate sound foundry practice. Thin cast-steel sections are sometimes substituted for heavy cast-iron casings and may even prove to be the more economical solution. Although every centrifugal pump has its own pressure and temperature limitations, an upper limit exists at which practical and economic reasons dictate the use of cast steel, whatever the pump design. Cast-iron casings are seldom used for pressures over 1,000 to 1,100 psi and temperatures over 350°F.

The repeated heating up and cooling off of pumps handling a hot liquid will aggravate minute imperfections in both iron and steel casings. No certainty exists that an originally tight casing of either material, will remain tight after the unit has been in prolonged service. Leaks in a cast-steel casing, however, can be corrected by welding, whereas welding may be impractical in cast-iron units.

The piping in high-temperature, high-pressure pumping systems is usually made of steel; steel casings are normally preferred, therefore, because it would not be desirable to have a casing weaker than the piping itself. Even pipe connections affect the selection of the casing material. Cast-iron casings are not recommended if these connections have raised face flanges.

As cast-iron loses tensile strength and becomes quite brittle at low temperatures, pumps handling liquids at very low temperatures (for example, brine) usually have casings of alloyed cast iron or cast steel.

Bronze is frequently used for pump casings if the liquid pumped is mildly corrosive (for example, sea or harbor water). Cast steel or even forged steel is used if the discharge pressure or the pumping temperature or a combination of both factors make cast iron unsuitable. Stainless steel is used if the pumped liquid is corrosive or excessively abrasive. Porcelain or glass casings are sometimes used for very special applications.

Impeller materials

Bronze impellers are usually preferred for handling normal liquids for the following reasons: (1) Bronze is easier to cast for complicated cored sections, (2) it is easier to machine, (3) it provides smoother surfaces, (4) it does not rust. However, bronze impellers should not be used with cast-iron casings if the liquid handled is a strong electrolyte. Such liquids require ferrous materials.

Heated bronze expands approximately 40 per cent more than steel. Since pumps are assembled at normal room temperatures, the original radial clearance between the hub of a bronze impeller and a steel shaft will increase in service at higher temperatures. The increased clearance can loosen the impeller on the shaft and introduces the possibility of leakage and erosion. It must also be remembered that the impeller assembly must be held axially by means of shaft nuts or shaft sleeves that are fixed on the shaft by means of external nuts. If the axial assembly is designed to be tight when cold, the unequal expansion of the impeller hub and of the shaft will cause "crushing" of the bronze impeller hub and may even set up severe stresses in the shaft itself. If the assembly is designed to be tight for operating temperatures, the impeller will be loose at low temperatures, and accurate axial alignment will not be possible until the pump becomes heated. Finally, the assembly cannot be kept tight at all operating temperatures without setting up excessive stresses. All these dangers are removed if the pump fittings provide for equal expansion of the shaft, impellers, and sleeves. Therefore, bronze is seldom used for pump fittings if the temperature of the pumped liquid exceeds 250°F.

The use of bronze impellers is also limited by the effect of peripheral speeds. The centrifugal stress exerted on an impeller and the resulting stretch at the impeller hub may become quite appreciable at the higher peripheral speeds of modern high-head pumps. For example, a typical 12-in. bronze or cast-iron impeller mounted on a 3-in. shaft and operating at 3,600 rpm will have a stretch of approximately 0.0011 in. If the pump also handles hot water, say at

250°F, a bronze impeller will undergo an additional temperature expansion difference of 0.0014 in., resulting in a total looseness of 0.0025 in. between the shaft and the impeller, which is excessive.

To avoid the cumulative effect of excessive thermal and centrifugal expansion, the empirical limit on the peripheral speed of bronze impellers handling hot liquids is approximately 160 fps, or a head of 375 ft per stage.

Materials for wearing rings, shafts, sleeves, and glands

Wearing rings are usually made of bronze, and for the same reasons that impellers are. However, cast-iron, cast-steel, stainless steel, or monel rings are sometimes used—whatever the impeller itself is made of—if hardness or other properties unobtainable with bronze are required.

Shafts in pumps requiring shaft sleeves are normally made of open-hearth steel. If high stresses are to be encountered, high tensile strength alloy steels may be preferred. If corrosive liquids are handled, some seepage may occur either through the pores of the impeller hub or through the joint between the impeller and the sleeves, thus requiring shafts of a noncorrosive

metal like stainless steel, phosphor bronze, or monel metal. Shafts in pumps without shaft sleeves are usually made of stainless steel, phosphor bronze, or monel metal, depending upon the liquid to be handled.

Most shaft sleeves are made of bronze. If bronze is not satisfactory because of its relatively low abrasion resistance, stainless steel is substituted. The growing popularity and reduced cost of the stainless alloys has caused stainless steel to supersede ordinary steel (formerly the most common substitute) and, for services like boiler feed, even bronze as sleeve material. A shaft sleeve that protects a shaft in a stuffing box must have a smooth finish if it is to function properly with the packing. Cast iron is thus rarely used for shaft sleeves.

Glands are normally made of bronze, although cast iron or steel may be used in all iron fitted pumps. If pumps handle hydrocarbons, iron or steel glands are bushed with bronze to avoid sparking, which might ignite flammable vapors.

Pump fittings

The expression "pump fittings" is used rather loosely to mean two entirely separate things. In the most accepted sense, it refers to the general construction features of the

TABLE 16.1   MATERIALS FOR VARIOUS FITTINGS

Materials for bearing housings, bearings, and other parts are not usually affected by the liquid handled.

Part	Part no.[1]	Standard fitting	All iron fitting	All bronze fitting
Casing	1	Cast iron	Cast iron	Bronze
Suction head	35	Cast iron	Cast iron	Bronze
Impeller	4	Bronze	Cast iron	Bronze
Impeller ring	12	Bronze	Cast iron or steel	Bronze
Casing ring	3	Bronze	Cast iron	Bronze
Diffuser		Cast iron or bronze	Cast iron	Bronze
Stage-piece	5	Cast iron or bronze	Cast iron	Bronze
Shaft (with sleeve)	2	Steel	Steel	Steel, bronze, or monel
Shaft (without sleeve)	2-A	Stainless steel or steel	Stainless steel or steel	Bronze or monel
Shaft sleeve	10	Bronze	Steel or stainless steel	Bronze
Gland	15	Bronze	Cast iron	Bronze

[1] Parts in this list and in Fig. 16.1–16.4 are numbered according to a proposed standard listing suggested to the Hydraulic Institute by Charles J. Tullo, Chief Engineer, Worthington Corporation. This proposed standard gives stationary parts odd numbers and rotating parts even numbers.

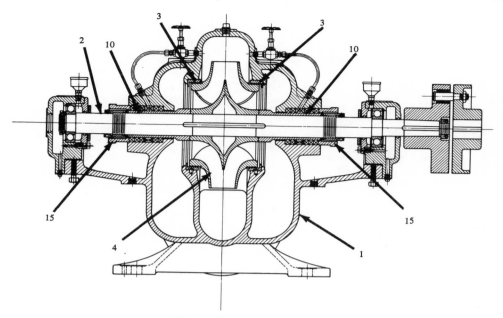

Fig. 16.1 Section of a double-suction, single-stage pump with shaft sleeves

Numbers refer to parts listed in Table 16.1.

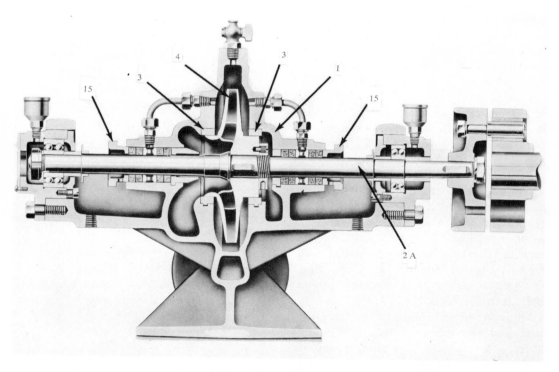

Fig. 16.2 Section of a single-suction, single-stage pump without shaft sleeves

Numbers refer to parts listed in Table 16.1.

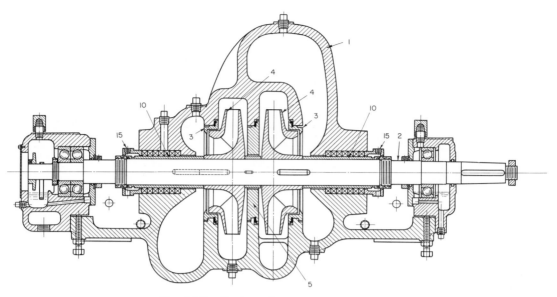

Fig. 16.3   Section of a two-stage pump

Numbers refer to parts listed in Table 16.1.

pump, for example, "ball bearing fitted pump," or to the combination of materials used in the pump, for example, "all iron fitted pump." In a different sense (as in the expression "underwriter fittings"), it may refer to various pieces of auxiliary equipment like valves, gages, or even tools. Table 16.1 indicates the materials commonly used for various types of fittings; Fig. 16.1–16.4 illustrate the materials used for different parts of different pumps.

Standard fitted pump

The so-called "standard fitted" centrifugal pump as defined by the Hydraulic Institute is bronze fitted. It has a cast-iron casing, steel shaft, bronze impeller, bronze wearing rings, and bronze shaft sleeves (if sleeves are used). Some manufacturers regularly furnish stainless steel shafts on pumps without shaft sleeves.

A number of centrifugal pump manufacturers prescribe specific materials for pumps designed for special services. A pump so constructed is usually termed a "standard pump." It is also sometimes erroneously referred to as a standard fitted pump, even

though its materials do not correspond to the Hydraulic Institute definition.

All bronze pump

If all parts of a centrifugal pump that come into contact with the pumped liquid are made of bronze, the pump is called an "all bronze pump."

All iron pump

If all parts of a centrifugal pump that come into contact with the pumped liquid are made of iron or ferrous metals, the pump is called an "all iron pump."

Acid-resisting pump

An acid-resisting pump is one in which all parts in direct contact with the pumped liquid are constructed of materials that will offer the maximum resistance to its corrosive action.

Salt water pumps

Centrifugal pumps handling salt or sea water may be built with standard fittings (cast-iron casings with bronze trim), with all iron or all bronze fittings, or with iron casings and stainless fittings.

Although thousands of standard fitted pumps are used for this purpose, such fittings are not suitable if the sea water is contaminated (for example, harbor water). Failures are usually caused by the galvanic action between the bronze parts and the cast-iron casing, which graphitizes and is ultimately a total loss. High-suction lift accelerates the galvanic action because of the release of oxygen from the water.

Although an all iron pump discourages this electrolytic action, it may occur nevertheless. A certain amount of iron dissolution may take place, leaving graphitized areas that act as cathodes to the untouched anodic parts of cast iron. The resulting galvanic action is self-accelerated. To avoid the graphitization (and poor resistance to cavitation) of cast iron, the impellers and other small pump parts may be made of stainless steel.

An all bronze pump usually offers the longest life for sea water handling (some have lasted more than 20 years). Certain harbor waters are contaminated by chemicals that make bronze unsuitable, however.

Material choice and pH values

The pH value of a liquid is a quantitative representation of its relative acidity or alkalinity. The value is based on the concentration of H+ (positive hydrogen) ions as opposed to OH− (negative hydroxyl) ions in the solution. It is calculated as follows:

$$pH = \log \frac{1}{H+ \text{ concentration}}$$

The lower the pH, obviously, the more acidic the solution.

A solution with a pH value of 7.0 is neutral; values above 7.0 indicating alkalinity and values below 7.0, acidity. As pH values are expressed logarithmically, it must be remembered that changes in pH represent more than a direct linear change. For instance, a solution having a pH of 5.0 is ten times more acidic than one with a pH of 6.0.

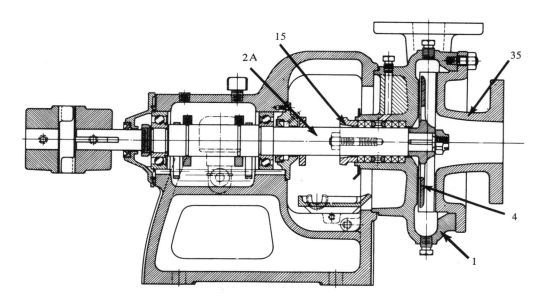

Fig. 16.4 Section of an end-suction, single-stage pump for light duty

Numbers refer to parts listed in Table 16.1.

The pH of a given solution varies somewhat with temperature changes, decreasing rather rapidly up-to 300°F and remaining fairly constant at higher temperatures. For instance, a solution with a pH of 8.5 at 70°F, will have a pH of about 7.0 at 300°F and 6.8 at 500°F (Fig. 16.5).

Although pH values are only one factor influencing the selection of pump materials, it may be stated that standard bronze fitted pumps should not be used for pH values below 6.00 or above 8.5 at pumping temperature. For values below 6.00, all bronze pumps or stainless steel fitted pumps should be used, and for values above 8.5, all iron or stainless steel fitted pumps are preferable.

Material choice and galvanic corrosion

If two dissimilar metals are used in close proximity in a pump handling an electrolytic liquid, severe galvanic corrosion may be expected because of the cell action between the two metals immersed in the electrolyte. The immersion of two connected but dissimilar metals in an electrolytic solution is actually the form of an electric battery cell. The electrochemical reaction causes an electric current and the flow of

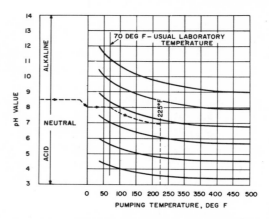

Fig. 16.5   Effect of temperature on pH values

A laboratory test for pH usually involves contact with atmosphere, and as a result, the value is about 0.5 high.

small metal particles from one metal to the other. In centrifugal pumps, these particles may become deposited on the second metal or washed away, depending on the flow velocity. The protected metal is the cathode and the corroded metal is the anode. Obviously, dissimilar metals should be avoided in pumps in which the pumped liquid is electrolytic.

A galvanic series is a tabulation of metals in order of their relative susceptibility to galvanic action. The series in Table 16.2 gives an approximate idea of the interrelation of the metals most commonly used in centrifugal pump construction. It does not include the various stainless steels used for pump parts, because the position of these alloys in the series has been known to change, depending on the exact nature of the electrolyte. It is safe to state, however, that with weak electrolytes they may be placed in the iron and chrome iron group and that the galvanic action within this group is almost negligible. The relative position of the listed metals is subject to some changes; however, changes across the blank spaces between adjoining groups of metals are unusual.

Material choice and structural features

Although materials are often chosen on the basis of their corrosion and abrasion

TABLE 16.2   GALVANIC SERIES OF METALS COMMONLY USED IN PUMP CONSTRUCTION

The series is not complete as it includes only those materials most commonly encountered in centrifugal pump construction. In addition, it does not include the various stainless steels used for pump impellers, wearing rings, or sleeves.

Corroded end (anodic)

Zinc

Iron
Chromium iron
Chromium-nickel iron

Tin
Lead

Brasses
Bronzes
Nickel-copper alloys
Copper

Protected end (cathodic)

resistance, the structural features of a pump part may play a definite role in the final selection. Some parts, for instance, may require extremely thin wall sections for which a material like cast iron would be unsuitable despite its corrosion resistance. Other parts like shaft sleeves require a high degree of polish, and only materials capable of receiving such a finish can be used.

The material used for pressed-on or shrunk-on parts, must be suitable for this method of mounting. For instance, bronze cannot be used if pump impellers have to be shrunk on the shaft, and cast steel or stainless steel must be substituted.

The strength of metal castings depends a great deal on the relative uniformity of their cross sections. The more uniform the cross sections, the stronger the casting, and the less the possibility of internal cracks, inclusions, and the like.

An example of the interrelation of structural features and material problems is shown in the construction of double-suction impellers (Fig. 16.6). The central hub of these impellers is usually cored out in order to avoid a heavy cross section that might result in porous and imperfect castings.

Material choice and the load factor

It is obvious that a selection of metals to provide the longest possible life for a temporary installation would be very unec021nomical. Thus, standard fitted pumps are frequently used for services in which corrosion or erosion will wear a pump out in a relatively short time, if this pump will no longer be needed after the service is per-

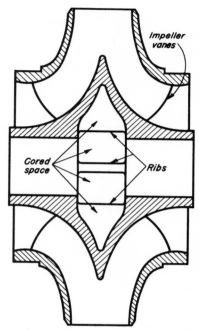

Fig. 16.6 Coring of impeller hub

formed. The same reasoning applies to installations in which pumps operate an extremely small percentage of the time, providing that contact with the pumped liquid during idle periods does not continue the disintegration process, or else that the pump can be drained and flushed out.

Plain common sense dictates that materials be chosen on the basis of optimum economic life, that is, for an initial cost and a cost of part replacement (including the necessary labor) that will yield the lowest over-all total investment during the expected life of the equipment. The materials chosen, therefore, may often be neither the cheapest nor the most expensive available.

TABLE 16.3 SELECTION OF BOILER FEED PUMP MATERIALS

Casing	Fittings	Max temp, deg F	Use
Cast iron	Bronze	250	pH neutral to 8.5, unless water is known to be corrosive
Cast iron	Stainless steel	350	Any pH, unless water is known to be excessively corrosive
Stainless steel	Stainless steel	Any temperature normally encountered	Any pH or if water is corrosive or previous trouble has been reported

If outstanding reliability is desired, on the other hand, the best materials are none too good, even if a pump is to operate only once every ten years. Although operation of the centrifugal pumps on board ships of the U.S. Navy is relatively infrequent in peacetime, for example, and not constant even in war, the most rigid material specifications are enforced because failure of any part of the equipment may prove fatal.

Material choice and metallurgical progress

Metallurgical advances have had a marked effect on centrifugal pumps designed for the pumpage of chemicals and the handling of boiler feedwater. The selection of materials for the former naturally depends upon the liquid used. These pumps comprise a very specialized field; detailed information should be obtained from manufacturers actively engaged in their production. As the selection of materials for boiler feed pumps has been the most affected by recent metallurgical developments, it will be more profitable to discuss some present trends in this field.

Most boiler feed pumps designed for pressures of 800 psi and above are now stainless steel fitted, regardless of temperature or feedwater analysis. Stainless steel fittings show signs of invading the lower pressure field, moreover, and are almost mandatory for all pressure conditions at temperatures above 250°F and for all installations in which feedwater pH is below 7.0 (neutral) or more than 8.5.

Stainless steel casings have been utilized to an increasingly great degree since 1944, especially after the findings of an investigation conducted by the Boiler Auxiliary Subcommittee of the Prime Movers Committee of the Edison Electric Institute. This investigation was initiated to determine the cause of the rapid pump deterioration being experienced in many high-pressure power plants, a condition that had reached alarming proportions in the early forties.

Among the findings made by this subcommittee was the fact that stainless steels containing 5 per cent chromium or more are immune to the attack of any boiler feedwater known.

Table 16.3 has been prepared as a general guide for the selection of boiler feed pump materials. It goes without saying that it represents an average of the recommendations in use today and is subject to some modifications.

The wide range of selection offered by corrosion-resistant stainless steels includes both straight chromium steels and the so-called austenitic chromium-nickel steels. The chromium content of the former may be 5, 9, or 13 per cent; the composition of the latter may range from the familiar 18 per cent chromium, 8 per cent nickel to 25 per cent chromium, 20 per cent nickel. The final selection depends both on design details and operating conditions. It is important to remember, however, that the coefficient of expansion of the austenitic steels approaches that of bronze and is 40 per cent greater than that of carbon steel or steels of lower chromium content. For pumps in which this difference in coefficient may lead to operational difficulties, the use of 18-8 steel or steels of higher chromium content is not recommended.

A certain tendency exists to select stainless steel for a pump casing on the basis of its weldability. Although certain stainless steels are more readily welded than others, it should be remembered that their use, in itself, practically eliminates any need for field welding. However, even the 5 per cent chromium alloy can be practically welded in the field. It must be heated to approximately 400°F prior to welding and to about 1,200 to 1,300°F for stress relieving; it must not be permitted to cool quickly and air harden. Similarly, 13 per cent chromium steel should be heated after welding to about 1,300°F, and 18-8 stainless to 900°F, for complete stress relieving. Of course, stress relieving is not necessary for small welds.

II
PUMP
PERFORMANCE

17 *Heads, Conditions of Service Performance Characteristics*

In selecting the most suitable centrifugal pump for a given application, the most important information to be given the manufacturer is the desired capacity and the head against which the pump will be required to operate while delivering the specified rate of flow.

Units of capacity

The standard unit of capacity for centrifugal pumps varies with the application of the pump as well as the design standards of the country where the pump is used—gallons per minute in the United States, imperial gallons per minute in the British Commonwealth, and cubic meters per hour in countries using the metric system. In the United States, units vary with the pump application as follows: million gallons per day, cubic feet per second, gallons per hour, barrels per day, barrels per hour, pounds per hour, and acre feet per day.

It is a simple matter to convert the various units into gallons per minute (gpm). The equivalents for most units are incorporated in Table 17.1. For a direct conversion chart, see Fig. 17.1.

The pump capacity required by an installation should be stated in gallons per minute at the pumping temperature; any desired or imposed variation in the range of capacities should also be clearly stated. The proper method of specifying required capacity in preparing an inquiry for centrifugal pumps is discussed in some detail in Chap. 23.

Heads

Pumping is the addition of kinetic and potential energy to a liquid for the purpose of moving it from one point to another. This energy will cause the liquid to do work, such as flow through a pipe or rise to a higher level. A centrifugal pump transforms mechanical energy from a rotating impeller into the kinetic and potential energy required. Although the centrifugal force developed depends on both the peripheral speed of the impeller and the density of the fluid, the amount of energy imparted per pound of fluid is independent of the fluid itself. Therefore, for a given machine operating at a certain speed and handling a definite volume, the mechanical energy applied and transferred to the fluid —in foot-pounds per pound of fluid—is the same for any fluid, regardless of density. The pump head, or energy in foot-pounds per pound, will therefore be expressed in feet. Barring viscosity effects, the head gen-

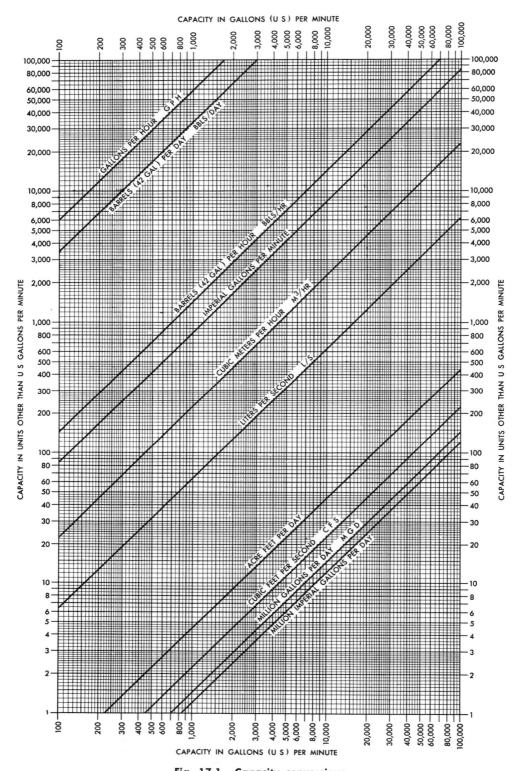

Fig. 17.1 Capacity conversions

For more accurate values, calculate from the equivalents shown in Table 17.1.

erated by a given pump at a certain speed and capacity will remain constant for all fluids. Thus, it is natural to speak of heads in centrifugal pumps in terms of feet of liquid.

Before discussing the various head terms involved in pumping systems it should be mentioned that (1) heads can be measured in various units, such as feet of liquid, pounds per square inch of pressure, inches of mercury, and others depending upon the application and the units of measurement of the country; (2) pressures and head readings can be in gage or absolute units; (3) the difference between gage and absolute units is affected by the existing atmospheric pressure and thus by the altitude; and (4) the pressure at any point in a system handling liquids must never be reduced below the vapor pressure of the liquid.

Conversion of pressures to feet of liquid

A column of cold water approximately 2.31 ft high will produce a pressure of 1 psi at its base. Thus for water at ordinary temperatures, any pressure calculated in pounds per square inch can be converted into an equivalent pressure in feet of water by multiplying by 2.31. For liquids other

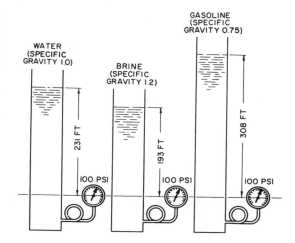

Fig. 17.2 Effect of fluid density on static head

Comparison of the heights of column of water, brine, and gasoline needed to effect a 100-psi pressure at datum line.

than water, the column of liquid equivalent to 1-psi pressure can be calculated by dividing 2.31 by the specific gravity of the liquid.

Figure 17.2 illustrates the effect that specific gravity has on the height of a column of various liquids for equal pressures. Thus a pump that must handle 1.2 specific gravity brine against 100-psi net pressure

TABLE 17.1 CAPACITY EQUIVALENTS

For conversion chart, see Fig. 17.1.

Various units	gpm
1 second-foot or cubic foot per second (cfs)	448.8
1,000,000 gallons per day (mgd)	694.4
1 imperial gallon per minute	1.201
1,000,000 imperial gallons per day	834.0
1 barrel (42 gal) per day (bbl/day)	0.0292
1 barrel per hour (bbl/hr)	0.700
1 acre-foot per day	226.3
1,000 pounds per hour (lb/hr)	2.00 [1]
1 cubic meter per hour (m³/hr)	4.403
1 liter per second (1/s)	15.851
1 metric ton per hour	4.403 [1]
1,000,000 liters per day = 1,000 cubic meters per day	183.5

[1] These equivalents are based on a specific gravity of 1 for water at 62°F for English units and a specific gravity of 1 for water at 15°C for metric units. They can be used with little error for cold water of any temperature between 32°F and 80°F. For specific gravity of water at various temperatures, see Fig. 17.6.

would be designed for a head of 193 ft. If the pump had to handle cold water against the same net pressure, the head would have to be 231 ft, whereas a pump handling 0.75 specific gravity liquid against the same net 100-psi pressure would require a head of 308 ft.

It is obvious that a pump designed to handle water but applied on brine service would develop a 231-ft head of brine or 120-psi pressure while if it was applied to pump 0.75 specific gravity gasoline it would develop a 231-ft head of gasoline or only 75-psi pressure.

The equivalents for the conversion of various pressure and head units other than feet into feet of liquid are indicated in Table 17.2. For quick conversion of pres-

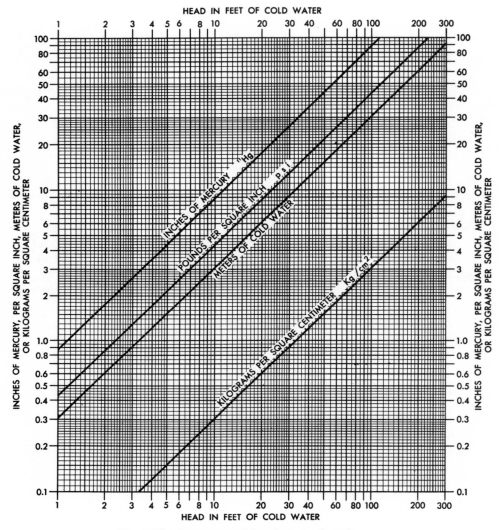

Fig. 17.3 Pressure and head conversion chart

Values are plotted for 62°F (18.7°C) water but can be used for water between 32°F and 80°F. For liquids other than cold water, divide the head by the specific gravity (62°F water = 1.0) of the liquid at the pumping temperature to get the head in feet. For more accurate values, calculate heads from the head equivalents in Table 17.2.

sures and heads into feet of liquid, see Fig. 17.3.

Gage and absolute units

Pressures and their corresponding heads can be expressed either in absolute units or gage units, for instance, 100 psig or 160 psi abs. In gage readings the pressure is given merely in relation to the atmospheric pressure, whereas absolute pressures are gage readings plus the existing atmospheric pressure. In other words, the pressure is referred to an absolute vacuum (Fig. 17.4).

To illustrate, assume a person standing part way up a hill, 25 ft from the bottom. The elevation or level at which he is standing would be his gage basis of measurement. Points below him would be negative (−) gage elevations, and points above him would be positive (+) gage elevations. Thus he would speak of a point 100 ft up as 100 ft gage elevation or of one 10 ft down as

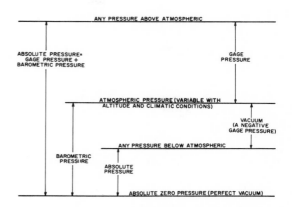

Fig. 17.4 Graphical illustration of atmospheric, gage, and absolute pressures

10 ft below gage level, corresponding to a vacuum in our problem. If he desired to express an elevation measured from the bottom of the hill (absolute datum level, corresponding to zero absolute pressure or a perfect vacuum), he would add 25 ft to his

TABLE 17.2 PRESSURE AND HEAD EQUIVALENTS

For conversion chart, see Fig. 17.3.

$$1 \text{ lb/sq in.} = \frac{2.310}{\text{specific gravity}^{1}} \text{ ft of liquid} = 2.310 \text{ ft of } 62\,^{\circ}\text{F water}$$

$$1 \text{ in. mercury } (32\,^{\circ}\text{F}) = \frac{1.134}{\text{specific gravity}^{1}} \text{ ft of liquid} = 1.134 \text{ ft of } 62\,^{\circ}\text{F water}$$

$$1 \text{ atmosphere}^{2} = \frac{33.95}{\text{specific gravity}^{1}} \text{ ft of liquid} = 33.95 \text{ ft of } 62\,^{\circ}\text{F water}$$

$$1 \text{ kilogram/sq cm} = 1 \text{ metric atmosphere}$$

$$= \frac{32.85}{\text{specific gravity}^{1}} \text{ ft of liquid} = 32.85 \text{ ft of } 62\,^{\circ}\text{F water}$$

$$= \frac{10.01}{\text{specific gravity}^{1}} \text{ m of liquid} = 10.01 \text{ m of } 15\,^{\circ}\text{C water}$$

$$1 \text{ meter} = 3.281 \text{ ft}$$

[1] These equivalents are based on a specific gravity of 1 for water at 62 °F for English units and a specific gravity of 1 for water at 15 °C for metric units. They can be used, with little error, for cold water of any temperature between 32 °F and 80 °F. For the actual specific gravity of water for temperatures to 220 °F, see Fig 17.6.

[2] Not used in conjunction with pumps.

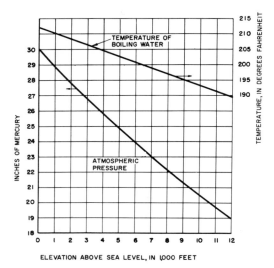

Fig. 17.5 Atmospheric pressures for altitudes up to 12,000 ft

gage reading so that the point 100 ft above him would be $100 + 25$, or 125 ft above the bottom of the hill (125 ft absolute elevation) whereas the point 10 ft below him would be $-10 + 25$, or 15 ft above the bottom of the hill (15 ft absolute elevation).

It is usually feasible to work in terms of gage pressure, but a complicated problem can occasionally be clarified by working entirely in terms of absolute pressure.

Effect of altitude on atmospheric pressure

For pumps installed at elevations above sea level, it must be remembered that there is a decrease in atmospheric pressure of about 1 in. of mercury per 1,000 ft of elevation. At an elevation of 4,000 ft, therefore, the atmospheric pressure is 4 in. of mercury (or about 4.5 ft of water) less than that at sea level, with the result that a centrifugal pump will operate satisfactorily for the same maximum capacities only if the suction lift is 4.5 ft less than that at sea level. This effect should not, however, lead to the confused notion that the net positive suction head required for a pump changes with elevation above sea level. It does not, but the available atmospheric pressure is

reduced. For barometric pressures at various altitudes, see Fig. 17.5.

Vapor pressure

The vapor pressure of a liquid at a given temperature is that pressure at which it will flash into vapor if heat is added to the liquid or, conversely, that pressure at which vapor at the given temperature will condense into liquid if heat is subtracted.

For homogeneous or single component liquids, such as water, the vapor pressure has a very definite value at any given temperature, and tables (such as steam tables) are available that give the vapor pressure of such liquids over a wide range of temperatures (see Table 22.1). Certain mixed liquids, however, such as gasoline, are made up of several components, each having its own vapor pressure, and partial vaporization may take place at various pressures and temperatures.

In figuring heads for pumps, it is important that pressures expressed in pounds per square inch or other pressure units be converted into feet of liquid at the pumping temperature. Care must be taken not to use conversion factors applying to other temperatures for such conversions. For example, the vapor pressure of 212°F water is 14.7 psia (standard barometric pressure at sea level). The equivalent head in feet of water is 33.9 ft of 62°F water. As 212°F water has a specific gravity of 0.959

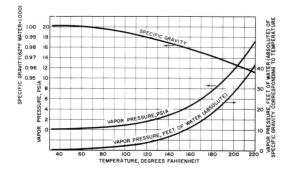

Fig. 17.6 Specific gravity, temperature, and vapor pressure relations for water

compared to a gravity of 1.0 for 62°F water, its equivalent head would be 33.9/0.959, or 35.4 ft (Fig. 17.6).

HEAD TERMS

In its elementary form, "head" denotes the distance at which the free surface of a body of water lies above some datum line; as such, it represents an energy or ability to do work. Energy can also exist as a pressure. Some consider that static head is the sum of the pressure head and the static head of elevation; however, these two factors are generally considered separately. In any pumping system, the liquid must be moved through pipes or conduits that offer certain resistances or, in other words, cause certain frictional losses. This energy dissipation, or head loss, is called a frictional head whereas the energy that has been converted into velocity energy is called velocity head. Thus, static heads, pressure heads, friction heads, and velocity heads may all be encountered in any system. When considering a pump by itself, "head" is a measure of the total energy imparted to the liquid at a certain operating speed and capacity.

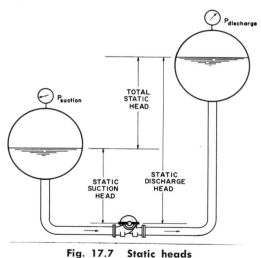

Fig. 17.7   Static heads

System with pump suction and discharge liquid levels under pressures other than atmospheric.

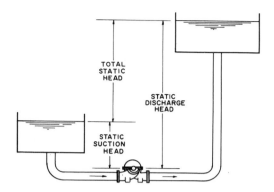

Fig. 17.8   Static heads

Discharge liquid level under atmospheric pressure and suction level above pump centerline.

System head

The total head of a system against which a pump must operate is made up of the following components:

1. Static head
2. Difference in pressures existing on the liquid
3. Friction head
4. Entrance and exit losses
5. Velocity head.

Static head

Static head refers to a difference in elevation. Thus the "total static head" of a system is the difference in elevation between the discharge liquid level and the suction liquid level (Fig. 17.7–17.9). The "static discharge head" is the difference in elevation between the discharge liquid level and the centerline of the pump. The "static suction head" is the difference in elevation between the suction liquid level and the centerline of the pump. If the static suction head is a negative value because the suction liquid level is below the pump centerline, it is usually spoken of as a "static suction lift."

If either the suction or discharge liquid level is under a pressure other than atmospheric, this pressure is sometimes considered as part of the static head, but it is

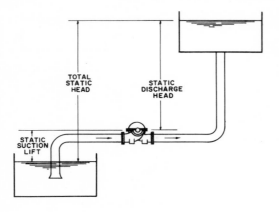

Fig. 17.9 Static heads

Discharge liquid level under atmospheric pressure and suction level below pump centerline.

often considered separately. The latter practice usually permits a clearer picture of the system. If the suction supply is taken from a closed vessel and the liquid level lies above the pump centerline, the difference in elevation of the suction liquid level and the pump centerline is commonly spoken of as "submergence" instead of "static suction head."

Friction head

Friction head is the equivalent head, expressed in feet of the liquid pumped, that is necessary to overcome the friction losses caused by the flow of the liquid through the piping, including all the fittings. The friction head varies with (1) the quantity of flow, (2) the size, type, and condition of the piping and fittings, and (3) the character of the liquid pumped.

Entrance and exit losses

Unless it comes from a main under pressure, such as a city water supply, the suction supply of a pump comes from some form of reservoir or intake chamber. The point of connection of the suction pipe to the wall of the intake chamber or the end of the suction pipe projecting into the intake chamber or reservoir is called the entrance

of the suction pipe. The frictional loss at this point is called the "entrance loss." The magnitude of this loss depends upon the design of the pipe entrance, a well-designed bellmouth providing the lowest possible loss.

Similarly, on the discharge side of the system where the discharge line terminates at some body of liquid, the end of the piping is called the exit. This exit is usually of the same size as the piping, and the velocity head of the liquid is entirely lost. The end of the discharge piping is sometimes a long taper so that the velocity can be effectively reduced and the energy recovered.

Some engineers consider entrance and exit losses as part of the suction and discharge pipe friction losses. Others prefer to consider them separately to make sure that they are not overlooked. This method has the additional advantage of clearly showing if either or both losses are excessive.

Velocity head

Velocity head is the kinetic energy in a liquid at any point, expressed in footpounds per pound of liquid, that is, in feet of the liquid in question. If the liquid is moving at a given velocity, the velocity head is equivalent to the distance the mass of water would have to fall in order to attain this velocity. Thus velocity head can be calculated by the equation:

$$h_v = \frac{V^2}{2g}$$

in which:

    h_v = the velocity head, in feet
    V = the liquid velocity, in feet per second
    g = the acceleration due to gravity, or 32.2 feet per second per second.

In determining the head existing in a pipe at any point, it is necessary to add the velocity head to the pressure gage reading, for the pressure gage can indicate only the pressure energy, whereas the actual head is the sum of the kinetic (velocity) and potential (pressure) energies. Thus, to determine the actual suction head or discharge head, it is necessary to add the velocity head to the gage reading.

If the suction and discharge pressures of a centrifugal pump are taken at points at which the velocities are the same, the velocity head component of each will be the same. The kinetic energy components of both the suction head and the discharge head will also be equal, and the total head can be determined by subtracting the suction gage reading from the discharge gage reading.

In high-head pumps, the kinetic energy is relatively small, but in low-head pumps, it is relatively high. Thus failure to consider the velocity head in determining heads in high-head pumps will not appreciably affect the results. For example, consider a pump handling 1,500 gpm with a 6-in. discharge and 8-in. suction. The discharge velocity head is 4.5 ft whereas the suction velocity head is 1.4 ft. If the suction gage showed an 8.6-ft head and the discharge head showed a 105.5 ft head, the true total head would be (105.5 + 4.5) less (8.6 + 1.4), or 100 ft, whereas the difference in gage readings would be 96.9 ft. Thus the error would be 3.1 per cent of the total head. Had this been a pump in which the discharge gage reading was 1,000 ft, the true total head would be 994.5 ft, whereas the difference in gage readings would be 991.4 ft. The error of 0.3 per cent is too small to be of any concern. If this were a pump in which the discharge head was 45.5 ft, however, the true total head would be 40 ft, whereas the difference in gage readings would be 36.9 ft, for an error of 7.8 per cent.

Whether or not the velocity head can be ignored depends upon the desired accuracy of head determination and upon the accuracy of the pressure readings that can be made. For the 1,000-ft head reading cited above, even with an accurate large scale gage it would be impossible for anyone to read the pressure within 10 ft, a basic error of 1 per cent.

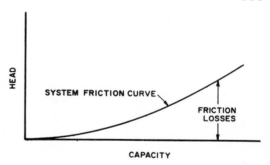

Fig. 17.10   System-friction curve

System friction curve

The friction-head loss in a system of pipes, valves, and fittings varies as a function (roughly as the square) of the capacity flow through the system. For the solution of pumping problems, it is often convenient to show the relation between capacity and friction-head loss through the system graphically. The resulting curve is called the "system friction curve," as shown in Fig. 17.10. The determinations of friction losses are usually rough approximations at best, for the roughness of the pipe is not known. As the friction loss will increase when the pipe tuberculates or otherwise deteriorates with age, it is usual to base the friction loss on constants that have been found from the average of pipe 10 or 15 years old, thus allowing for friction losses in excess of those that will be obtained when the pipe is new. As a result, the pump is generally designed for excess head and delivers over-capacity when installed in a new system or in one that has not suffered from pipe deterioration. (For a complete treatment of friction loss calculations, see Chap. 18.)

System-head curve

The friction-head losses, pressure differences, and static heads of any system can be graphically related (Fig. 17.11). The resulting curve is called the "system-head curve." For systems with varying static heads or pressure differences, it is possible to construct curves for minimum and maximum

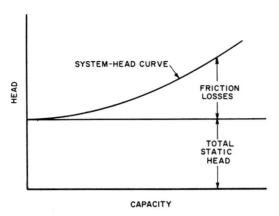

Fig. 17.11   System-head curve

static heads or pressure differentials. The capacity that a pump will be able to deliver under varying conditions can be predicted by superimposing such system-head curves on a pump head-capacity curve (see Fig. 23.3).

DEFINITIONS

Explanation of the head terms used with centrifugal pumps should be applicable to all installations although one or more elements of the total head are usually not involved (because they have zero values). Except as otherwise noted, the definitions given here are based on the current Standards of the Hydraulic Institute.

Suction head and suction lift

As now defined, the total suction head (h_s) is the static head on the pump suction line above the pump centerline minus all friction head losses for the capacity being considered (including entrance loss in the suction piping) plus any pressure (a vacuum being a negative pressure) existing in the suction supply. Rather than express the suction head as a negative value, the term "suction lift" is normally used when the suction head is negative and when the pump takes its suction from an open tank under atmospheric pressure. As the suction lift is a negative suction head measured

below atmospheric pressure, the total suction lift (symbol also h_s) is the sum of the static suction lift measured to the pump centerline and the friction head losses as defined above. (It is sometimes advantageous to express both suction and discharge heads in absolute pressure, but usually it is more suitable to measure them above or below atmospheric pressure.) A gage on the suction line to a pump, when corrected to the pump centerline, measures the total suction head above atmospheric pressure minus the velocity head at the point of attachment. As suction lift is a negative suction head, a vacuum gage will indicate the sum of the total suction lift and velocity head at the point of attachment.

The three most common suction supply conditions are illustrated in Fig. 17.12.

System I involves a suction supply under a pressure other than atmospheric and located above pump centerline; it includes all the components of suction head (h_s). If h_s is to be expressed as a gage reading and P_s is a partial vacuum, the vacuum expressed in feet of liquid would constitute a negative pressure head and carry a minus (−) sign. If the pressure P_s is expressed in absolute pressure values, h_s will also be in absolute pressure values.

A very common installation, II, involves a suction supply under atmospheric pressure located above the pump centerline. As the suction head (expressed as a gage value) has a P_s value of zero, the P_s value can be dropped from the formula.

System III, the most common installation for pumps handling water, involves a suction supply under atmospheric pressure located below the pump centerline. It is optional whether the suction head is expressed as a negative suction head or in positive values as a suction lift. As the source of supply is below the pump centerline (which is the datum line), S is a negative value. It should be noted that the suction lift formula is the same as that for suction head except that both sides have been multiplied by −1. A gage attached to the pump suction flange, when corrected to the pump centerline, will register a partial vacuum or negative pressure. To determine the suction head, it is therefore necessary to add the velocity head to this negative pressure *algebraically*, or, if it is desired to work in terms of a vacuum, the velocity head must be subtracted from the vacuum to obtain the suction lift. For example, if the gage attached to the suction of a pump having a 6-in. suction and pumping at a capacity of 1,000 gpm of cold water showed a vacuum of 6 in. of mercury (equal to 6.8 ft of water), the velocity head at the gage attachment would be 2.0 ft of water, and the suction head would be $-6.8 + 2.0$, or -4.8 ft of water, or the suction lift would be $6.8 - 2.0$, or 4.8 ft of water.

As most centrifugal pump troubles occur on the suction side of the pump, it is a very important part of pump selection to supply complete information on suction conditions, including all operational variations. For some complex problems, it is often necessary to superimpose the variation in total suction head graphically on the suction head limitations of the pump being considered in order to make sure the pump will be suitable.

NPSH

In the pumping of liquids, the pressure at any point in the suction line must never be reduced to the vapor pressure of the liquid. The available energy that can be utilized to get the liquid through the suction piping and suction waterway of the pump into the impeller is thus the total suction head less the vapor pressure of the liquid at the pumping temperature. The available head—measured at the suction opening of the pump—has been named "net positive suction head." It is usually indicated by its initials, NPSH.

Both suction head and vapor pressure should be expressed in feet of liquid being handled and must both be expressed either in gage or absolute pressure units. A pump handling 62°F water (vapor pressure of 0.6 ft) at sea level with a total suction lift of 0 ft has an NPSH of $33.9 - 0.6$, or 33.3 ft, whereas one operating with 15-ft total suction lift has an NPSH of $33.9 - 0.6 - 15$, or 18.3 ft.

A pump operating on suction lift will handle a certain maximum capacity of cold water without cavitation. The NPSH or amount of energy available at the suction

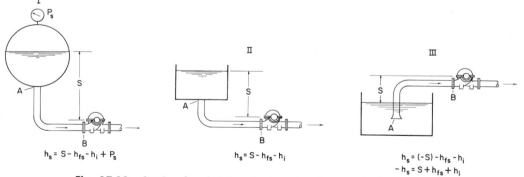

Fig. 17.12   Suction head determination for three typical examples

KEY:
P_s = pressure other than atmospheric
S = static head
h_s = suction head
h_{fs} = total friction loss from A to B
h_i = entrance loss at A
$-h_s$ = suction lift.

The gage reading at B corrected to pump centerline equals the suction head minus velocity head at B.

nozzle of such a pump is the atmospheric pressure minus the sum of the suction lift and the vapor pressure of the water. To handle this same capacity with any other liquid, the same amount of energy must be available at the suction nozzle. Thus, for a liquid at its boiling point (in other words, under a pressure equivalent to the vapor pressure corresponding to its temperature) this energy has to exist entirely as a positive head. If the liquid is below its boiling point, the suction head required is reduced by the difference between the pressure existing in the liquid and the vapor pressure corresponding to the temperature.

It is necessary to differentiate between *available* NPSH and *required* NPSH. The former, which is a characteristic of the system in which a centrifugal pump works, represents the difference between the existing absolute suction head and the vapor pressure at the prevailing temperature. The required NPSH, which is a function of the pump design, represents the minimum required margin between the suction head and vapor pressure at a given capacity.

Both the available and required NPSH vary with capacity (Fig. 17.13). With a given static pressure or elevation difference at the suction side of a centrifugal pump, the available NPSH is reduced with increasing capacities by the friction losses in the suction piping. On the other hand, the required NPSH, being a function of the velocities in the pump suction passages and at the inlet of the impeller, increases basically as the square of the capacity.

A great many factors—for example, eye diameter, suction area of the impeller, shape and number of impeller vanes, area between these vanes, shaft and impeller hub diameter, impeller specific speed, the shape of the suction passages—enter in some form or another into the determination of the required NPSH.

Suction limitations for given capacities in a specific pump were formerly expressed in terms of permissible suction lift at sea level. This calculation required consider-

able work if the pump was to handle water at higher temperatures, if it was to handle a liquid other than water, or if it was installed at a location above sea level. There is a marked tendency to indicate the capacity-suction limitation of all centrifugal pumps in a NPSH-capacity form.

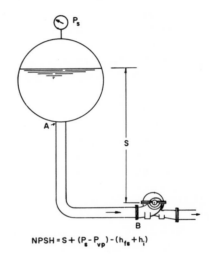

$$NPSH = S + (P_s - P_{vp}) - (h_{fs} + h_i)$$

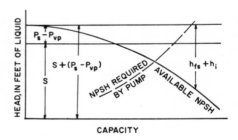

Fig. 17.13 Determination of available positive suction head (NPSH) at pump centerline and relationship of required NPSH and pump capacity

KEY:
S = static head
P_s = pressure value above or below atmospheric
P_{vp} = vapor pressure of liquid
h_{fs} = friction loss from A to B
h_i = entrance loss at A.

All heads and pressures must be expressed in feet of liquid at the pumping temperature with the proper algebraic sign. While P_s and P_{vp} can either be in gage or absolute values, they must both be measured under the same conditions.

Specifying suction conditions

The importance of accurately advising a manufacturer of the actual suction conditions for a centrifugal pump cannot be overemphasized. A pump will be unable to meet its design capacity conditions unless the suction head can provide enough energy to get the liquid into the pump as previously discussed. If a cold nonvolatile liquid is to be handled, it is necessary to know whether there will be suction head or suction lift, and if the latter, what maximum lift can be expected. If the liquid is to be hot or under a pressure corresponding to or near its vapor pressure, the pump must be installed with head on suction, and the available submergence must be indicated. For liquids other than water, information on the pumping temperature and vapor pressure is also necessary. All expected or probable variations in suction conditions should also be specified.

Discharge head

The discharge head (h_d) of a centrifugal pump is the head measured at the discharge nozzle. It is the algebraic sum of the static head, the friction head losses for the capacity being considered, the exit loss at the end of the discharge line, and the terminal head or pressure. It can be expressed with absolute or gage readings in feet of liquid.

Established practice expresses the discharge and suction heads of a horizontal pump with the pump centerline as datum. Usually, discharge and suction heads of a vertical pump are given with the centerline of the discharge as datum. Both heads can be given with other elevations as datum, but it is then necessary to indicate the datum at which they are measured. This practice is often necessary because the exact elevations of the pump centerline or discharge centerline have not been determined prior to the purchase of a pump. When the reading of a gage at the pump discharge has been corrected to the pump centerline, it will indicate the discharge head minus the velocity head at the point of attachment.

Typical discharge systems

Some typical discharge systems are illustrated in Fig. 17.14.

System I shows a system of pump delivery to an elevated tank in which a pressure other than atmospheric exists; it therefore includes all the components of discharge head.

System II is similar to I except that atmospheric pressure exists on the discharge-liquid level (typical of pumps delivering to open reservoirs and elevated tanks). If the discharge head is to be expressed as a gage reading, P_d equals zero and is therefore not shown in the formula. Should it be

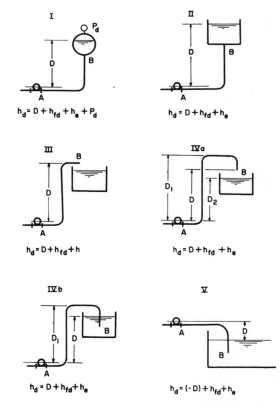

Fig. 17.14 Determination of discharge heads for six typical discharge layouts

KEY:
P_d = pressure deviation from atmospheric
h_e = exit loss at B
h_{fd} = friction loss from A to B (including any siphon losses)
h_{vd} = velocity head at A.

necessary to express the discharge head in absolute values, the atmospheric pressure expressed in feet of liquid must be added to the discharge head expressed as a gage reading.

Although system III illustrates an overhead tank, it applies to all conditions of "overboard discharge." The actual useful static head (the distance from the pump centerline to the discharge water level) is less than the actual static discharge head, D. It is possible to recover all or part of this difference by incorporating a siphon leg on the discharge. Although systems IVb and V would theoretically be the most efficient, it is often desirable not to use a sealed discharge. One reason is to prevent the possibility of back siphonage when the pump is stopped.

In systems IVa and IVb, the effectiveness of the siphon will depend both upon the length of the leg and the design of the piping. Design differences can make the recovery vary from 0 to 100 per cent. For example, if the pipe in IVa was very large in relation to the capacity, the pipe would not run full and the actual static discharge head would consequently become the distance to the actual water level in the loop of the piping. All systems using a siphon leg must be investigated carefully to see what percentage of recovery can be expected and what loss is to be included in the friction loss (h_{fd}).

As the absolute pressure at any point in a siphon must exceed the vapor pressure of the liquid, it is theoretically possible to employ a siphon leg nearly 34 ft long with airfree cold water at sea level. The water being handled is usually not airfree and a reduction in pressure to below atmospheric causes separation of this air, reducing the effectiveness of the siphon. Water siphons more than 20 ft high are rarely encountered. It is even questionable if many of those under 20 ft are 100 per cent effective.

Siphon design must provide for the washing out or removal of entrapped air when operation begins, so that the siphon will be established. Unless the air in the loop can be evacuated, the pump will have to operate during the starting period against a maximum static component, D_1 (IVb). In some condenser circulating installations, this condition results in a starting head much higher than the normal operating head, and special consideration has to be given to the head-capacity curve of the pump. The same effect can occur in system I if P_d is a negative pressure that is not established until after pumping has commenced (as in a barometric condenser).

In systems with variable discharge head, it is usually advantageous to establish the head at various capacities and prepare a graph showing the variation with capacity. When this graph is related to the suction head, the resulting chart will indicate the system head.

The proper method of specifying discharge heads in preparing an inquiry for centrifugal pumps is discussed in some detail in Chap. 23.

Total head

The total head, H, of a centrifugal pump is the energy imparted to the liquid by the pump, that is, the difference between the discharge head and the suction head. As a suction lift is a negative suction head, the total head is the sum of the discharge head and the suction lift. If the discharge head and the suction head are not determined independently, the total head can be calculated (Fig. 17.15) by determining the algebraic sum of the static head from supply level to discharge level, H_{st}, plus all friction losses for the capacity being considered, h_f, plus the entrance, h_i, and exit, h_e, losses plus the terminal pressure, P_d, minus the suction supply pressure, P_s. For complex systems involving both vacuums and pressure, it is often easier to convert all the vacuums and pressures into absolute pressure values of the liquid being handled, expressed in feet. (To convert psi to ft of liquid, multiply by 2.31 and divide by the specific gravity of the liquid at pumping temperature.)

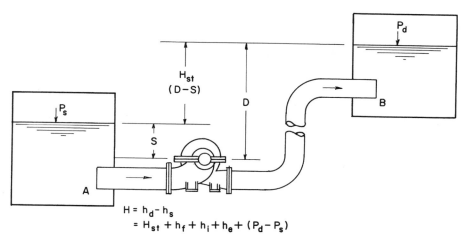

$$H = h_d - h_s$$
$$= H_{st} + h_f + h_i + h_e + (P_d - P_s)$$

Fig. 17.15 Determination of total head

As measured by gages attached to the pump suction and discharge openings, the total head is the discharge head (the sum of the discharge gage reading corrected to the pump centerline and the velocity head at the point of attachment of the discharge gage) minus the suction head (the sum of the suction gage reading corrected to the pump centerline and the velocity head at the point of attachment of the suction gage). As the plus and minus signs of the various elements are easily reversed, and as there are numerous precautions to be considered in taking gage readings, it is advantageous in any test to follow the instructions in the Test Code of the Hydraulic Institute.

Outmoded terminology

Total dynamic head, dynamic suction head, dynamic suction lift, and dynamic discharge head are outmoded terms. Total dynamic head referred to what is now called total head; dynamic suction head, dynamic suction lift, and dynamic discharge head were defined as the heads measured by a gage corrected to the pump centerline, and thus did not include the velocity head element.

Misunderstandings arose if the size of the pump suction and discharge openings were not specified, resulting in different head values for pumps working under identical conditions if their openings were not the same. Furthermore, in determining the total dynamic head from the dynamic discharge head and dynamic suction head it was necessary to correct for any difference in velocity head. The present method of specifying heads is more satisfactory than the dynamic head method.

Head terms for vertical wet-pit pumps

Vertical wet-pit pumps can be either the volute or the turbine type, the latter covering both propeller and vertical turbine pumps, which were formerly called deep-well pumps. The special hydraulic and mechanical problems of vertical turbine pumps have caused them to become virtually independent of the regular centrifugal pump field with different practices and terminology.

Both volute and propeller wet-pit pumps have been handled primarily by engineers in the regular centrifugal pump field. With these two types, total head is the discharge head measured at the centerline of the discharge nozzle, with velocity head included, plus the static distance to the suction water level. Thus, the loss in the suction bell and further losses of the suction strainer and suction piping, if either is furnished, as well as the losses in the column pipe and

TABLE 17.3   HEAD CALCULATIONS FOR PUMPS IN FIG. 17.16

All suction and discharge heads are as corrected to centerline of pumps.
For explanation of head symbols, see Fig. 17.12–17.15.

	I	II	III	IV	V	VI	VII
Liquid	Water	Water	Gasoline	Brine	Water	Water	Water
Temperature, in deg F	62	62	70	32	212	100	185
Specific gravity	1.0	1.0	0.73	1.2	0.959	0.995	0.970
Altitude	Sea level	4,000 ft	Sea level	Sea level	Sea level	Sea level	Sea level
Barometric pressure, psi absolute	14.7	12.65	14.7	14.7	14.7	14.7	14.7
P_s, gage	0	0	0	0	0	28.1 in. Hg vacuum	0
P_s, psi absolute	14.7	12.65	14.7	14.7	14.7	0.9 −	14.7
P_s, in ft of liquid, gage	0	0	0	0	0	−32.0	0
P_s, in ft of liquid, absolute	34.0	29.2	46.5	28.3	35.4	2.2	35.0 −
P_{vp}, psi absolute	0.275	0.275	6.0	0.07	14.7	0.9 +	8.38
P_{vp}, in ft of liquid, absolute	0.6 +	0.6 +	19.0	0.1 +	35.4	2.2	20.0 −
$h_{fs} - h_i$, in ft	2.0	2.0	2.0	2.0	2.0	2.0	2.0
S, in ft	−13.0	−8.2	−7.1	−7.8	20.4	20.4	5.4
NPSH, in ft of liquid = $S - (h_{fs} + h_i) + P_s - P_{vp}$	18.4	18.4	18.4	18.4	18.4	18.4	18.4
h_{sr}, in ft of liquid, gage = $S - (h_{fs} + h_i) + P_s$	−15.0	−10.2	−9.1	−9.8	18.4	−13.6	3.4
h_{vsr}, in ft of liquid	1.0 +	1.0 +	1.0 +	1.0 +	1.0 +	1.0 +	1.0 +
$h_{sg} = h_s - h_{vsr}$, in ft, gage	−16.0 ft	−11.2 ft	−10.1 ft	−10.8 ft	17.4 ft	−14.6 ft	2.4 ft
h_{sg}, pressure gage	14.1 in. Hg vacuum	9.9 in. Hg vacuum	6.5 in. Hg vacuum	11.4 in. Hg vacuum	7.2 psi	12.8 in. Hg vacuum	1.0 − psi
$h_d = H + h_{sr}$, in ft of liquid	65.0	69.8	70.9	69.2	98.4	65.4	83.4
h_{vd}	3.9 +	3.9 +	3.9 +	3.9 +	3.9 +	3.9 +	3.9 +
$h_{dg} = h_d - h_{vd}$, in ft of liquid	61.1	65.9	67.0	65.3	94.5	61.5	79.5
h_{dg}, pressure gage	26.4	28.5	21.2	33.9	39.2	26.5	33.4

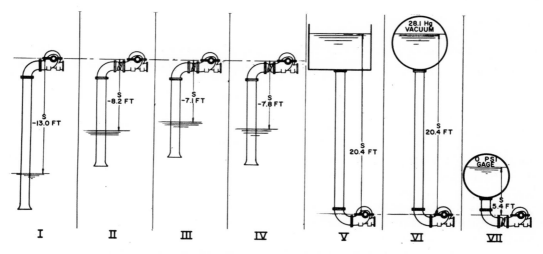

Fig. 17.16 Seven installations of duplicate pumps referred to in Table 17.3

Pumps have 8-in. discharge and 10-in. suction and all operate at 2,000 gpm, 80 ft total head, and 18.4 ft NPSH.

elbow in propeller pumps are charged to the pump.

The following head terminology is used by the National Association of Vertical Turbine Pump Manufacturers for vertical turbine pump applications.

1. *Laboratory head*—discharge pressure by gage in feet plus static vertical distance to suction water level in a test setup using the minimum length of column and shafting for a laboratory test.

2. *Total head*—discharge pressure by gage in feet plus distance to suction water level. (In case of a closed suction, the total head is discharge pressure plus distance to centerline of suction gage minus suction pressure, in feet.)

3. *Dynamic laboratory head*—laboratory head as defined above plus velocity head at the point of the discharge gage attachment.

4. *Total dynamic head*—total head as defined above plus velocity head at the point of the discharge gage attachment minus the velocity head at the point of the suction gage attachment, in case of a closed suction. (What was formerly called field pumping head in vertical turbine pump terminology is now called total head.) In most verti-

cal turbine pumps, the velocity head is a very small portion of the head developed by the pump, and its omission is of little importance.

EXAMPLES OF HEAD CALCULATIONS

The head calculations of seven different installations, with the same hydraulic conditions (gallons per minute, total head, and NPSH), are shown in Fig. 17.16 and Table 17.3. This illustration and table show the effect on the physical installation of a pump, for various applications, if the same NPSH is to be available at the pump.

Installations I and II have the same conditions except for altitude. The reduction in barometric pressure at 4,000 ft elevation makes it necessary to locate the pump 4.8 ft lower to obtain the same NPSH.

Installation III illustrates how a liquid with a high vapor pressure forces a reduction in possible suction lift. In this case the liquid is considerably lighter than water, and, if it had a vapor pressure equal to water at 60°F, it would have been possible to have a static suction lift of 25.2 ft. The effect of the specific gravity of the liquid on suction conditions is illustrated more clearly in installation IV. Here, because brine has a specific gravity of 1.2, the atmospheric pressure corresponds to only 28.3 ft of liquid instead of 34.0 ft as with cold water (in installation I). As a result, in order to obtain the same 18.4 ft of NPSH, the value of S, the static component of the suction head, can only be -7.8 ft instead of -13.0 ft as with cold water. (Usually in installations handling gasoline and brine, the pumps are located so that the liquid level is above the pump. Many such installations have long suction pipes with considerable friction head loss, so that the same suction conditions indicated here could logically result.)

Installation V is similar to many boiler feed pump installations. When the liquid handled is at a temperature corresponding to its boiling point at the suction pressure, the suction head available to overcome friction and provide the required NPSH must be entirely static as $P_s - P_{vp}$ is zero. This is also demonstrated in installation VI, which shows a typical condition encountered in condensate or hotwell pumps. Condensate pumps serving surface condensers are generally located on the floor, just slightly below the liquid level in the hotwell, and pumps of a special design requiring a very low NPSH have to be used.

A comparison of installations V and VII shows how the reduction in temperature below that corresponding to suction pressure affects the required suction conditions.

PUMP CHARACTERISTIC CURVES

Unlike positive displacement pumps, a centrifugal pump operating at constant speed can deliver any capacity from zero to a maximum value dependent upon the pump size, design, and suction conditions. The total head developed by the pump, the power required to drive it, and the resulting efficiency vary with the capacity. The interrelations of capacity, head, power, and efficiency are called the pump characteristics. These interrelations are best shown graphically, and the resulting graph is called the characteristic curves of the pump. The head, power, and efficiency are usually plotted against capacity at a constant speed, as shown in Fig. 17.17. It is possible for special problems, however, to plot any three components against any fourth component. When variable-speed drivers are used, a fifth component, the operating pump speed expressed in rpm, is involved. Where suction conditions may be critical, the limit of suction-lift–capacity curve or required NPSH-capacity curve is often shown. Many other relationships can be shown on the same graph as required for specialized studies, for example, specific speed plotted against capacity.

The curve H–Q in Fig. 17.17, showing the relationship between capacity and total head, is called the head-capacity curve. Pumps are often classified on the basis of the shape of their head-capacity curves as described below.

The curve P–Q in Fig. 17.17, showing the relation between power input and pump capacity, is the power-capacity curve, but is generally referred to as the power curve, the brake horsepower curve, or the bhp curve.

The curve η–Q in Fig. 17.17, showing the relation between efficiency and capacity, is

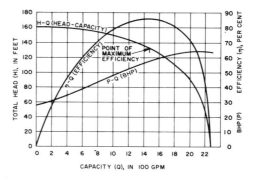

Fig. 17.17 Typical centrifugal pump characteristics

Double-suction, single-stage volute pump with 8-in. suction and 6-in. discharge at 1,760 rpm.

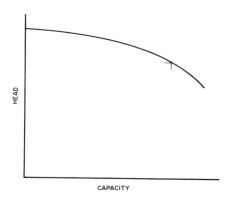

Fig. 17.18   Rising head-capacity curve

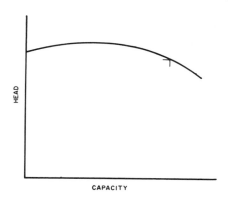

Fig. 17.19   Drooping head-capacity curve

properly called the efficiency-capacity curve, but is commonly referred to as the efficiency curve.

Usually the graph of a pump characteristic is made for a capacity range from zero to the maximum operating capacity of the unit. The scales on the graph for head, efficiency, and brake horsepower (bhp) all have the same zero line at the base of the graph (Fig. 17.17).

In some cases, the curve is made for a limited range in capacity. In other cases, to permit clearer presentation, the head, efficiency, and power scales are so selected that their zero lines do not coincide, and sometimes these scales are so enlarged that their full range cannot be shown on the graph.

Classification of head-capacity curve shapes

Pump head-capacity curves are commonly classified as follows:

1. *Rising characteristic*—or rising head-capacity characteristic, meaning a curve in which the head rises continuously as the capacity is decreased (Fig. 17.18).

2. *Drooping characteristic*—or drooping head-capacity characteristic, indicating cases in which the head-capacity developed at shutoff is less than that developed at some other capacities. This is also known as a looping curve (Fig. 17.19).

3. *Steep characteristic*—a rising head-capacity characteristic in which there is a large increase in head between that developed at design capacity and that developed at shutoff. It is sometimes applied to a limited portion of the curve; for example, a pump may have a steep characteristic between 100 per cent and 50 per cent of the design capacity (Fig. 17.20).

4. *Flat characteristic*—a head-capacity characteristic in which the head varies only slightly with capacity from shutoff to design capacity. The characteristic might also be either drooping or rising. All drooping curves have a portion where the head developed is approximately constant for a range in capacity, called the flat portion of the

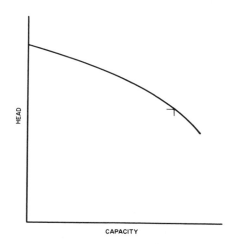

Fig. 17.20   Steep head-capacity curve

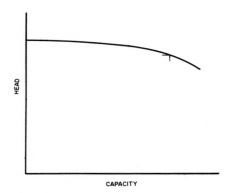

Fig. 17.21   Flat head-capacity curve

curve. Other curves are sometimes qualified as flat, either for their full range or for a limited portion of their range (Fig. 17.21).

5. *Stable characteristic*—a head-capacity characteristic in which only one capacity can be obtained at any one head. Basically this has to be a rising characteristic ·(Fig. 17.18 and 17.20).

6. *Unstable characteristic*—a head-capacity characteristic in which the same head is developed at two or more capacities. (Fig. 17.19 and 17.22). The successful application of any pump depends as much upon the intrinsic characteristics of the system on which it is operated as upon the head-capacity characteristic. Most pumping systems permit the use of pumps with moderately unstable characteristics.

Classification of bhp curve shapes

Power-capacity (bhp) curves are also classified according to shape. Figure 17.23 illustrates a pump characteristic with a bhp curve that flattens out and decreases as the capacity increases beyond the maximum efficiency point. This is called a non-overloading curve. When the bhp curve continues to increase with an increase in capacity, as in Fig. 17.24, the pump is said to have an overloading curve. The shape

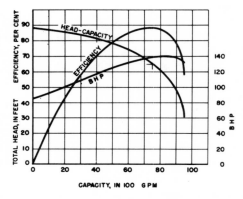

Fig. 17.23   Characteristics of a pump with a non-overloading power curve with reduction in head

of the bhp curve varies with the specific speed type. As a result, the bhp curve may have a very low value at shutoff (see Fig. 17.23 and 17.24), it may have a high value at shutoff (Fig. 17.25), or any value in between. Whereas in Fig. 17.24 the bhp curve is an overloading curve with a decrease in head and *increase* in capacity, the bhp curve in Fig. 17.25 is an overloading curve with an increase in head and *decrease* in capacity.

Pumps with non-overloading power curves are advantageous because the driver is not overloaded under any operating conditions, but they are not obtainable in all specific speed types of pumps. The actual

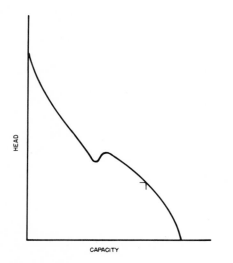

Fig. 17.22   Unstable head-capacity curve

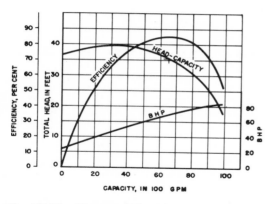

Fig. 17.24 Characteristics of a pump with an overloading power curve with a reduction in head

$$whp = QH \frac{\text{specific gravity}}{3,960}$$

in which:

whp = water horsepower

Q = pump capacity, in gallons per minute

H = total head, in feet.

The power required to drive the pump is regularly determined in horsepower and is called the bhp input to the pump. The ratio of the whp output to the bhp input is the pump efficiency. The relation between bhp, capacity, head, and efficiency is therefore:

$$bhp = \frac{QH}{\text{efficiency}} \frac{\text{specific gravity}}{3,960}$$

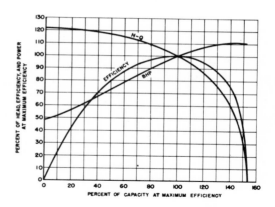

Fig. 17.26 Type characteristic, or 100 per cent curve

range of operating conditions encountered in the operation of a pump determines the range in power requirements, and the driver size should be selected for the power to be encountered.

Mathematical relations of head, capacity, efficiency, and brake horsepower

The useful work done by a pump is the weight of liquid pumped in a period of time multiplied by the head developed by the pump and is generally expressed in terms of horsepower, called water horsepower (whp)—liquid horsepower would be more correct. It can be determined by the relation:

Type characteristics

If the operating conditions of a pump at the design speed, that is, the capacity, head, efficiency, and power input at which the efficiency curve reaches its maximum, are taken as the 100 per cent standard of comparison, the head-capacity, power-capacity, and efficiency-capacity curves can all be plotted in terms of the percentage of their respective values at the capacity at maximum efficiency. Such a set of curves represents the type characteristic or 100

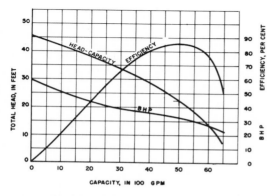

Fig. 17.25 Characteristics of a pump with an overloading power curve with an increase in head

per cent curve of the pump. Figure 17.26 shows the type characteristic of the pump whose performance is shown in Fig. 17.17.

CENTRIFUGAL PUMP CHARACTER-ISTICS RELATIONS

Certain relations exist that allow the performance of a centrifugal pump to be predicted for a speed other than that for which the pump characteristic is known. Certain relations also exist that allow prediction of the performance of a pump if the impeller is reduced in diameter (within a limit dependent upon the impeller design) from the characteristics obtained at the larger diameter.

When the speed is changed: (1) The capacity for a given point on the pump characteristics varies as the speed, *and at the same time,* (2) the head varies as the square of the speed, and (3) the brake horsepower varies as the cube of the speed. These relations take the form of equations as follows:

$$Q = Q_1(n/n_1)$$

$$H = H_1(n/n_1)^2$$

$$P = P_1(n/n_1)^3$$

or:

$$\frac{n}{n_1} = \frac{Q}{Q_1} = \sqrt{\frac{H}{H_1}} = \sqrt[3]{\frac{P}{P_1}}$$

in which:

n = new speed desired, in revolutions per minute

Q = capacity, in gallons per minute at desired speed n

H = head, in feet, at desired speed n for capacity Q

P = brake horsepower, at desired speed n at H and Q

n_1 = a speed, in revolutions per minute, at which the characteristics are known

Q_1 = a capacity, at speed n_1

H_1 = head, at capacity Q_1 at speed n_1

P_1 = brake horsepower, at speed n_1 at H_1 and Q_1.

For example, a pump is tested at 1,800 rpm and gives the following results:

Capacity, in gpm	Head, in ft	Power, in bhp	Efficiency, per cent
4,000	157	189.5	83.7
3,500	183.5	185	87.6
3,000	200.5	174.5	87.0
2,000	221	142.3	78.4
1,000	228.5	107	54.0
0	230	76.5	0

To obtain the performance of this pump at 1,600 rpm, the first set of values is corrected to 1,600 rpm, as follows:

$$Q = 4,000 \ (1,600/1,800) = 3,556 \text{ gpm}$$
$$H = 157 \ (1,600/1,800)^2 = 124 \text{ ft}$$
$$P = 189.5 \ (1,600/1,800)^3 = 133.1 \text{ bhp.}$$

Changing the other sets of values yields the following (Fig. 17.27):

Capacity, in gpm	Head, in ft	Power, in bhp
3,556	124	133.1
3,110	145	129.9
2,667	158.3	122.5
1,777	174.6	100
890	180.6	75.2
0	181.8	53.7

The capacity and head figures for these various points can be calculated on a slide rule with one setting. In this case 1.8 on the C scale would be set over 1.6 on the D scale, and the new capacities would be read on the D scale opposite the 1,800-rpm capacities on the C scale. The new heads would be read on the A scale opposite the 1,800-rpm heads on the B scale. Although it is possible to obtain the cube of a ratio on a slide rule, errors are often made in this step. Except for shutoff (zero capacity) the bhp can be calculated from the new head and capacity (at 1,600 rpm) using the same efficiency as for the corresponding head-capacity at 1,800 rpm. Thus the bhp for the first point can be calculated as $(3,556 \times 124)/(3,960 \times 0.837)$ or 133.1. The shutoff horsepower can only be obtained by using the cube of the speed ratio, as both the capacity and the efficiency are zero.

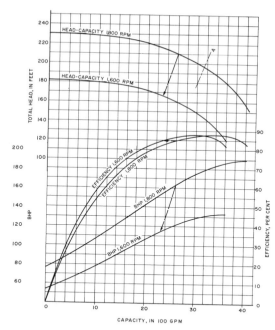

Fig. 17.27 Effects of speed change on pump characteristics

varies as the impeller diameter, *and at the same time*, (2) the head varies as the square of the impeller diameter, and (3) the horsepower varies as the cube of the impeller diameter. Expressed as equations, these are:

$$Q = Q_1(D/D_1)$$
$$H = H_1(D/D_1)^2$$
$$P = P_1(D/D_1)^3$$

or:

$$\frac{D}{D_1} = \frac{Q}{Q_1} = \sqrt{\frac{H}{H_1}} = \sqrt[3]{\frac{P}{P_1}}$$

in which:

D_1 = original diameter, in inches
D = cut-down diameter, in inches
Q_1 = capacity with D_1 impeller
Q = corresponding capacity with D impeller
H_1 = head with D_1 impeller at Q_1
H = corresponding head with D impeller at Q
P_1 = bhp with D_1 impeller at Q_1 and H_1'
P = bhp with D impeller at Q and H.

Referring back to the tabulation of values of the pump tested at 1,800 rpm (with an impeller diameter of 14.75 in.), if the impeller is reduced to 14 in. in diameter, the first set of values is corrected as follows:

$$Q = 4{,}000\ (14/14.75) = 3{,}797 \text{ gpm}$$
$$H = 157\ (14/14.75)^2 = 141.5 \text{ ft}$$
$$P = 189.5\ (14/14.75)^3 = 162.0 \text{ bhp}$$

or:

$$(3{,}797 \text{ gpm} \times 141.5 \text{ ft})/(3{,}960 \times .837)$$
$$= 162.0 \text{ bhp}.$$

The other sets of values yield the following (Fig. 17.28):

Capacity, in gpm	Head, in ft	Power, in bhp	Efficiency, per cent
3,797	141.5	162	83.7
3,322	165.3	158.2	87.6
2,847	180.7	149.2	87.0
1,898	199	121.6	78.4
949	206	91.5	54.0
0	207.4	65.4	0

These relations for a change in speed can be used safely for moderate speed changes. They may not be accurate in large changes in speed, particularly in increases in speed.

The diameter of an average impeller can be cut down on a lathe by 20 per cent of its original maximum value without adverse effect. Cutting it down to less than 80 per cent will generally result in a much lower efficiency. This 20 per cent limit is approximate, as some impeller designs can be cut more than this, whereas others cannot be cut more than a small percentage without adverse effect. Any change in diameter will affect the proportions of the impeller, and some variations from the theoretical results should be expected when tested.

If an impeller is cut in diameter, it is found that, at the same speed, the characteristics of the pump will have a definite relation to its original characteristics.

These relations are: (1) The capacity for a given point in the pump characteristic

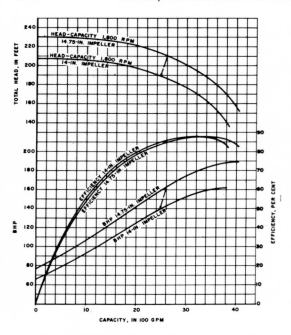

Fig. 17.28 Effects of change in impeller diameter on pump characteristics

These relationships are used most commonly to determine the change in speed, the change in diameter of the impeller, or the combination of both that is necessary to produce a head-capacity curve passing through a given point. For example, suppose a pump has to meet the condition of 3,000 gpm at 180 ft total head. Since this falls below the head-capacity curve of the 14.75-in. impeller at 1,800 rpm, the desired head-capacity is obtained by reducing the speed or reducing the diameter of the impeller.

If the pump, which is to give 3,000 gpm at 180 ft, were speeded up, or the impeller diameter increased so that point on the characteristic became 3,100 gpm, the head, at the same time, would have become 180 (3,100/3,000)² or 192.2 ft. Similarly, if 3,200 gpm were obtained by a further increase in speed or impeller diameter, the head would be 180(3,200/3,000)² or 204.8 ft. Plotted as shown on Fig. 17.27 these values form a section of a curve (A). This intersects the 1,800 rpm (14.75-in. D₂) head-capacity curve

at 3,135 gpm and 196.5 ft, indicating the desired point on that characteristic. To obtain 3,000 gpm and 180 ft, the required speed can be determined by calculation of 1,800(3,000/3,135) or by 1,800(180/196.5)½ both of which give 1,723 rpm.

If no speed change was desired, it would have been necessary to change the diameter to 14.75 (3,000/3,135) or 14.12 in. Had the new driver run at 1,760 rpm, the 14.75-in. diameter impeller would have given 3,065 gpm and 187.9-ft head requiring, in addition, a cut in the impeller diameter to 14.44 in. In all three cases a new curve that would pass through 3,000 gpm at 180 ft can be plotted by stepping down a number of the 1,800 rpm and 14.75-in. diameter curve points, the capacities being reduced by the ratio of 3,000/3,135, whereas the corresponding heads are reduced by the ratio of (3,000/3,135)², and the corresponding bhp reduced by the ratio of (3,000/3,135)³. The bhp can also be calculated from the resulting capacity, head, and the pump efficiency at the point in question.

DESIGN CONSTANTS

The designing of centrifugal pumps is not an exact science because of the many interrelated factors whose combined effect cannot be accurately foreseen and thus must be determined experimentally. The development of centrifugal pumps has been largely a result of the accumulation of data on the performance of both specific designs in service and of experimental designs, the result of research and experiences in other hydraulic fields, and the application of this information to the development of new designs. In analyzing data, centrifugal pump designers use various constants, formulas, and relations, two of which are of interest to users of centrifugal pumps: (1) Model pump relations; and (2) specific speed.

Pumps are analyzed and compared basically at their so-called design conditions;

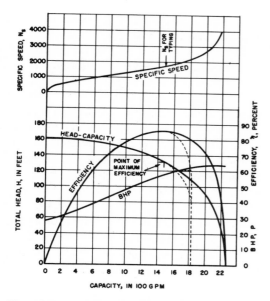

Fig. 17.29 Typical pump characteristic curve with auxiliary specific speed curve

Double-suction, single-stage 6-in. pump, operating at 1,760-rpm constant speed.

that is, at the head and capacity condition at rated speed at which maximum efficiency is obtained. Thus, for the pump whose characteristics are shown in Fig. 17.29, the design conditions would be 1,480 gpm and 132 ft total head at 1,760 rpm.

Model pumps

A model pump has the design features of a full-size unit on a smaller scale. To meet the requirements of a strict model, all linear dimensions of the model must be in the same proportion as the corresponding dimensions of the full-size pump. The theoretical relationship of the performance of a model pump can be easily visualized by considering two pumps identically proportioned, with one having twice the linear dimensions of the other.

The impeller of the smaller pump will be half the diameter of the larger and will, therefore, have to run at twice the rotative speed of the larger for the same peripheral velocity and equal design head. The areas through the waterways of the smaller pump

will be one-half squared or one-quarter the areas of the larger pump. Thus, at equal velocities, the capacity of the smaller pump will be one-quarter that of the larger pump. Therefore, it is apparent that for the same design head the interrelationship of exactly similar pumps would be theoretically:

$$f = \frac{L_a}{L_b} = \frac{n_b}{n_a} = \sqrt{\frac{Q_a}{Q_b}}$$

in which:

f = the ratio or factor of the two pumps

L_a and L_b = comparable dimensions of the two pumps

n_a and n_b = the rotative speeds of the two pumps

Q_a and Q_b = the capacities of the two pumps at comparable points on their characteristic curves.

The above equation is based on the assumption that the two pumps are proportional in every way and that the same relative degree of smoothness is obtained in the two pumps. This is difficult to attain, as the actual smoothness of castings is approximately the same, regardless of size. Thus, the relative internal smoothness of a larger pump is greater than that of a smaller pump. This is reflected in the head losses in the pump waterways; the larger pump should produce a higher head than the smaller pump for points of similar capacity. Inasmuch as part of the liquid pumped leaks through the wearing rings, this loss may not be in exact proportion in both sizes of pumps, thus affecting the net quantity delivered. Part of the power input goes into mechanical losses (bearings and stuffing boxes), that are roughly but not exactly proportional to the pump sizes, resulting in a third discrepancy. Good mechanical design (especially in the production of a commercial line of pumps) precludes making the shaft, casing thickness, or thickness of impeller vanes of two pumps in exact proportion to their size factor. A comparison of the largest and smallest pumps of a closely homologous line of a commercial design, therefore, will show some difference in performance. The magnitude of this difference will depend upon the size factor and the actual physical sizes of the two pumps. Centrifugal pump designers are careful, when making model pumps, to use a size that will be close enough to the full-size pump so that the results of the model will permit a reasonably close prediction of the performance of the full-size pump.

Model pumps have been used to prove within a reasonable degree of accuracy the performance of the full-size unit for almost every case involving special large-capacity pumps.

Specific speed

An analysis of the performance of a projected centrifugal pump would be difficult without the progress achieved in the science of hydrodynamics in the four centuries of its existence. This progress may be directly credited to the almost universal application of model study, which precludes the necessity of experimenting upon full size commercial constructions that are too expensive and least convenient for securing the necessary information. Sir Isaac Newton evolved the theory of dynamical similarity, in 1687, thereby introducing the mathematical background for model investigations.

The application of the Newtonian principle of dynamical similarity has since given rise to the wide use of models in hydraulic machinery, as well as in other fields of science, and to an extensive knowledge of the relative performance of models and prototypes.

One such application of the principle of model and prototype relationship has enabled engineers to predict the performance of centrifugal pumps on the basis of the behavior of other machines, smaller or larger in size, operating over a wide range

of design conditions, but modeled from and similar to each other.

The principle of dynamical similarity expresses the fact that two pumps geometrically similar to each other will have similar performance characteristics. In order to afford some basis of comparison among various types of centrifugal machines, it became necessary to evolve a concept which would link the three main factors of these performance characteristics—capacity, head, and rotative speed—into a single term. The term "specific speed" is such a concept. The mathematical analysis used to establish the relationship between the specific speed of a pump and its operating characteristics does not enter the scope of this book. In its basic form, the specific speed is a non-dimensional index number which is *numerically* equal to the rotative speed at which an exact theoretical model centrifugal machine would have to operate in order to deliver one unit of capacity against one unit of total head. It is mathematically expressed as:

$$N_s = \frac{n\sqrt{Q}}{(gH)^{3/4}}$$

in which:

N_s = specific speed
n = rotative speed
Q = capacity
H = head (head per stage for a multistage pump)
g = gravitational constant, 32.2 feet per second per second at sea level.

In order for this relation to remain dimensionless, when using English units, the rotative speed would have to be expressed in revolutions per second, the capacity in cubic feet per second and the head in foot-pounds per pound or foot. However, since specific speed is used only as an index or type number, certain liberties are permissible in selecting the units used. Thus, the gravitational constant, g, is dropped out of the relation, leaving:

$$N_s = \frac{n\sqrt{Q}}{H^{3/4}}$$

The rotative speed is expressed in revolutions per minute. For some time, two units of capacity, gallons per minute and cubic feet per second, were used in the US to determine specific speed, but the gallons per minute basis has been accepted as standard by the Hydraulic Institute and is now the approved basis. The unit of head is one foot.

The formula for the specific speed of a pump remains unchanged whether a single- or a double-suction impeller is used. It is customary, therefore, when listing a definite value of specific speed, to mention what type of impeller is in question.

Operating specific speed

The customary presentation of the performance characteristics of a centrifugal pump consists in plotting its head, power consumption, and efficiency as ordinates against the pump capacity as abscissa, at a constant rotative speed. Since the pump speed, capacity, and head all enter into the concept of specific speed, it is also possible to calculate the specific speed for any given operating condition of head and capacity and to plot this operating specific speed against the pump capacity. For example, Fig. 17.29 shows the performance characteristics of a 6-in. pump operating at 1,760 rpm whose maximum efficiency is at 1,480 gpm and 132-ft head. The operating specific speed curve has been added to the usual performance curve. The operating specific speed is 0 at zero flow and increases with the capacity until it reaches infinity at the maximum capacity of 2,270 gpm and zero head.

Type specific speed

The type specific speed, by definition, is that operating specific speed that gives the maximum efficiency for a particular pump and is the number that identifies the pump type. It should be noted that this index number is independent of the rotative speed at which the pump is operated, since any change in speed carries with it a change in capacity in a direct proportion and a change in head varying as the square of the speed.

The normal range in specific speeds encountered in single-suction impeller designs is from 500 to 15,000. Basically, the lower the specific speed type, the higher the head per stage that can be developed by the pump.

Normally, the conditions of service for which a pump is sold are relatively close to the maximum efficiency point, and the specific speed determined from the conditions of service will be a close indication of the pump type. For example, the *true* type specific speed of the pump whose characteristics are illustrated in Fig. 17.29 is 1,740. This pump would normally be applied for a range of conditions between 1,300-gpm and 140-ft total head and 1,600-gpm and 125-ft total head. The corresponding range of specific speeds is 1,590 to 1,890, which varies about 10 per cent from the actual type specific speed.

Graphical solution for specific speed

Whereas the calculation of the formula for the specific speed of any given pump can be made with relative ease on a slide rule (since a number to the three-quarter power is equivalent to the square root of the number multiplied by the square root of its square root) even technically educated men can easily make errors in such calculations. In order to avoid these calculations, it is practical to solve the formula for the specific speed by using a chart such as that in Fig. 17.30. To determine the specific speed, knowing the capacity, head, and speed, project vertically from the capacity scale to the head line, then horizontally to the rpm line, and finally vertically again down to the specific speed scale. Thus, for example, if it is desired to establish the specific speed of a pump designed

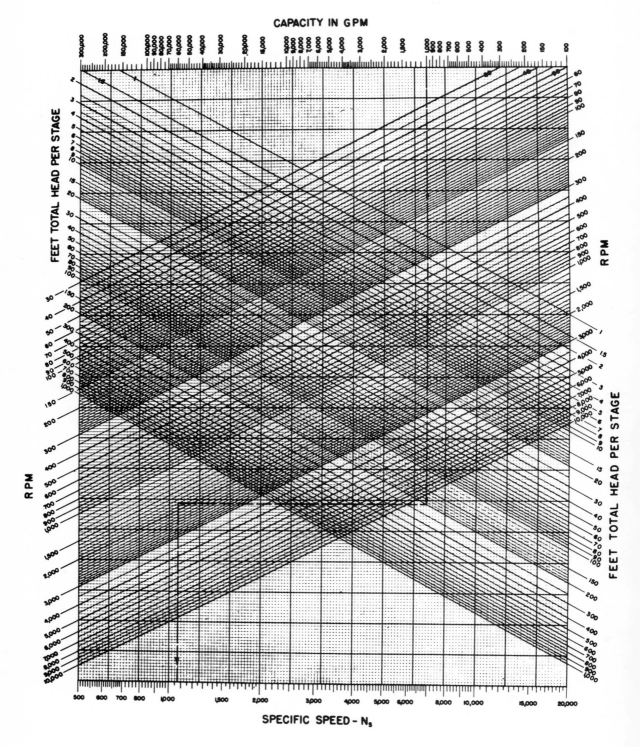

Fig. 17.30   Graphical computation of specific speed

to deliver 1,000 gpm against a total head of 200 ft when operating at 1,800 rpm, one projects vertically from 1,000 gpm to the intersection with the diagonal 200-ft line, then horizontally to the 1,800-rpm line, and vertically to the specific speed scale, reading $N_s = 1,070$. Reverse these steps to determine the capacity when specific speed, rpm, and head are known. To determine the rpm when the capacity, head, and specific speed are known, project from the capacity scale to the head line, then project horizontally to a point directly over the specific speed, and note the speed in rpm corresponding to this point.

Significance of type specific speed

One of the most important applications of the specific speed concept is the fact that all sizes of pumps can be indexed by the rotative speed of their unit capacity-unit head model. Thus, the specific speed concept can be used in such a manner that for homologous designs, the performance of any impeller of the series can be predicted from the knowledge of the performance of any other impeller of the series. Because the physical characteristics and the general outline of impeller profiles are intimately connected to their respective type specific

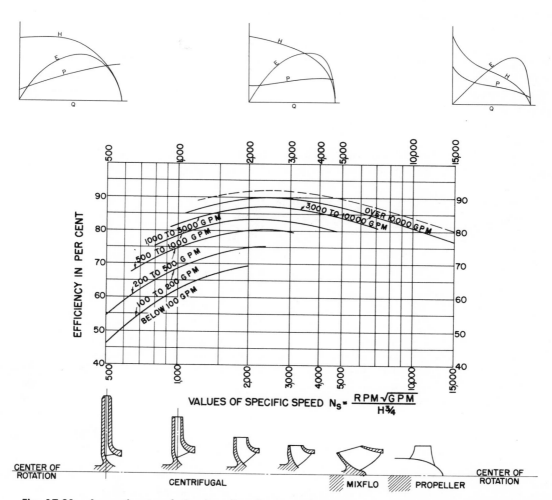

Fig. 17.31 Approximate relative impeller shapes and efficiency variations with specific speed

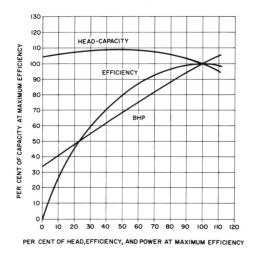

Fig. 17.32   Type characteristics for $N_s = 600$ single-suction impeller

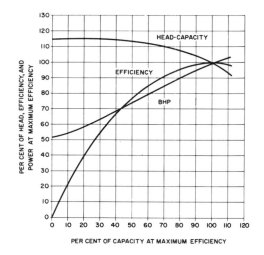

Fig. 17.33   Type characteristics for $N_s = 1,550$ single-suction impeller

speeds, the value of the latter will immediately describe the approximate impeller shape in question. As an illustration of this statement, Fig. 17.31 represents a few typical impeller outlines tied down to their type specific speeds.

Figure 17.31 also indicates the maximum range of efficiencies obtainable from pump

impellers of different type specific speeds. Low type specific speed impellers have a lower maximum efficiency than medium type specific speed impellers because the former have considerably more disk area for a given set of operating conditions and, therefore, a greater loss in disk-horsepower. High specific speed types also have a lower

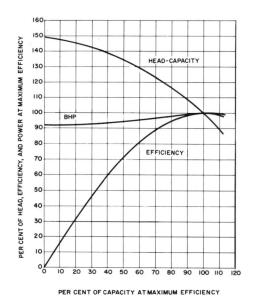

Fig. 17.34   Type characteristics for $N_s = 4,000$ single-suction impeller

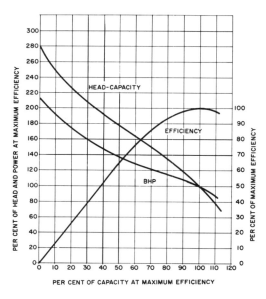

Fig. 17.35   Type characteristics for $N_s = 10,000$ single-suction impeller

maximum efficiency than medium specific speed types because, although they have still further reduced areas and therefore still lower disk-horsepower losses, they present poor flow conditions from inlet to discharge.

The specific speed of a given pump will also definitely be reflected in the shape of the pump characteristic curves, and, whereas some variations in the shape of these curves can be obtained by changes in the design of the impeller and casing waterways, the variation that can be obtained without adversely affecting the pump efficiency is relatively small. Approximate type characteristics for four single-suction impeller types are shown in Fig. 17.32–17.35. Figure 17.36 shows the variation of head with the specific speed for shutoff, 25, 50, 75, and 110 per cent capacity. Figure 17.37 shows the variation of power with specific speed for the same capacities, while Fig. 17.38 shows the variation of efficiency with specific speed for these same capacities. The values shown in Fig. 17.32–17.38 are for more or less normal impeller-casing designs and combinations. Variations in the shape of the curve will be found, depending upon the individual design of the pump.

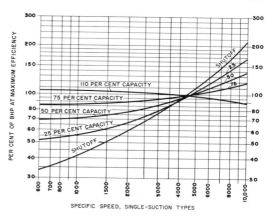

Fig. 17.37 **Variation in power values at shutoff, 25, 50, 75, and 110 per cent capacity with specific speed**

The variation in the shape of the type characteristics between a single-suction impeller with shaft through the eye and an overhung single-suction impeller is small, therefore Fig. 17.32–17.38 can be applied to either type. A double-suction impeller will have a type characteristic approximating that of a single-suction impeller having a specific speed 70.7 per cent or $(1/\sqrt{2})$ of that of a double-suction impeller.

Approximating specific speed type from impeller outline

Prior to the general adoption of specific speed as a type indicator, the ratio of outside diameter (D_2) to suction eye diameter (D_1), or the reciprocal of this relationship,

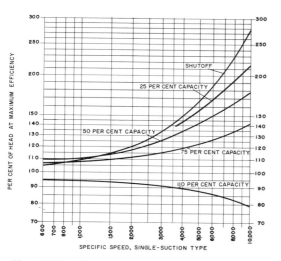

Fig. 17.36 **Variation in head values at shutoff, 25, 50, 75, and 110 per cent capacity with specific speed**

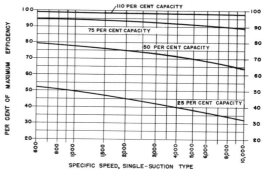

Fig. 17.38 **Variation in efficiency values at shutoff, 25, 50, 75, and 110 per cent capacity with specific speed**

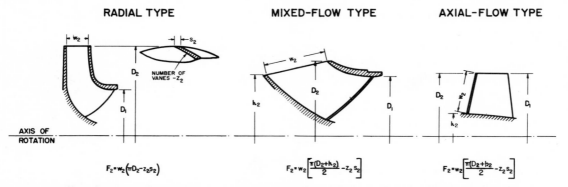

Fig. 17.39   **Dimensional symbols for impellers and formula for determining discharge area F_2**

was generally used for that purpose (Fig. 17.39). An approximate relation of the D_2/D_1 ratio to N_s for single-suction impellers is shown in Fig. 17.40. These values are necessarily approximate, as a true curve would be a fairly wide band. One reason for this can easily be seen if one considers an impeller for a given set of conditions. The velocity of the liquid would have to be approximately the same whether the impeller had no hub or a hub extending into the eye. Thus an impeller with a hub extending into the eye would necessarily have a larger D_1 for the same capacity and head or the same D_2. Multistage pumps require a large shaft because of the power involved, so that the impellers of such pumps would have relatively large hubs. The impeller would have an abnormally large D_1 or a smaller D_2/D_1 ratio than normal for its specific speed type.

This method of identifying the specific speed fails for axial-flow impellers. Axial-flow impellers fall into a 9,000 to 20,000 specific-speed range. Their output and, therefore, their specific speed depends on the angle and length of the vanes as well as the number of vanes. To predict the characteristics of axial-flow impellers, a designer would require very detailed information on the impeller and other pump parts. No simplified guide can be offered for general use for this type of pump.

Head and capacity constants

Two design constants can be used to approximate the performance of a centrifugal pump. One expresses the relation between the impeller peripheral speed and the total head. The second relates the radial discharge velocity from the impeller (and, therefore, the capacity) and the total head. The formulas for these constants are:

$$\phi = \frac{u_2}{\sqrt{2gH}}$$

and:

$$K_{cr} = \frac{c_2}{\sqrt{2gH}}$$

in which:

u_2 = peripheral velocity, in feet per second

g = gravitational constant, 32.2 feet per second per second

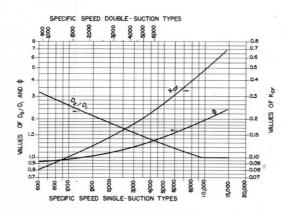

Fig. 17.40 **Variation with specific speed of D_2/D_1 ratio, Φ, and K_{cr} constants**

c_2 = radial discharge velocity, in feet per second

H = total head (per stage), in feet.

These relationships can be converted into terms of impeller dimensions that can be measured:

$$u_2 = \frac{D_2 n \pi}{12 \times 60} = \frac{D_2 n}{229}$$

and:

$$c_2 = \frac{144Q}{7.48 \times 60 \times F_2} = \frac{Q}{3.117 F_2}$$

in which:

D_2 = outside diameter of impeller, in inches

n = speed, in revolutions per minute

Q = capacity, in gallons per minute

F_2 = discharge area of impeller, in square inches (Fig. 17.39).

(In high-speed Francis vanes, mixed-flow, and axial-flow impellers, the effective discharge diameter is not D_2 but a geometric mean: $\sqrt{(D_2{}^2 + h_2{}^2)/2}$. When there is little difference between D_2 and h_2, an arithmetical mean is commonly used. However for simplicity, constants presented here are calculated on the basis of D_2, not the effective diameter that most designers would use in calculating their constants.)

The equations for the design constants can now be replaced by:

$$\phi = \frac{D_2 n}{229 \sqrt{2gH}} = \frac{D_2 n}{1{,}840 \sqrt{H}}$$

and:

$$K_{cr} = \frac{Q}{3.117 F_2 \sqrt{2gH}} = \frac{Q}{25 F_2 \sqrt{H}}$$

These equations can be further transformed to give head and capacity values directly:

$$H = \left(\frac{D_2 n}{1{,}840 \phi}\right)^2$$

and:

$$Q = 25 K_{cr} F_2 \sqrt{H}$$

Both ϕ and K_{cr} vary with the specific speed type and, to some extent, with the individual impeller and casing design. They are also affected by the physical size of the pump. Figure 17.40 shows a ϕ–N_s relation and a K_{cr}–N_s relation that are representative for normal pump design. Like the D_2/D_1–N_s curve, these curves show average values; the true values for an individual design will vary somewhat from those shown.

APPROXIMATING CHARACTERISTICS FROM PHYSICAL MEASUREMENTS

Generally, a centrifugal pump user who wishes to determine the performance characteristics of a given pump in his possession has a large reservoir of information from which these data can be determined:

1. A copy of the order on which the pump was purchased.

2. The nameplate of the pump. This nameplate generally carries the pump shop serial number, the manufacturer's type designation, and the rated conditions of service, including the operating speed.

3. The pump driver nameplate. If this nameplate is missing, a direct measurement on the driver will give the operating speed.

If the pump make and serial number are known, it is a simple matter to get the desired information from the manufacturer. If the make, but not the serial or other identifying number or letter is known, most manufacturers can identify the pump type and impeller design, if given the following: (1) The nozzle sizes; (2) a sketch showing the external appearance and dimensions of the pump; and (3) the major impeller dimensions.

It is seldom that a pump is without both pump and driver nameplates and that no record of the purchase is available, leaving the physical presence of the pump itself as the only thing certain. But even in this case it is possible to carry out certain measurements and calculations to obtain the desired information with some degree of accuracy.

Two separate phases exist in the problem of estimating the performance of a centrifugal pump when nothing is known except the physical dimensions and proportions. The first phase of the problem is the theoretical aspect, which gives a reasonable approximation of the pump head-capacity curve at any given operating speed and the expected power consumption of the pump. This phase requires the application of the data presented on the preceding pages and in Fig. 17.32–17.40 for its solution.

The second phase of the problem concerns the practical aspects of the application of the pump to a particular service. Is the physical design of the pump suitable for the power and the speed selected? Is the casing design suitable for the operating pressure? Will the pump operate satisfactorily under the suction conditions contemplated?

Maximum speed limitation

Present-day practice in the United States places a limit on pump speed by limiting the maximum specific speed type for any combination of total head and suction conditions. Before the widespread use of the specific speed concept, the maximum rotative speed of an impeller was often determined by limiting the peripheral velocity of its suction eye (D_1) to a certain value, depending on the suction conditions. This velocity can be established exactly, as in the case of the peripheral velocity of the outside impeller diameter:

$$Suction\ eye\ velocity = \frac{D_1 n}{229}$$

A reasonable maximum peripheral velocity for the suction eye is shown in Fig. 17.41 for various suction conditions.

Expected power consumption

Determining the possible head and capacity of a pump would be of little value if the power required to drive the pump could not be predicted. Figure 17.31 shows the approximate maximum efficiency for both single- and double-suction pumps that can be obtained with present-day designs. How close an existing pump would approach these values would depend on the individual pump. Between 80 and 100 per cent of the efficiencies shown in Fig. 17.31 can be expected if the pump is of the single-stage type and not more than 30 years old. Multistage pumps, with abrupt crossover passages from one stage to the next, should be expected to be less efficient by two to three percentage points.

Determination of safe power input

The shaft of a centrifugal pump is subject to both bending and torsional stresses. Usually, its smallest diameter is at the coupling and this section is subjected primarily to torsional stress only. Generally, centrifugal pump shafts are designed not to exceed a torsional stress at the coupling of 7,000 psi for safe maximum continuous loading. Therefore, in predicting pump performance, a check should be made to see if the maximum expected power at the selected speed keeps the torsional stress

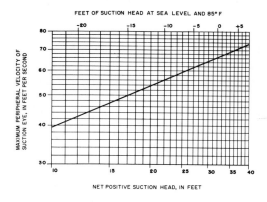

Fig. 17.41 Approximate limit for peripheral velocity of suction eye and suction head

within the recommended 7,000-psi limit. If not, the pump is not mechanically safe for operation at the selected speed. The relation between transmitted horsepower, speed, shaft diameter, and permissible torsional stress is given by the formula:

$$hp = \frac{Snd^3}{321,000}$$

in which:

hp = horsepower
S = permissible stress, in pounds per square inch
n = speed, in revolutions per minute
d = shaft diameter at coupling, in inches.

Determination of safe operating pressure

Determination of the safe operating pressure of a given pump requires a very detailed study. Some idea of a possible maximum can be obtained by examining the size and drilling of the discharge nozzle flange. This is an indefinite limit, especially in the United States where the "125-lb flange," which is good up to 175-psi hydraulic operating pressure, is generally used for all pressures below that value. Many pump designs using such flanges, however, are not good for operating pressures that high.

It is more satisfactory to make a check of safe pressures at the bolting of the casing or casing heads. In actual design studies, maximum safe bolt stresses are calculated with full knowledge of the areas subjected to internal pressures, of the maximum expected hydraulic pressure, and of the forces required to compress joint gaskets. A good approximation can be obtained by limiting the bolting stress at the root of the threads to 5,000 psi; thus, safe working pressure can be computed as follows:

$$swp = \frac{A_r n_b 5,000}{A}$$

in which:

swp = safe working pressure, in pounds per square inch
A_r = root area of bolt, in square inches
n_b = number of bolts
A = area subjected to hydraulic pressure, in square inches.

For purposes of analysis, assume a modern double-suction, single-stage centrifugal pump with an 8-in. discharge and 10-in. suction, both with 125-lb flanges. Various relevant dimensions (see Fig. 17.39) are given below for the impeller:

$$D_2 = 12\frac{1}{4} \text{ in.}$$
$$D_1 = 6\frac{1}{8} \text{ in.}$$
$$W_2 = 2\frac{1}{16} \text{ in.}$$
$$z_2 = 7$$
$$s_2 = \frac{3}{8} \text{ in.}$$

The shaft diameter is $1\frac{9}{16}$ in. at the coupling, increasing to a maximum at the impeller. The area of the horizontal split is 200 sq. in. The pump casing is held together by 21 $\frac{3}{4}$-in. studs and bolts. The intended drive is a 60-cycle induction motor.

$$D_2/D_1 = 12.25/6.125 = 2.0$$

Referring to Fig. 17.40, a D_2/D_1 ratio of 2.0 indicates an impeller type with specific speed $N_s = 1,700$ if single-suction or $N_s = 2,400$ if double-suction. The head and capacity constants are:

$$\phi = 1.06$$
$$K_{cr} = 0.13$$

Assuming that the pump will be applied to an installation involving a 15-ft suction lift at sea level, handling cold water, the NPSH will be, roughly, 17 ft. From Fig. 17.41, it can be established that the maximum safe peripheral velocity at the impeller suction eye is about 50 fps. Solving for the rotative speed gives:

$$n = \frac{50 \times 229}{6.125} = 1,860 \text{ rpm}$$

Therefore, 1,750 rpm will be the maximum possible rotative speed with a 60-cycle motor.

The pump head can now be calculated at 1,750 rpm:

$$H = \left(\frac{12.25 \times 1,750}{1,840 \times 1.06}\right)^2 = 120.5 \text{ ft}$$

Using the formula in Fig. 17.39, F_2, the discharge area of the impeller, can be calculated:

$$F_2 = 2.0625[(12.25\pi) - (7 \times 0.375)]$$

$$= 74.2 \text{ sq in.}$$

The pump capacity is estimated:

$$Q = 25 \times 0.13 \times 74.2\sqrt{120.5} = 2,650 \text{ gpm}$$

When operated at 1,750 rpm, this pump will deliver 2,650 gpm against a total head of 120.5 ft at its best efficiency point. Its specific speed can be recalculated:

$$N_s = \frac{1,750\sqrt{2,650}}{(120.5)^{3/4}} = 2,480 \text{ (double-suction)}$$

Figure 17.31 shows that, for this specific speed and capacity, the maximum efficiency would be somewhat over 85 per cent. If 85 per cent is used, the power consumption would be:

$$hp = \frac{2,650 \times 120.5}{3,960 \times 0.85} = 95 \text{ hp}$$

Working from these values of capacity, head, efficiency, and horsepower, and using the percentages shown in Fig. 17.36–17.38 (for a single-suction N_s of $2,480 \times 0.707 =$

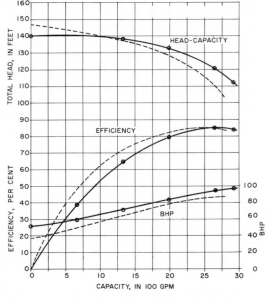

Fig. 17.42 Comparison of characteristics predicted from impeller measurements and actual characteristics determined by test

Dash lines are predicted curves, and solid lines are actual test curves.

1,743), the points in Table 17.4 can be obtained. Plotting these points, the approximate curve shown in Fig. 17.42 is obtained. (The actual test curve of the pump has been superimposed on the same graph for comparison.)

To determine if the pump shaft is suitable for this application:

$$\text{Safe horsepower} = \frac{7,000 \times 1,750 \times 1.562^3}{321,000}$$

$$= 145 \text{ hp}$$

TABLE 17.4   CHARACTERISTICS PREDICTED FROM IMPELLER MEASUREMENTS

Capacity		Head		Power		Efficiency	
Per cent	gpm	Per cent	ft	Per cent	bhp	Per cent of max	Actual per cent
0	0	116.5	140	54	51.2	0	0
25	662			62.5	59.4	45.7	38.8
50	1,325	115	138	75	71.2	75.6	64.3
75	1,987	110	132.5	88	83.6	93.8	79.6
100	2,650	100	120.5	100	95	100	85
110	2,915	93	112	103	97.8	99	84

The shaft is obviously safe for 1,750-rpm operation.

It remains to check the pump for safe operating pressure. The ¾-in. bolts and studs holding the two halves of the casing together have an area of 0.302 sq in. at the root of the threads.

$$\text{Safe working pressure} = \frac{5,000 \times 0.302 \times 31}{200}$$

$$= 158.5 \text{ psi}$$

This is approximately 2.6 times the expected shutoff head of 140/2.31 or 60.5 psi, when operating at 1,750 rpm. The casing will be strong enough for the intended operation. (The actual pump design is good for 175-psi operating pressure.)

The foregoing example has demonstrated that it is possible to approximate the performance of centrifugal pumps other than those of the axial-flow impeller type. Nevertheless, this should be used with caution and only in the absence of a more reliable method.

Specific speed and suction condition limitations

A final and very important application of the specific speed concept is its effect on the permissible and recommended suction conditions. In the early 1930's, the Hydraulic Institute sponsored studies of centrifugal pumps in which cavitation troubles had been experienced. It was found that to avoid difficulties, for any given total head and suction lift condition, the specific speed of a pump should be below a certain value. These studies finally permitted the plotting of a series of charts prepared by the Hydraulic Institute to show the recommended limitations. Two of these charts (Fig. 17.43 and 17.44) apply to single-stage, double-suction pumps and to single-stage, single-suction pumps with a shaft through the eye of the impeller. A third chart (Fig. 17.45) indicates recommended specific speed limitations for single-

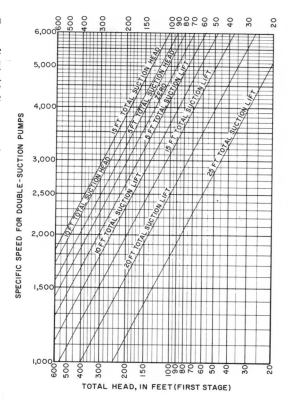

Fig. 17.43 Hydraulic Institute specific speed limit chart for double-suction, single-stage pumps with shaft through eye of impeller

Reprinted from the Standards of the Hydraulic Institute, *Tenth Edition. Revised in 1960. Copyright 1955 by the Hydraulic Institute, 122 East 42nd Street, New York 17, N. Y.*

suction overhung-impeller pumps. A fourth chart (Fig. 17.46) applies to single-suction, mixed-flow and axial-flow pumps. All of these charts are based on 85°F water at sea level and are commonly known among centrifugal pump engineers and users as "specific speed limit charts."

When using these charts it must be realized that pumps built for the limit allowed are not necessarily the best design for the intended service and that a lower specific speed type might be more economical. It must also be realized that the individual pump design limits its application for both maximum head and for suction condition limitations. For example, the

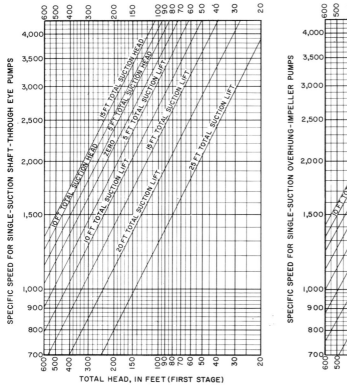

Fig. 17.44 Hydraulic Institute specific speed limit chart for single-suction, single-stage pumps with shaft through eye of impeller

Revised in 1960. Copyright 1955 by the Hydraulic Institute.

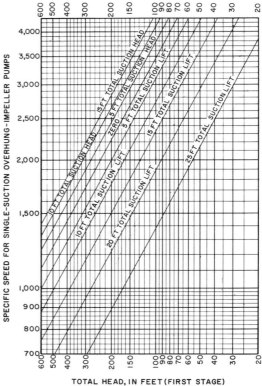

Fig. 17.45 Hydraulic Institute specific speed limit chart for single-suction, overhung-impeller pumps

Revised in 1960. Copyright 1955 by the Hydraulic Institute.

maximum recommended specific speed for a double-suction, single-stage pump is 1,970 for a 200-ft total head and a 15-ft suction lift. It does not follow that all double-suction, single-stage pumps of 1,970 specific speed type are suitable for operation at speeds which will cause them to develop a 200-ft total head (at maximum efficiency); nor that the pump, if suitable for operation at a 200-ft total head is suitable for operation with a 15-ft suction lift; nor that a pump of this type operating against a 200-ft total head would on test be found capable of operating on only a 15-ft maximum suction lift. These charts are intended to indicate only the maximum rotative speed for which experience has shown a centrifugal

pump can be designed with assurance of reasonable and proper operation for the combination of operating conditions.

Nothing on these charts suggests that the specific speed indicated corresponds to the point of maximum efficiency. However, pumps are normally applied for conditions near their maximum efficiency points. Thus, even though the service conditions do not correspond exactly with the design conditions, the specific speed value is generally sufficiently close to the specific speed of the design conditions.

Suction specific speed

Just as pump specific speed is an index number indicative of pump type, the

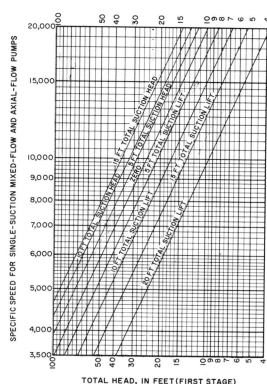

Fig. 17.46 Hydraulic Institute specific speed limit chart for single-suction, mixed- and axial-flow pumps

Revised in 1960. Copyright 1955 by the Hydraulic Institute.

parameter known as "suction specific speed" is essentially an index number descriptive of the suction characteristics of a given impeller. It is defined as:

$$S = \frac{n\sqrt{Q}}{h_s^{3/4}}$$

in which:

S = suction specific speed

n = rotative speed, in revolutions per minute

Q = flow, in gallons per minute

h_s = NPSH required for satisfactory operation, in feet.

Figure 17.47 has been provided for easy conversion of h_s to $h_s^{3/4}$. Note that for single-suction impellers, Q is the total flow.

When a double-suction impeller is involved, Q should be taken as one-half of the total flow.

In using this formula, the proper selection of the S value will determine the accuracy of the required NPSH so calculated. The Hydraulic Institute Standards recommended NPSH curves for single-suction boiler feed pump applications are based upon an S of about 7,900 at the capacity corresponding to the point of best efficiency (Fig. 17.48). This is a quite conservative limitation, values of 9,000 are quite common. As a matter of fact, centrifugal boiler feed pumps producing S values up to 10,000 and 11,000 are known to be in successful operation. Recommended Hydraulic Institute values for double-suction pumps are shown in Fig. 17.49. They are based on a value of $S = 6,660$.

Figure 17.50 shows a chart for determining required NPSH values for boiler feed pumps at speeds of 3,575 to 15,000 rpm, based on an S value of 8,000. Figure 17.51 gives a K factor which can be used to convert results obtained from Fig. 17.50 for other S values from 7,000 to 10,000. This K

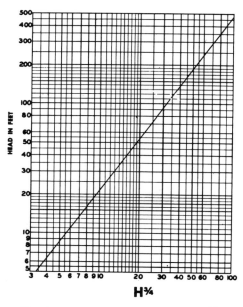

Fig. 17.47 Conversion of h_s to $h_s^{3/4}$

factor is based upon the following derivation, in which subscript a refers to values of $S = 8,000$, and subscript b refers to a selected value of S.

$$h_{S_a}^{3/4} = \frac{n\sqrt{Q}}{8,000} \quad \text{and} \quad h_{S_b}^{3/4} = \frac{n\sqrt{Q}}{S_b}$$

Since n and Q are constant:

$$S_b h_{S_b}^{3/4} = 8,000 h_{S_a}^{3/4}$$

$$h_{S_b}^{3/4} = h_{S_a}^{3/4}\left(\frac{8,000}{S_b}\right)$$

$$h_{S_b} = h_{S_a}\left(\frac{8,000}{S_b}\right)^{4/3}$$

Let

$$K = \left(\frac{8,000}{S_b}\right)^{4/3}$$

$$h_{S_b'} = K h_{S_a}$$

This is the K given in Fig. 17.51.

Although the S value at the point of best efficiency is fixed for a particular design, several methods for attaining NPSH improvement are available to the pump manufacturers. The NPSH required by a pump at equivalent operating points on its curve varies approximately as the square of a speed change. Therefore, selection of operating speeds can be a very important factor in determining NPSH requirements. With high-speed pumps (over 4,000 rpm), for example, an adequate NPSH produced through static elevation alone frequently is found to be prohibitive since the required height is so great. For this reason, most such high-speed boiler feed pumps operating in open feedwater cycles (see Fig. 22.2) require booster pumps to produce an adequate NPSH.

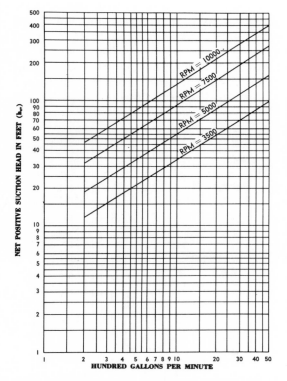

Fig. 17.48 Hydraulic Institute recommended values of NPSH for single-suction, multistage boiler feed pumps

Revised 1958. Copyright 1955 by the Hydraulic Institute.

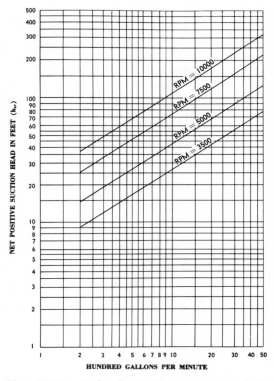

Fig. 17.49 Hydraulic Institute recommended values of NPSH for double-suction, multistage boiler feed pumps

Revised 1958. Copyright 1955 by the Hydraulic Institute.

The fact that NPSH varies as the square of a speed change can be demonstrated as follows:

$$S = \frac{n_1 \sqrt{Q_1}}{h_{S1}^{3/4}} = \frac{n_2 \sqrt{Q_2}}{h_{S2}^{3/4}}$$

$$\left(\frac{h_{S2}}{h_{S1}}\right)^{3/4} = \frac{n_2 \sqrt{Q_2}}{n_1 \sqrt{Q_1}}$$

but:

$$\frac{Q_2}{Q_1} = \frac{n_2}{n_1}$$

therefore:

$$\left(\frac{h_{S2}}{h_{S1}}\right)^{3/4} = \frac{n_2^{3/2}}{n_1^{3/2}}$$

and:

$$\frac{h_{S2}}{h_{S1}} = \left(\frac{n_2}{n_1}\right)^2$$

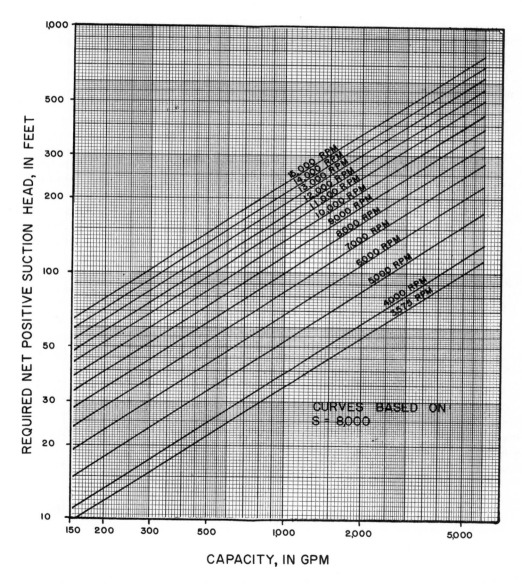

Fig. 17.50 Chart of required NPSH for boiler feed pumps operating at speeds of 3,575 to 15,000 rpm

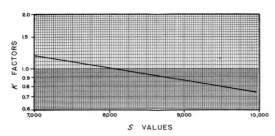

Fig. 17.51   Correction factor for Fig. 17.50 showing multiplier for NPSH based on S values differing from 8,000

Effect of suction conditions on pump characteristics

The suction limitation of centrifugal pumps is determined by the fact that the impeller cannot impart energy to the liquid until the liquid is in the impeller between the vanes. Thus, the energy necessary to overcome the frictional losses up to the entrance of the suction vane ends of the impeller and the energy necessary to create the velocity required at this point have to come from some outside source. Furthermore, sufficient additional energy must be available in excess of these and other requirements so that the absolute pressure at all points is above the vapor pressure of the liquid, to prevent its flashing into vapor.

Figure 17.29 shows the characteristics of a 6-in. pump. If it is operated in a system in which the available external energy on the suction side could only force 1,400 gpm into the impeller, and if the total head of the system with this capacity is 100 ft, the pump will work to pump over 1,900 gpm against this head. As there is insufficient suction head to get more than 1,400 gpm into the impeller, however, the pressure at that point would be reduced below the vapor pressure of the liquid and part of the liquid would flash into vapor.

If there is not sufficient available NPSH to permit a pump to develop its normal characteristics, cavitation will result and the pump will "work in the break." Thus, the characteristics of a centrifugal pump will vary with the available NPSH. For the

specific pump shown in Fig. 17.29, the characteristics in solid lines are for 0 suction lift (32 ft NPSH) whereas with a 20-ft suction lift (12 ft NPSH approximately) the pump follows the 0 suction lift characteristics out to 1,500 gpm when cavitation starts, evidenced by the pump producing less head. Some increase in capacity results with further reduction in head until 1,820 gpm is reached, when further reduction in head causes no increase in capacity. Thus the pump characteristics with a 20-ft suction lift would be as shown by the solid lines out to 1,500-gpm capacity, and then by the broken line.

The pump illustrated in Fig. 17.29 is of a fairly low specific speed type. In higher specific speed types like the high-speed francis screw vane and mixed-flow impeller designs, the operation at reduced NPSH also reduces the head developed at or near shutoff. With high specific speed types this reduction in head is even more pronounced. Unlike the low specific speed types, the higher specific speed types may deliver (with reduced NPSH), with lower total heads, up to a maximum capacity, and then, as the total head is further reduced, the capacity may be reduced below this maximum, reversing the head-capacity curve on itself.

Usually cavitation is to be avoided. However, one type of pump, the condensate pump operating on non-throttled systems, is especially designed for such operation. Figure 17.52 shows the normal head-capacity curve, with sufficient NPSH to prevent cavitation, and the system head-capacity curve; Fig. 17.53 shows the layout of the system on the suction side. If the amount of steam being condensed is equivalent to 52½ gpm, the level in the hotwell will be that which gives 12-in. NPSH at the suction nozzle, so that the pump is operating in the break at 52½-gpm capacity and a 54.2-ft total head, as dictated by the system-head curve. If the amount of steam increases to equal 71 gpm, the liquid level in the hotwell will build up until it is 18 in. plus the

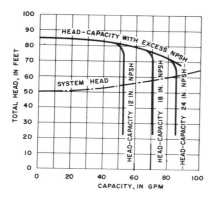

Fig. 17.52 Characteristics of a condensate pump operating on a submergence-controlled system

friction losses above the pump centerline, and the pump will be delivering 71 gpm against a 57.8-ft total head [the intersection of the head-capacity (18 in.) and the system-head curves]. A regular impeller design on such service would be noisy and would show evidence of cavitation by damage to the vanes after a short time. For hotwell or condensate service, special impeller designs have been developed with larger suction areas (to operate on low NPSH) and with special suction vanes to give quiet operation and long life even though cavitating all the time they are in operation.

In handling liquids containing dissolved gases, the pressure at or near the suction vanes of the impeller may be reduced sufficiently so that the dissolved gases are liberated and the pump is actually handling a gas-and-liquid mixture. If the amount of gas liberated is not excessive, the only effect may be a reduction in capacity output and efficiency. This separation of gas from liquid is often mistaken for cavitation; it is not. If both cavitation and gas separation occur in a pump, the cushioning effect of the gas often quiets the cavitation noise. The cushioning effect has sometimes been used to quiet noisy cavitating pumps by bleeding air into the suction. Although it serves as a temporary expedient, the most economical solution should be replacement of the impeller by a design suitable for the

suction conditions or a redesign of the pumping system so that the pump has sufficient NPSH to operate on its normal characteristics.

Effect of specific gravity on pump characteristics

The only effect that specific gravity of liquids with viscosities equal to water has on the operation of a pump is to vary the power required to drive it. The capacity and head (measured in feet of liquid) are the same as for water and so is the efficiency. The power input for any capacity is that required with cold water multiplied by the specific gravity.

Effect of viscosity on pump characteristics

Two of the major losses in a centrifugal pump are through fluid friction and disk friction. These losses vary with the viscosity of the liquid so that the head-capacity output, as well as the mechanical input, differ from the values produced when handling water.

It is not practical to present here a complete discussion on viscosity as a property of liquids and on the effect of viscosity on flow of liquids. The reader can find such a discussion in textbooks on fluid mechanics.

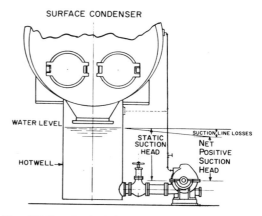

Fig. 17.53 Typical hook-up for submergence-controlled condensate pump

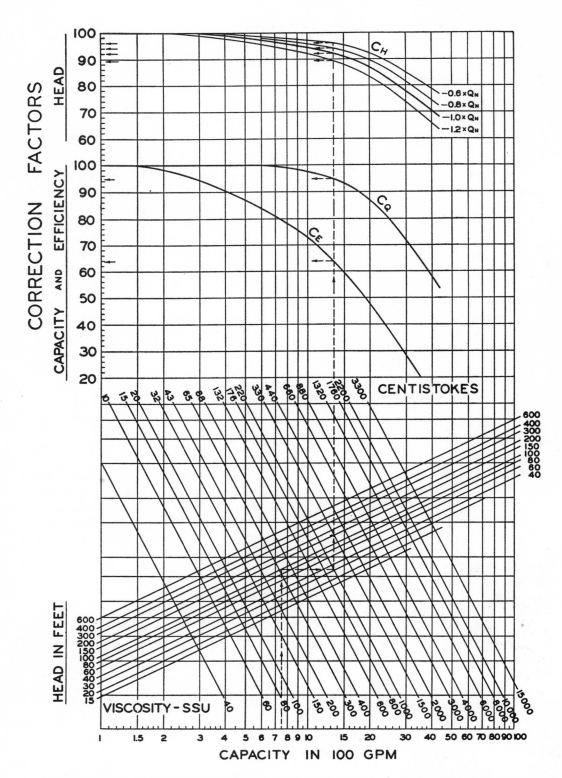

Fig. 17.54 Hydraulic Institute correction chart for pumps handling viscous liquids
Revised 1958. Copyright 1955 by the Hydraulic Institute.

It is necessary, however, to know the three different units that may be encountered describing the viscosity of a specific liquid:

1. Saybolt seconds universal, or ssu
2. Centistokes—defining the kinematic viscosity
3. Centipoises—defining the absolute viscosity.

Data for the conversion from one to another of these units and relations between the viscosity and temperature of a number of liquids are given in the Data Section of this book.

Considerable experimental testing has been done in determining the effect of liquid viscosity on the performance of different centrifugal pumps. Even with extensive data on the effect of viscosity, it is difficult to predict accurately the performance of a pump when handling a viscous fluid from its performance when handling cold water. The Hydraulic Institute has published the chart shown in Fig. 17.54 which permits approximating the characteristics on uniform liquids (not paper stock, slurries, or the like) of conventional 2-in. to 8-in. discharge single-stage centrifugal pumps, not of the mixed-flow or axial-flow types. This chart can also be used for multistage pumps if the correction factors are selected on the basis of the head per stage, and provided the losses (which result in heating the liquid) do not cause sufficient increase in the temperature to change the viscosity of the liquid appreciably. The correction factors in Fig. 17.54 are selected for the head (per stage) and capacity at which the pump gives maximum efficiency on cold water. For example, a pump whose maximum efficiency capacity $(1.0 \times Q_n)$ was 750 gpm at a 100 ft total head on water would, on a 1,000-ssu viscosity liquid, have the following characteristics: (1) A reduction of capacity to 95 per cent of its corresponding water capacity, (2) a reduction in the head produced at these reduced capacities to 96, 94, 92, and 89 per cent of the cold water heads at 60, 80, 100, and 120 per cent of

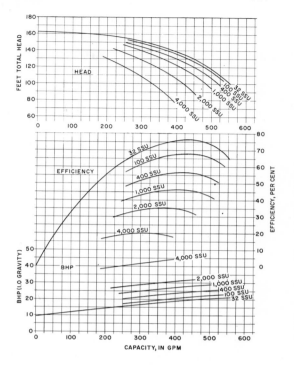

Fig. 17.55 Predicted characteristics for a centrifugal pump for liquids of various viscosities

normal capacity respectively, and (3) a reduction in the efficiency to 63.5 per cent of that produced on water for the corresponding capacities. The power required to drive is determined by calculating by the formula:

$$bhp = \frac{QH \times \text{specific gravity}}{3,960\eta}$$

in which:

bhp = brake horsepower
Q = capacity, in gallons per minute (corrected for viscosity)
H = total head, in feet (corrected for viscosity)
η = efficiency (corrected for viscosity).

Applying these corrective factors to a pump whose cold water characteristics are identified in Fig. 17.55 by 32 ssu, the approximate performance for 100-, 400-, 1,000-, 2,000- and 4,000-ssu liquids have been developed, the values for bhp being calculated on basis of 1.0 specific gravity. Whereas the pump produced a maximum efficiency of 76 per cent when pumping 440 gpm of cold water against a 132-ft total head, it would be expected to produce a maximum efficiency of only 19.7 per cent when pumping 321 gpm of a 4,000-ssu liquid against a 102.5-ft total head.

In applying regular cold-water pumps for use in pumping viscous liquids, care must be taken to make sure that the shaft design is strong enough for the required power, which may be considerably in excess of the cold-water brake horsepower, even though the specific gravity of the liquid may be less than that of water.

Effect of liquid characteristics on pump suction performance

The statement made previously that the NPSH required by a centrifugal pump for satisfactory operation is independent of the liquid vapor pressure at the pumping temperature is not one hundred per cent true. It is actually an oversimplification used to illustrate the definition that NPSH is a measurement of the energy in the liquid at the pump suction over the datum line of its vapor pressure.

Laboratory tests run on pumps handling a wide variety of liquids and over a range of operating temperatures have shown that the NPSH required for a given capacity and with a given pump apparently vary appreciably. For example, the required NPSH when handling some hydrocarbons is frequently much less than that required when the pump handles cold water. Even when pumping water, there is definite evidence that required NPSH decreases as the water temperature increases. Altogether, the reduction in the required NPSH appears to be a function of the vapor pressure and of the specific gravity of the particular liquid handled by the pump.

It is obviously desirable to correlate the results of all these separate and individual tests, and, as a result, various charts have

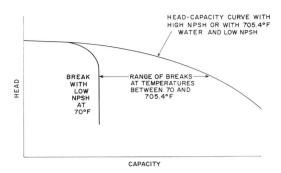

Fig. 17.56 Effect of temperature on maximum capacity at fixed pump speed and with fixed NPSH

been suggested to provide correction factors that could be used to translate values of required NPSH on cold water into NPSH values for liquids with various other characteristics, such as composition, vapor pressure, and gravity. At present, however, there is insufficient correlation among the many tests cited in the technical literature. The reader is therefore cautioned to refer to the latest available information. Presently, two papers seem to merit special attention:

1. "Cavitation and NPSH Requirements of Various Liquids" by Victor Salemann, *Journal of Basic Engineering* (Transactions of ASME), June 1959.
2. "Cavitation in Centrifugal Pumps with Liquids other than Water" by A. J. Stepanoff, ASME paper 59-A-158, presented at ASME Annual Meeting, Atlantic City, November 29–December 4, 1959.

The required NPSH may be reduced from the values obtained when testing a pump on cold water because mild and partial cavitation can occur in a pump without causing extremely unfavorable effects. Thus, if satisfactory operation is defined to mean that the loss in hydraulic performance is negligible, that no serious mechanical difficulty such as vibration occurs, and that no objectionable noise or damage to metal through cavitation takes place, then some partial cavitation can be permitted.

When a pump handles hot water, the higher the temperature, the less volume will be occupied by the steam into which a small amount of water will flash if NPSH is reduced until cavitation just begins. But, in addition, the flashing of some water into steam absorbs a certain amount of heat from the rest of the water, which is then cooled in this process. Its vapor pressure is reduced, automatically increasing the available NPSH and limiting further flashing. The cavitation process is therefore self-regulated to a great extent and a slight NPSH reduction will have no major ill effects. The degree of these effects is further and further minimized, the higher the water temperature. Conversely, the higher the temperature, the more NPSH reduction can be permitted for the same degree of effect on the pump performance.

These phenomena can also be demonstrated by strictly logical considerations. Consider the effect of temperature on the performance of a centrifugal pump operating at a given speed and supplied with a fixed NPSH when handling: (1) Cold water at 70°F; and (2) hot water at 705.4°F (critical temperature for water).

In the case of cold water at 70°F, the break in the head-capacity curve will occur at some capacity determined by the geometry of the impeller and by the speed at which the pump is operating, as indicated in Fig. 17.56.

When it comes to handling water at 705.4°F, which is the critical temperature above which no evaporation takes place (as long as the pressure is maintained at 3,206.2 psia), the volume occupied by water is the same as that occupied by steam. Under these conditions, a reduction in the available NPSH below that required when handling cold water can have no appreciable effect on the performance characteristics of the pump since true cavitation cannot take place. This statement is predicated on the fact that a change from water to steam takes place without a change in the volume occupied by either fluid. Nor is there a difference between the density of

TABLE 17.5 EFFECT OF TEMPERATURE ON NPSH AT DESIGN FLOW

Pump N_s = 1,600; N = 3,585 rpm.

Fluid	Temperature, deg F	NPSH, in ft (min accuracy ± 0.5 ft)	Decrease in NPSH
Water	70	12.3	0.0
	250	11.0	1.3
	300	8.6	3.7
Butane	35	9.8	2.5
	55	8.8	3.5
	90	3.5	8.8
Butane +3 per cent propane by weight	35	9.5	2.8
	55	7.8	4.5
	90	1.6	10.7
Benzene	180	12.4	(−0.1)
	230	9.7	2.6
Kerosene degasified	70	12.4	(−0.1)
Gasoline (Reid)	70	13.3	(−1.0)
Freon-1	85	10.2	2.1
	120	8.4	3.9
Water [1]	70	12.0	0
	290	9.5	2.5
	325	6.0	6.0
	410 [2]	2.0	10.0

[1] Pump N_s = 1,200; N = 3,585 rpm.
[2] Extrapolated from lower capacity.

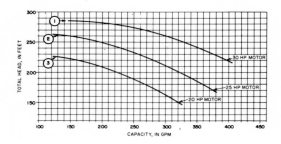

Fig. 17.57 Rating curve of 2½-in. motor-mounted pump

Pump equipped with different impellers that load up several sizes of motors.

water and that of steam. Expressed in terms of foot-pounds of energy per pound of fluid and plotted against volume, the pump performance (its head-capacity curve) remains unchanged by a reduction in NPSH. That is, the head-capacity curve of a pump handling 705.4°F water with less than the NPSH required for cold water will coincide with the head-capacity curve with cold water and ample NPSH (Fig. 17.56).

Having established the two limits of performance at 70°F and 705.4°F, the relationship between the location of the break and

the pumping temperature (or vapor pressure) can be assumed to be a continuous function. Therefore, all "breaks" at temperatures between these two limits must take place between the two indicated limits of capacity. Expressed differently, the required NPSH decreases as temperature increases, falling theoretically to zero at 705.4°F.

Examples of test results on three different pumps handling various liquids at varying temperatures are given in Table 17.5. They illustrate the wide range in the effect of different liquid characteristics on the suction performance of centrifugal pumps and the need of very complete information in order to correct for this effect.

RATING CURVES AND CHARTS

Rating curves and rating charts were originally intended for pump salesmen to use for making pump selections. They are now also common in bulletins and other sales literature.

A rating curve for a centrifugal pump of specific design shows in a condensed form,

TABLE 17.6 PORTION OF MODERN PUMP RATING CHART

gpm		Total head, in feet								
		20	25	30	35	40	45	50	55	60
200	Size-type	2½-CF-1	2½-CF-1	2½-CF-1	2½-CF-1	2½-CF-1	2½-CF-1	2½-CF-1	2½-CF-1	2½-CF-1
	hp	1.6	2.0	2.4	2.8	3.25	3.75	4.2	4.65	5.05
	rpm	1,010	1,090	1,170	1,240	1,310	1,380	1,440	1,500	1,550
	Size-type		3-CF-1	3-CF-1	3-CF-1	3-CF-1	3-CF-1	3-CF-1	3-CF-1	3-CF-1
	hp		2.25	2.6	3.0	3.3	3.6	4.1	4.6	5.0
	rpm		830	890	945	1,000	1,048	1,095	1,140	1,185
225	Size-type	2½-CF-1	2½-CF-1	2½-CF-1	2½-CF-1	2½-CF-1	2½-CF-1	2½-CF-1	2½-CF-1	2½-CF-1
	hp	1.9	2.3	2.7	3.2	3.65	4.15	4.7	5.1	5.55
	rpm	1,055	1,130	1,205	1,280	1,345	1,410	1,470	1,530	1,585
	Size-type		3-CF-1	3-CF-1	3-CF-1	3-CF-1	3-CF-1	3-CF-1	3-CF-1	3-CF-1
	hp		2.6	2.9	3.3	3.6	4.0	4.6	5.0	5.5
	rpm		850	907	962	1,015	1,065	1,113	1,155	1,205
250	Size-type	2½-CF-1	2½-CF-1	2½-CF-1	2½-CF-1	2½-CF-1	2½-CF-1	2½-CF-1	2½-CF-1	2½-CF-1
	hp	2.2	2.6	3.1	3.6	4.1	4.6	5.1	5.6	6.2
	rpm	1,095	1,170	1,245	1,315	1,385	1,450	1,510	1,565	1,620
	Size-type		3-CF-1	3-CF-1	3-CF-1	3-CF-1	3-CF-1	3-CF-1	3-CF-1	3-CF-1
	hp		2.9	3.2	3.6	4.0	4.4	4.9	5.5	5.9
	rpm		870	927	980	1,035	1,085	1,130	1,175	1,220

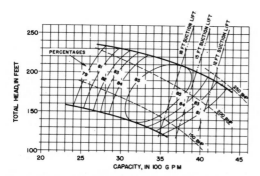

Fig. 17.58 Rating curve of 10-in. double-suction single-stage pump

Revealing the wide range that can be covered by an impeller of single design by machining it to proper diameter for a particular service.

the possible range of applications of that pump, either for a range in speed or for a range in impeller diameter. Some small centrifugal pumps for belt drive are manufactured in lots for stock sale, and the most efficient operating speed for each installation is obtained by selecting a proper pulley ratio to give the head and capacity condition desired. A curve showing the head, capacity, and brake horsepower for such a pump at a number of speeds, could be utilized for determining the speed necessary and the power involved, but these rating curves are now rarely used. Instead, a table (Table 17.6) is generally more convenient and permits showing a number of pump sizes on the same sheet.

A few lines of pumps, notably small motor-driven stock units, are made with several different impeller diameters, of the same or different patterns, that load up various sizes of motors. A rating curve for this type is shown in Fig. 17.57. With such a line of pumps, a pump with a 25-hp motor and an impeller that would approximate the results shown on curve (2) would be furnished if the desired head condition fell anywhere within the zone between the head-capacity curves (2) and (3) in Fig. 17.57. Thus, for some customers' requirements, such as 250 gpm, 200 ft total head, the pump supplied would produce more

capacity or head than required. When intalled, this unit would either give excess capacity or excess pressure, depending upon the system, unless throttling were employed to increase the frictional head artificially.

For pumps which are built-to-order with an impeller pattern and diameter individually selected for the prevalent service condition, a curve showing the range in conditions that can be met by a given impeller design or by several impeller designs for a given speed is used. These are generally complicated in appearance because the efficiency that can be obtained varies with the diameter of the impeller. This variation in efficiency is covered either by iso-efficiency curves, as shown in Fig. 17.58 or by figures on the curves of similar points ($Q/Q_1 = \sqrt{H/H_1}$ relation) or lines approximating that relation as shown in Fig. 17.59. For their proper use, rating curves must also show the limits of suction lift at sea level (Fig. 17.58) or the required minimum NPSH shown in Fig. 17.59.

A different chart is required for each motor speed for which the particular pump may be offered. For unusual conditions of

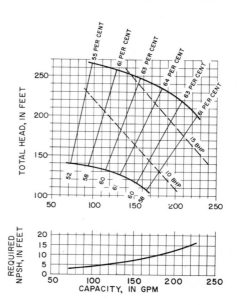

Fig. 17.59 Rating curve of 2-in. discharge, 3,500-rpm pump

driver speeds not covered by a curve, the use of standard relations for speed changes permits determining what the pump will do.

It should be noted that the required NPSH-capacity relation for a given impeller design at a given speed varies with the diameter to which the impeller is machined. With low specific speed types, the variation over the usable range of impeller diameter is small, and a single curve, as in Fig. 17.59, can be used to apply to all diameters even though it is slightly conservative for the maximum impeller diameters. When the required NPSH-capacity relation for an impeller design varies considerably over the useful range of impeller cutdowns, the use of a single curve to show required NPSH is not practical. Then the required NPSH can best be shown by diagonal lines like the suction lift limit lines on Fig. 17.58, but labeled with NPSH values, or with both suction lift and required NPSH values.

If the suction lift limitation (at sea level and with 62°F water) for any head-capacity point of a pump is known, it is a simple matter to determine the required NPSH at that point, since the NPSH is the atmospheric pressure minus vapor pressure minus suction lift. As the atmospheric pressure at sea level is 14.7 psia or 33.9 ft of 62°F water and the vapor pressure of 62°F water is 0.26 psia or 0.6 ft of 62°F water, standard atmospheric pressure minus the vapor pressure is 33.3 ft of water. Thus 33.3 ft minus the indicated suction lift limit gives the required NPSH. Similarly 33.3 ft minus the required NPSH gives permissible suction lift at sea level with 62°F water.

18 *System-Head Curves*

A centrifugal pump must be suitable for operation with the system in which it is used. In order to select a suitable pump, the characteristics of the system must be considered. It is usually easy to determine the characteristics of the system, but, occasionally, a complicated system that requires analysis of each of its parts is encountered. It is not possible to provide a detailed analysis of every type of problem that may be encountered when using centrifugal pumps. However, the following discussion of a few typical examples will acquaint the reader with the general method of solving such problems.

The total operating head for a given capacity through a system is the algebraic sum of the static head from supply level to discharge level (H_{st}); the terminal pressure minus the suction pressure ($P_d - P_s$); all friction losses at this capacity (h_f); and the entrance and exit losses (h_i and h_e). These values are expressed in feet of the liquid being handled (see Fig. 17.15).

Ideally, the simplest system would have only one static head. In an actual system, there would also be some friction losses. If there is no static head component and no difference in pressure on the suction and discharge liquid levels (Fig. 18.1), the head would be entirely frictional.

Causes of friction

The characteristics of the flow of liquid in a pipe vary with the velocity. When the velocity is very low, the flow is viscous. Under these conditions, the effect is that of concentric cylinders of the liquid shearing past each other in an orderly fashion. The greatest velocity is at the center of the pipe; the velocity falls to zero at the pipe walls. With water, viscous flow occurs when the average velocity is very low. As a result, viscous flow with water is rarely encountered in normal applications. As the average velocity of the liquid is increased, the flow becomes turbulent. Under turbulent flow conditions, the axial velocity measured across the pipe diameter is more uniform than in viscous flow; the flow is viscous in an area adjacent to the pipe walls. The average velocity at which the flow changes from viscous to turbulent is not absolute; there is a critical range in which the character of the flow may be of either type.

The flow of any liquid is accompanied by two types of friction: Internal friction caused by the rubbing of the fluid particles against one another; and external friction caused by the rubbing of the fluid particles against the pipe walls or against the static layer of liquid adhering to the walls. Energy must be expended to overcome this friction.

223

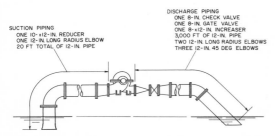

SUCTION PIPING
ONE 10-x12-IN. REDUCER
ONE 12-IN LONG RADIUS ELBOW
20 FT TOTAL OF 12-IN. PIPE

DISCHARGE PIPING
ONE 8-IN. CHECK VALVE
ONE 8-IN. GATE VALVE
ONE 8-x12-IN. INCREASER
3,000 FT OF 12-IN. PIPE
TWO 12-IN. LONG RADIUS ELBOWS
THREE 12-IN. 45 DEG ELBOWS

Fig. 18.1   Simple pumpage system with head that is entirely friction

If the flow is turbulent, the friction developed is partly dependent upon the roughness of the walls. Because the interior surfaces of pipes of the same material are practically the same irrespective of diameter, small pipes are relatively rougher than large ones. Thus, for equal velocities, the larger the pipe, the smaller will be the friction loss. The roughness of the pipe wall also depends upon the material from which the pipe is made and, after the pipe has been in service, upon any change that occurs at the inner surface.

Numerous pipe friction experiments and studies have been made, and a great number of tables and charts are available. Williams' and Hazen's tables are one of the earlier standards for water and have been found particularly reliable for cast-iron pipes of 3-in. or larger diameter. These tables are based on an empirical formula that can be modified to the following form:

$$h_f = 10.45(Q/C)^{1.852}(L/d)^{4.87}$$

in which:

h_f = loss in head for length L, in feet of water

Q = flow, in gallons per minute

d = inside diameter of the pipe, in inches

C = coefficient of pipe smoothness

L = length of pipe, in feet.

The coefficient C is an index of the smoothness of the interior pipe surface (the smoother the pipe interior, the higher the C value), and the selection of the proper value of this coefficient will determine the accuracy of the friction head loss calcu-

lated for any problem. For new unlined cast-iron pipe, a C of 130 is the common value, but some new pipes in which the friction head losses indicate C values of 140 or higher have been encountered. Pipes coated on the interior to give a smoother surface naturally have a higher C value. There are also records of pipes made of rolled metal and very smooth cement that have C values of 145 to 150 or higher.

Most pipes deteriorate with age, and thus the C value becomes lower. This decrease in C value or increase in friction head loss depends upon the material of the pipe, the pipe coating used (if any), and the character of the water. Therefore, any C value selected for an old pipe represents a pure guess. When it is necessary to ascertain friction head losses in such pipe, a test should be made, if possible, to find the friction loss at some known capacity so that the coefficient can be approximated. If such a test cannot be made, some guide to indicate the average C change with age is desirable. Figure 18.2 shows in chart form the coefficients that might be expected for cast-iron pipes handling soft, clear, unfiltered water.

Pipes carrying water that has been filtered but not chemically treated have been found to deteriorate less rapidly than pipes handling unfiltered water. Chemically treated waters have sometimes been found to produce more corrosion in the pipe than untreated water. Brackish water usually results in increased tuberculation. Some moderately hard waters have been found to cause a slow rate of deterioration. On the other hand, they are also capable of depositing calcium carbonate on the interior of the pipe, thus both reducing its size and increasing its roughness. Smooth cement and cement-lined pipes have been found to maintain a high C value for many years.

All these possibilities make it difficult to select, with any assurance of accuracy, the proper coefficient, so that any guide should be used with reservations. In important studies it is often possible to locate a simi-

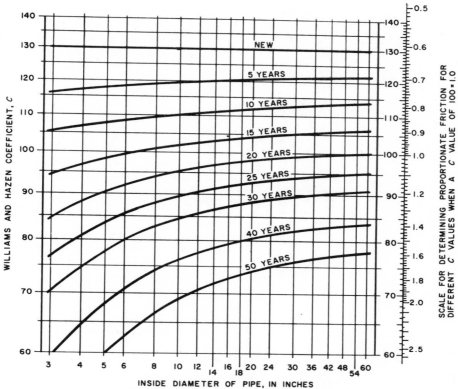

Fig. 18.2 Change in Williams and Hazen coefficient C with years of service, for cast-iron pipes handling soft, clear unfiltered water

Estimating friction loss

When selecting pumping machinery, it is particularly desirable to consider the friction head loss that may occur when the installation is new, as well as that which may result some years after it has been in service. Most charts and tables based on the Williams and Hazen formula have been made for a C value of 100, which is approximately the value expected for pipe that is 15 years old. A C value of 100 is commonly used as a design value; no calculation is made to determine the friction when the installation is new. This practice tends to distort the problem of compensat-

ing for friction losses, and may result in trouble when centrifugal pumps are used. For example, in a new installation in which most of the pumping head is friction, the actual friction head would be lower than that allowed for in the selection of the pump. As a result, the pump would deliver more capacity at some reduced head that would equal the system head. The increased capacity would depend upon the pump characteristics and the increase in system head with capacity, but might be 15 to 25 per cent more than the capacity for which the pump was selected. Operation at this increased capacity might cause the pump to require more power, thus overloading the driver. If the installation was such that the available net positive suction head (NPSH) at the design capacity exceeded only slightly the NPSH required

TABLE 18.1 VELOCITY AND FRICTION HEAD LOSS IN OLD PIPING

Friction values apply to cast-iron pipes after fifteen years service handling average water. Based on Williams and Hazen's formula with C = 100.

gpm	3-in. ID pipe v	3-in. f	4-in. ID pipe v	4-in. f	5-in. ID pipe v	5-in. f	6-in. ID pipe v	6-in. f	8-in. ID pipe v	8-in. f	10-in. ID pipe v	10-in. f	12-in. ID pipe v¹	12-in. f²	14-in. ID pipe v	14-in. f	16-in. ID pipe v	16-in. f	18-in. ID pipe v	18-in. f	20-in. ID pipe v	20-in. f
30	1.36	.534	.77	.131																		
40	1.81	.910	1.02	.224																		
50	2.27	1.38	1.28	.338	.82	.114																
60	2.72	1.92	1.53	.475	.98	.160																
70	3.18	2.56	1.79	.631	1.14	.213	.79	.088														
80	3.63	3.28	2.04	.808	1.31	.273	.91	.112														
90	4.08	4.08	2.30	1.01	1.47	.339	1.02	.139														
100	4.54	4.96	2.55	1.22	1.63	.412	1.14	.170														
125	5.68	7.50	3.19	1.85	2.04	.623	1.42	.256														
150	6.81	10.5	3.83	2.59	2.47	.874	1.70	.360	.96	.089												
175	7.95	14.0	4.47	3.44	2.86	1.16	1.99	.478	1.12	.118												
200	9.08	17.9	5.10	4.41	3.27	1.49	2.27	.613	1.28	.151												
225	10.2	22.3	5.74	5.48	3.68	1.85	2.55	.762	1.44	.188												
250	11.3	27.1	6.38	6.67	4.08	2.25	2.84	.926	1.60	.228	1.02	.077										
275	12.5	32.3	7.02	7.96	4.50	2.68	3.12	1.11	1.76	.272	1.12	.092										
300	13.6	37.9	7.65	9.34	4.90	3.13	3.41	1.30	1.91	.320	1.23	.108	.99	.059								
350	15.9	50.4	8.93	12.4	5.72	4.20	3.97	1.73	2.23	.425	1.43	.144	1.13	.076								
400	18.2	64.6	10.2	15.9	6.54	5.38	4.54	2.21	2.55	.545	1.63	.184	1.28	.094								
450			11.5	19.8	7.36	6.68	5.10	2.75	2.87	.678	1.84	.228	1.42	.114	.94	.044						
500			12.8	24.1	8.18	8.12	5.68	3.34	3.19	.823	2.04	.278	1.56	.136	1.04	.054						
550			14.0	28.7	8.99	9.69	6.24	3.99	3.51	.982	2.24	.331	1.70	.160	1.15	.064						
600			15.3	33.7	9.81	11.4	6.81	4.68	3.82	1.15	2.45	.389	1.84	.186	1.25	.076	.96	.039				
650			16.6	39.1	10.6	13.2	7.38	5.43	4.15	1.34	2.65	.452	1.99	.214	1.36	.088	1.04	.046				
700			17.9	44.9	11.4	15.1	7.94	6.23	4.47	1.53	2.86	.518	2.13	.242	1.46	.100	1.12	.052				
750					12.3	17.2	8.51	7.08	4.78	1.74	3.06	.589	2.27	.273	1.56	.114	1.20	.060	.95	.034		
800					13.1	19.4	9.08	7.98	5.10	1.97	3.26	.666	2.55	.339	1.67	.129	1.28	.067	1.01	.038		
900					14.7	24.1	10.2	9.92	5.74	2.44	3.67	.825	2.83	.412	1.88	.160	1.44	.084	1.13	.047		
1,000					16.3	29.3	11.4	12.1	6.38	2.97	4.08	1.00	3.12	.492	2.08	.195	1.60	.102	1.26	.057	1.02	.034
1,100					18.0	35.0	12.5	14.4	7.02	3.55	4.50	1.20	3.40	.578	2.29	.232	1.76	.121	1.39	.068	1.12	.041
1,200							13.6	16.9	7.66	4.17	4.90	1.41			2.50	.273	1.92	.143	1.51	.080	1.23	.048

Flow	12-in. ID pipe		14-in. ID pipe		16-in. ID pipe		18-in. ID pipe		20-in. ID pipe		24-in. ID pipe		6-in. ID pipe		8-in. ID pipe		10-in. ID pipe		Flow
	V[1]	hf[2]	V	hf	V	hf	V	hf	V	hf	V	hf	V	hf	V	hf	V	hf	
1,300	3.69	.671	2.71	.316	2.08	.165	1.64	.093	1.33	.056			14.8	19.6	8.30	4.83	5.31	1.63	1,300
1,400	3.97	.770	2.92	.363	2.24	.190	1.76	.107	1.43	.064			15.9	22.5	8.93	5.54	5.72	1.87	1,400
1,500	4.25	.875	3.12	.413	2.40	.215	1.89	.121	1.53	.073	1.06	.030	17.0	25.5	9.55	6.30	6.12	2.13	1,500
1,600	4.54	.985	3.33	.465	2.55	.243	2.02	.137	1.63	.082	1.13	.034	18.2	28.8	10.2	7.10	6.53	2.39	1,600
1,800	5.11	1.22	3.75	.578	2.87	.302	2.27	.170	1.84	.102	1.28	.042			11.5	8.83	7.35	2.98	1,800
2,000	5.67	1.49	4.17	.703	3.19	.367	2.52	.207	2.04	.124	1.42	.051			12.8	10.7	8.17	3.62	2,000
2,500	7.09	2.25	5.21	1.06	3.99	.555	3.15	.312	2.55	.187	1.77	.077			16.0	16.2	10.2	5.48	2,500
3,000	8.51	3.16	6.25	1.49	4.78	.778	3.78	.438	3.06	.262	2.13	.108			19.1	22.8	12.3	7.67	3,000
3,500	9.93	4.20	7.29	1.98	5.59	1.04	4.41	.583	3.57	.349	2.48	.143					14.3	10.2	3,500
4,000	11.3	5.38	8.33	2.54	6.39	1.33	5.04	.746	4.08	.447	2.83	.184					16.3	13.1	4,000
4,500	12.8	6.68	9.38	3.15	7.18	1.65	5.67	.928	4.59	.555	3.19	.228					18.4	16.3	4,500
5,000	14.2	8.13	10.4	3.83	7.98	2.00	6.30	1.13	5.10	.675	3.54	.278							5,000
5,500	15.6	9.70	11.5	4.58	8.78	2.39	6.93	1.35	5.61	.806	3.90	.332							5,500
6,000	17.0	11.4	12.5	5.38	9.68	2.81	7.56	1.58	6.12	.947	4.25	.390							6,000
6,500	18.4	13.2	13.6	6.24	10.4	3.26	8.19	1.83	6.73	1.10	4.61	.452							6,500
7,000	19.9	15.2	14.6	7.16	11.2	3.74	8.82	2.11	7.15	1.26	4.96	.518							7,000
7,500			15.6	8.13	12.0	4.24	9.45	2.39	7.66	1.43	5.32	.589							7,500
8,000			16.7	9.16	12.8	4.79	10.1	2.69	8.17	1.61	5.66	.664							8,000
9,000			18.8	11.4	14.4	5.95	11.3	3.39	9.18	2.01	6.38	.825							9,000
10,000					16.0	7.24	12.6	4.07	10.2	2.44	7.09	1.00							10,000
11,000					17.6	8.63	13.9	4.86	11.2	2.91	7.80	1.20							11,000
12,000					19.2	10.1	15.1	5.71	12.3	3.42	8.51	1.41							12,000
13,000							16.4	6.62	13.3	3.96	9.12	1.63							13,000
14,000							17.6	7.59	14.3	4.54	9.93	1.87							14,000
15,000							18.9	8.63	15.3	5.27	10.6	2.13							15,000
16,000									16.3	5.82	11.3	2.40							16,000
18,000									18.4	7.24	12.8	2.98							18,000
20,000											14.2	3.62							20,000
25,000											17.7	5.48							25,000

[1] Velocity, in feet per second.
[2] Friction head loss, in feet of water per 100 feet of pipe.

by the pump for this capacity, the resulting increase in pump capacity with the lower operating head would result in cavitation.

Table 18.1 shows pipe friction losses for 3-in. to 24-in. inside diameter pipes based on Williams' and Hazen's formula with C of 100. The usual values of C for new pipe are:

Smooth, unlined cast iron	130
Asphalted cast iron	140
Cement asbestos	130–140
Very smooth cement or cement-lined cast iron	130–140
Ordinary cement	110–120
Drawn steel or wrought iron	130–140
Riveted steel	90–110

Conversion factors for changing friction values based on $C = 100$ to other values are indicated on the right-hand side of Fig. 18.2. For example, with a flow of 700 gpm through a 6-in. pipe, the friction head loss is 6.23 ft per 100 ft of pipe with $C = 100$. For $C = 130$, the conversion factor is 0.613; therefore, the friction head loss will be 6.23×0.613 or 3.83 ft per 100 ft of pipe.

Steel and wrought-iron pipes are used extensively in sizes up to 8 in. and larger with cold water. In drainage and irrigation work, steel pipe is used almost exclusively with larger sizes fabricated of steel plate. Cast-iron pipe is now seldom used for water lines in sizes less than 3 in. Most long water lines are cast iron, but the cement asbestos type is also used. Steel pipe is made with the same outside diameter for a number of different weights or wall thicknesses. Therefore, the inside diameter will not be the same as the nominal diameter, and the friction losses for a given capacity must be corrected for such differences. With steel and wrought-iron pipe, it is more difficult to predict the change in friction head that will result when the pipe becomes older than it is to predict the changes in cast-iron pipes. In some situations, the pipe decreases in area due to tuberculation, while in other situations the pipe corrodes and the film is washed away. With smaller steel piping, it is best to consider what the fric-

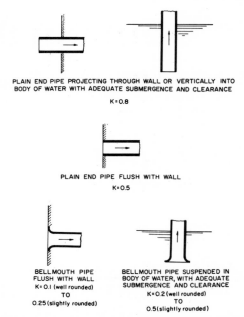

Fig. 18.3 Various types of piping connections and their K values

tion head will probably be when the pipe is new and to make allowance for an increase in the loss based on local conditions. If a basis of comparison is not available, an increase in friction of 25 per cent with age would be a reasonable allowance.

A friction table for the flow of water through new steel or wrought-iron pipe is shown in Table 18.2. Drawn brass or copper pipe would have a somewhat lower loss. This table is based on the Darcy formula:

$$h_f = f(L/D)(V^2/2g)$$

in which:

h_f = loss in head, in feet of liquid
L = length of pipe, in feet
D = inside diameter of the pipe, in feet
V = velocity, in feet per second
g = acceleration due to gravity (32.2 ft per second per second)
f = conversion factor dependent upon the relative roughness of the pipe, the velocity of the liquid, the size of the pipe, and the viscosity of the liquid.

This is a more precise formula than that of Williams and Hazen and is generally used in determining friction values when liquids other than water are being handled.[1]

Friction loss in valves and fittings

When liquid flows through valves, elbows, tees, and other fittings, there will be a frictional loss. Irrespective of the pipe size, these losses in fittings and valves can be expressed as percentages of the velocity head and may be calculated by the formula:

$$h_f = K(V^2/2g)$$

in which:

h_f = loss, in feet of liquid

K = constant (dependent upon the fitting design)

V = velocity, in feet per second

g = acceleration due to gravity (32.2 ft per second per second).

Values of K for common fittings, valves, and other resistances to flow have been determined experimentally.

Type of resistance	K value
Globe valve	10
Angle valve	5
Fully open swing check valve	1.5–2.5
Close return bend	2.2
Standard tee acting as elbow	1.8
Standard elbow	0.9
Long-sweep elbow	0.6
45-deg elbow	0.4
Fully open gate valve	0.2

The K values for various types of entrances are shown in Fig. 18.3. The K values for sudden enlargements and sudden contractions are shown in Fig. 18.4 and 18.5.

There is such a wide variation in the design of check valves that it is impossible to give any general values of K. In the swing-type valve, the disk is opened by the force of the flowing liquid. Thus, at low velocities when the disk is not fully open, the flow is throttled and the loss measured in terms of the velocity head is greater than at higher velocities. Very few data have

[1] Lewis F. Moody, "Friction Factors for Pipe Flow." Transactions of ASME, November 1944.

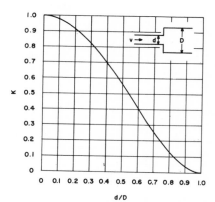

Fig. 18.4 Head loss in sudden enlargement of pipe

Based on assumption that the difference in velocity head is lost.

been published on the loss in swing check valves. One manufacturer provides a chart for equivalent length of pipe that gives values of $K = 2.0$, approximately, for valves of 3-in. to 24-in. size. Assumptions of a K value of 2.5 for 1-in. valves, 2.0 for 2-in. valves and 1.5 for 10-in. and larger valves should give reasonable friction head allowances.

A check valve using a disk hinged slightly above its center has become quite popular. Published data on two sizes of this type of valve indicate a K value of about 0.3 at

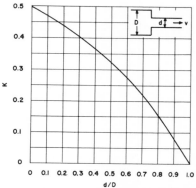

Fig. 18.5 Head loss in sudden contraction of pipe

Average of data obtained from various sources.

TABLE 18.2   VELOCITY AND FRICTION HEAD LOSS IN NEW PIPING

Friction values apply to Schedule 40 (standard weight) steel pipe carrying water.

gpm	1-in. pipe (1.049-in. ID) v	f	1¼-in. pipe (1.380-in. ID) v	f	1½-in. pipe (1.610-in. ID) v	f	2-in. pipe (2.067-in. ID) v	f	2½-in. pipe (2.469-in. ID) v	f	3-in. pipe (3.068-in. ID) v	f	3½-in. pipe (3.548-in. ID) v¹	f²	4-in. pipe (4.026-in. ID) v	f	5-in. pipe (5.047-in. ID) v	f	6-in. pipe (6.065-in. ID) v	f	8-in. pipe (7.981-in. ID) v	f
1	.37	.11																				
2	.74	.39	.43	.10																		
3	1.11	.82	.64	.21	.47	.10																
4	1.49	1.37	.86	.36	.63	.17																
5	1.86	2.08	1.07	.54	.79	.26																
6	2.23	2.83	1.28	.76	.95	.35	.57	.10														
8	2.97	4.88	1.72	1.29	1.26	.61	.76	.17														
10	3.71	7.12	2.14	1.95	1.57	.90	.96	.26	.67	.11												
15	5.56	15.0	3.21	4.06	2.36	1.87	1.43	.54	1.00	.23												
20	7.41	25.6	4.28	6.80	3.15	3.12	1.91	.92	1.34	.38	.87	.13										
25			5.35	10.3	3.94	4.70	2.38	1.39	1.67	.58	1.08	.20	.81	.10								
30			6.43	14.4	4.72	6.60	2.86	1.92	2.00	.81	1.30	.28	.97	.13								
40					6.30	11.2	3.82	3.35	2.68	1.36	1.73	.47	1.30	.23	1.01	.12						
50					7.87	16.6	4.77	5.00	3.34	2.06	2.16	.72	1.62	.34	1.26	.18						
60							5.72	7.00	4.02	2.85	2.60	.99	1.94	.48	1.51	.25						
70							6.68	9.40	4.68	3.80	3.03	1.33	2.27	.63	1.76	.34	1.12	.11				
80							7.62	11.9	5.35	4.95	3.46	1.72	2.59	.82	2.01	.43	1.28	.15				
90							8.60	14.7	6.02	6.05	3.89	2.13	2.91	1.00	2.27	.54	1.44	.18				
100							9.56	18.7	6.70	7.47	4.83	2.58	3.24	1.24	2.52	.65	1.60	.22	1.11	.09		
125									8.37	11.1	5.41	3.90	4.05	1.82	3.15	1.00	2.00	.33	1.39	.13		
150									10.0	15.4	6.50	5.44	4.86	2.55	3.78	1.39	2.40	.47	1.66	.18		
175									11.7	20.8	7.58	7.30	5.66	3.40	4.40	1.90	2.80	.62	1.94	.24		
200											8.66	9.18	6.48	4.35	5.04	2.40	3.20	.80	2.22	.31		
225											9.75	11.6	7.30	5.44	5.66	2.98	3.60	.97	2.50	.38	1.44	.10
250											10.8	14.0	8.10	6.59	6.29	3.68	4.00	1.19	2.77	.47	1.60	.12

For each pipe: V = velocity, in feet per second[1]; F = friction head loss, in feet of water per 100 feet of pipe[2].

Flow	3-in. pipe (3.068-in. ID) V	F	3½-in. pipe (3.548-in. ID) V	F	4-in. pipe (4.026-in. ID) V	F	5-in. pipe (5.047-in. ID) V	F	6-in. pipe (6.065-in. ID) V	F	8-in. pipe (7.981-in. ID) V	F	10-in. pipe (10.020-in. ID) V	F
275	11.9	16.9	8.91	7.90	6.92	4.35	4.40	1.43	3.05	.56	1.76	.15		
300	13.0	19.6	9.72	9.30	7.55	5.04	4.80	1.65	3.32	.66	1.92	.17		
350			11.3	12.2	8.80	6.85	5.60	2.21	3.88	.88	2.24	.23		
400			13.0	15.9	10.1	8.67	6.40	2.89	4.44	1.12	2.56	.29	1.62	.10
450			14.6	20.0	11.3	10.9	7.20	3.56	4.99	1.40	2.88	.37	1.82	.12
500					12.6	13.3	8.00	4.36	5.54	1.72	3.20	.45	2.03	.15
550					13.9	16.0	8.80	5.17	6.10	2.06	3.52	.55	2.23	.18
600					15.1	19.1	9.60	6.16	6.65	2.42	3.84	.63	2.44	.21
650							10.4	7.22	7.20	2.78	4.16	.73	2.64	.24
700							11.2	8.29	7.75	3.25	4.47	.85	2.84	.28
750							12.0	9.40	8.31	3.63	4.80	.97	3.04	.31
800							12.8	10.3	8.87	4.11	5.11	1.11	3.25	.35
900							14.4	13.0	9.96	5.12	5.75	1.33	3.65	.44
1,000							16.0	15.8	11.1	6.17	6.40	1.64	4.06	.55
1,100							17.6	19.0	12.2	7.45	7.04	1.98	4.46	.64
1,200									13.3	8.73	7.67	2.36	4.87	.75
1,300									14.4	10.2	8.31	2.71	5.27	.88
1,400									15.5	11.9	8.95	3.10	5.68	1.02
1,500									16.7	13.2	9.60	3.49	6.09	1.18
1,600									17.8	15.0	10.2	3.92	6.49	1.31
1,800									20.0	18.5	11.5	4.99	7.30	1.60
2,000											12.8	5.96	8.11	1.97
2,500											16.0	9.00	10.2	2.95
3,000											19.2	12.5	12.2	4.15
3,500											22.4	16.6	14.2	5.60
4,000													16.2	6.90
4,500													18.3	8.80
5,000													20.3	10.8
5,500													22.3	13.0
6,000													24.4	15.3

[1] Velocity, in feet per second.
[2] Friction head loss, in feet of water per 100 feet of pipe.

all velocities. Fully open butterfly valves should have K values in this range.

In many municipal or other important installations of large or fairly large size, a combined check and stop valve design made on the principle of the plug cock, with the plug rotated by an external mechanism, is used. This special valve has a straight full-size passage when fully open, and should have a loss no greater than a section of pipe of the same length.

A flap valve is a form of a check valve used on the end of a pipe. The flap is quite light in most designs. Some special designs with the flap partially counterweighted have also been made. Flap valves have very low losses even at low velocities when the

disk is not raised very high by the flow. In most designs, a loss of 0.2 ft of water irrespective of velocity should be ample allowance. The exit loss equal to the velocity head at the valve must also be added. Losses in multiported check and foot valves vary too widely to make any assumption. This type of construction is now rarely used so that it is likely to be encountered only in existing installations in which the loss can be determined by test. While designs of foot valves with strainers vary widely, a K value of 5 to 15 might be expected.

The value of K varies with the design of any valve or fitting and, in the case of elbows in which part of the loss is due to the bend and part due to the length of pipe

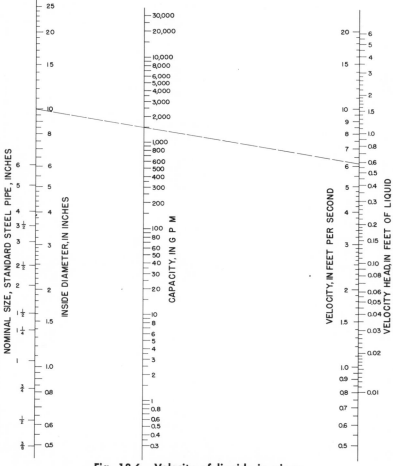

Fig. 18.6   Velocity of liquids in pipes

involved, the value of K varies with the smoothness of the walls. Thus, calculated friction values are approximations, not definite values. The approximate value of velocity head for any capacity in any size pipe up to 24-in. pipe can be quickly obtained (Fig. 18.6). This value when multiplied by the appropriate K value gives the head loss in the fitting. For example, with a flow of 1,500 gpm through 10-in. pipe, the velocity will be 6.1 ft per second, for which the corresponding velocity head is 0.58 ft.

The loss in any valve, fitting or other resistance can be expressed as the loss in a length of pipe of the same size as the fitting. The total friction loss involved can be determined for the total length of piping, plus the equivalent lengths of all the valves, fittings, and other resistances. Using the K value of the fitting, the resistance of any fitting expressed in equivalent length of pipe can be approximated from the chart in Fig. 18.7. For example, the loss in a fully open 10-in. gate valve ($K = 0.2$) would be

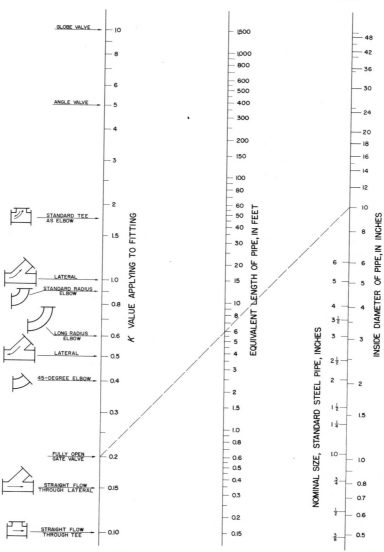

Fig. 18.7 Friction losses in fittings expressed in equivalent lengths of straight pipe

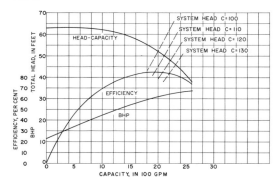

Fig. 18.8   System head for installation shown in Fig. 18.1 for various roughness factors using an 8-in. pump

approximately the same as the loss in 6 ft of 10-in. pipe. It must be remembered that any such conversion may result in a somewhat different value for the head loss in any fitting when compared with the value obtained by the velocity head method. The magnitude of the difference depends upon the Williams and Hazen C value or equivalent used in determining the friction head loss per unit length of pipe.

There is very little loss in taper reducers because a liquid can be accelerated with little loss. For long reducers there will be a greater loss because of length. In such cases, determine the loss as for a pipe with a diameter equal to the average diameter of the reducer. There is greater friction loss in increasers. A taper increaser up to about a 15-deg included angle (Fig. 18.4, $[D - d]/L$ equals 0.266 or less) will result in the water following the taper. A taper with over 60-deg included angle (Fig. 18.4, $[D - d]/L$ equals 1.15 or more) will have about the same loss as that which would be determined for a sudden expansion. For tapers between 15-deg and 60-deg included angle, calculate the loss as one-half the loss determined for a sudden expansion.

This subject of losses in fittings has been discussed primarily for systems handling water. In general, these methods of determining the head loss through fittings apply as well to systems handling other noncompressible liquids.

Determining friction head

Using frictional values for $C = 100$ for the pipe and figuring the losses in fittings and valves in the K-times-velocity-head basis, the head for the system shown in Fig. 18.1 for a flow of 2,000 gpm would be determined as follows:

Entrance loss (12-in. bell not well rounded)–K = 0.5	0.25 ft
12-in. long radius elbow–K = 0.2	0.10
20 ft of 12-in. pipe (at 1.49 ft loss per 100 ft of pipe)	0.30
10-in. × 12-in. reducer	0.10
8-in. gate valve–K = 0.2	0.51
8-in. swing check valve–K = 1.8	4.56
8-in. × 12-in. increaser–K = 0.30	0.76
3,000 ft of 12-in. pipe (at 1.49 ft loss per 100 ft of pipe)	44.70
Two 12-in. long-radius elbows–K = 0.2	0.20
Three 45-deg elbows–K = 0.2	0.30
Exit loss (12-in. pipe–1 velocity head)	0.50
	———
Total losses	52.28 ft
	say, 52.3 ft

Of this 52.3-ft loss, 45 ft is loss in the pipe and will vary according to the C value of the pipe. The remaining 7.3 ft is the allowance for loss in valves and fittings and will vary only slightly with the age of the pipe.

By computing the values for various other capacities, we would be able to graph the relation of the system head to the capacity (Fig. 18.8). Without further analysis it appears that for a flow of 2,000 gpm, a good selection would be a pump with the same characteristics as that in Fig. 18.8.

If the pipe were new, there would have been less friction loss. If the condition of the pipe was such that $C = 130$, the pipe friction loss would be 61.5 per cent of the friction loss when $C = 100$. For example, at 2,000 gpm the pipe friction loss would have been 45 times 0.615 or 27.7 ft. The total friction head, including friction losses in the fittings, would be 27.7 plus 7.3 or 35.0 ft. The resulting system-head curve labeled $C = 130$ is shown in Fig. 18.8. Thus, the pump delivers a greater amount of liquid (2,320 gpm) with less head (46-ft total head) and less efficiency (82.5 per cent).

If it was known that the water would cause a very slow increase in friction with increasing age of pipe (reaching a value of $C = 110$ in 15 years or longer) or if power costs were so high that the friction was to be kept low by periodic cleaning of the pipe, then it would be advisable to select the pump for a lower head. Using the system head for $C = 110$ as the maximum to be encountered and at which 2,000 gpm capacity is desired, the pump would have to be selected for 45-ft total head. If the same pump was used with a smaller impeller for 2,000 gpm, a 45-ft head would yield an initial efficiency of 82.5 per cent for the system when the piping had a C value of 130, and would reach an efficiency of about 84.5 per cent when the C value had fallen to 110. If it was desired to obtain greater economy over the entire operating range, a larger pump would have to be used. The larger pump would have a 10-in. discharge and a 10-in. gate valve. A 10-in. check valve could be used with smaller friction losses. A graph of the relation of the system head to the capacity for this system is shown in Fig. 18.9.

In circulating pumps for surface condensers the head is composed entirely, or almost entirely, of friction losses. These systems are generally complicated because of the need for less water in winter, when the water temperature is low, than in sum-

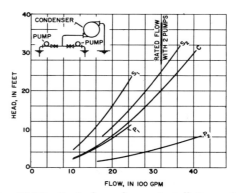

Fig. 18.10 Typical condenser installation with two pumps in parallel showing operating conditions for one- and two-pump operation

> KEY:
> S_1 = System head—one pump running $(C + P_1)$
> S_2 = System head—two pumps running $(C + P_2)$
> C = Friction losses in condenser and common piping
> P_1 = Friction losses in individual pump piping—one pump running
> P_2 = Friction losses in individual pump piping—two pumps running.

mer when the water temperature is high. Generally, two pumps of equal capacity are used. Both are run during the summer to give the required large capacity and one is run in winter, when less capacity is needed. In this installation, the head would be made up of losses through the piping and fittings carrying the capacity handled by each pump and losses through the piping, fittings, and condenser that carry the combined flow. Thus, the system-head curve for the operation of one pump would not be the same as the system-head curve for the operation of both pumps. The losses for the individual pumps are shown by curve P_1 in Fig. 18.10. If both pumps are running, twice the capacity flow through the condenser yields the same loss in the individual piping for each pump. This is shown by curve P_2. The loss in the condenser and in the piping and fittings in which the flow is the same is shown in curve C. At any capacity, the system head with one pump running (S_1) is the head

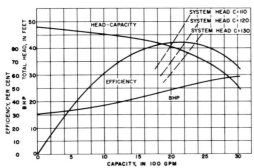

Fig. 18.9 System head for installation shown in Fig. 18.1 using a 10-in. pump

Performance is improved.

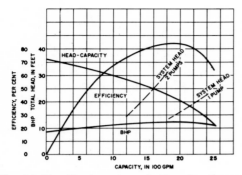

Fig. 18.11 Ideal pump selection for the system shown in Fig. 18.10 if one- and two-pump operation are equally important

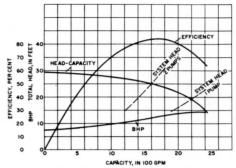

Fig. 18.12 Results of smaller pump for installation in the system in Fig. 18.10

This pump shows poor efficiency with single-pump operation.

shown in curve P_1 plus that shown in curve C. With two pumps running (S_2), it is the head shown in curve P_2 plus that shown in curve C. If the loss in the individual piping is low, curves S_1 and S_2 are so close that only S_2 is constructed and the discrepancy between S_1 and S_2 is ignored.

The system illustrated in Fig. 18.10 indicates no static head, and is based on the assumption that the full siphon head is recovered. Although siphons up to 25 ft or more are feasible, full recovery is rarely obtained. Also, in this system, unless the piping and condenser waterways are primed before the pumps are started, the pumps will have to fill the piping and condenser before the siphon can be established. Thus, in the starting cycle, a static head equal to the siphon leg will be encountered just before the siphon is established. The maximum starting head can be determined by adding the siphon leg as a static component to curves S_1 and S_2. It is often impossible to obtain a pump that will deliver sufficient capacity to establish the siphon without impairing the results obtained when the siphon has been established. Modern practice is to provide priming equipment, so that the siphon loop can be evacuated, and the siphon established without the necessity of a high starting head.

It is desirable that good efficiency be obtained when two pumps are running as

well as when one pump is running. The system head when both pumps are running can be plotted against the capacity handled by each pump (Fig. 18.11 and 18.12). The problem is to select a pump for this installation that will give good efficiency at 1,600 gpm and 25-ft head as well as at 2,100 gpm and 18-ft head resulting from the intersection of the pump head-capacity curve and the system curve S_1. This generally requires the use of a larger pump than would be used if the pump were selected only for the 1,600 gpm 25-ft head condition. The selection of a pump with 82 per cent efficiency at both operating conditions is ideal (Fig. 18.11). If a smaller pump had been selected, more actual capacity would have been obtained with one pump operating but only 74.5 per cent pump efficiency would have been obtained (Fig. 18.12). If, in this installation, both pumps were normally operated all the time and two units had been installed instead of one larger unit (in order to permit operation at reduced capacity if one unit is out of service), the selection shown in Fig. 18.12 would have been preferable to that shown in Fig. 18.11 because the efficiency at 1,600 gpm and a 25-ft head is greater and the first cost would be lower.

When there is a static head, or its equivalent in pressure, or both, included in the head, the system head is the sum of these

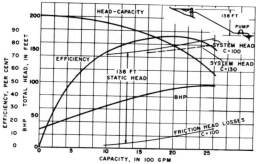

Fig. 18.13 Characteristics of pump whose total head is mostly static

Error in calculating friction element or change in friction with age has little effect on selection.

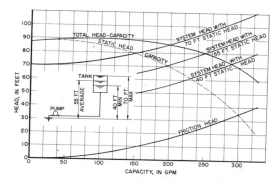

Fig. 18.14 Characteristics of pump installation with a variable static head

Friction head variation with capacity must be considered in the selection.

components plus the friction head. Thus, for the system in Fig. 18.13, the friction head losses have been determined in curve form and added to the static head, giving the system head indicated. If the pipe had been new and had a coefficient of $C = 130$, very little increase in capacity would have resulted. Consequently, the possible error in determining the friction loss becomes less important as the percentage of the friction head in the total operating head is reduced.

Effect of variable static head

At a constant speed, the head developed by a centrifugal pump varies with the capacity delivered by the pump. Thus, if a pump is to be used in a system in which there is a variation in static head, the capacity delivered through the system will also vary. The purchaser of a pump for such an installation, will often calculate the friction at rated capacity, add it to the average static head, and state the sum as the design head. In addition, he will add the same friction head to the maximum and minimum static heads and give the resulting heads as the maximum and minimum operating heads that the pump will encounter. Thus, the change in friction with change in capacity is neglected, and the manufacturer is handicapped in selecting

a suitable pump. For example, in the installation shown in Fig. 18.14 some purchasers would specify that the pump had to operate over a head range of 65 to 95 ft, giving a rated capacity of 250 gpm at 80-ft head. The pump in Fig. 18.14 would appear to be unsatisfactory, although it will actually deliver 79 per cent rated capacity at the maximum static head and 115 per cent rated capacity at the minimum static head.

It is often good to know what capacity will be delivered by a pump operating on a system in terms of static head (Fig. 18.14). For any capacity, the static head will be the total head of the pump minus the friction loss at that capacity. This can be graphed (Fig. 18.14) to show, for example, that the flow to the tank will be 244 gpm when the static head is 57 ft.

Determining pump delivery

If a plant or community is located at some distance from its source of water supply, the demand for water very often increases over a period of time. It ultimately becomes uneconomical to continue to use the existing pipe line because of the frictional head loss with increased capacity. If the original line is in good condition, the usual solution is to install a second line, in parallel with the existing line, that will allow economical pumpage of the desired increased rated capacity. If two pipes are operating in parallel, the friction head loss in each branch must be the same. The proper approach to this problem is to plot the relation of capacity to friction head loss for each line and then to determine the relation of combined capacity to friction head loss for the two by adding together the capacity of each line, when the head losses are the same, at a number of points. For example, Fig. 18.15 shows the head loss for a 10-in. line and the head loss for a paralleling 12-in. line in the form of a

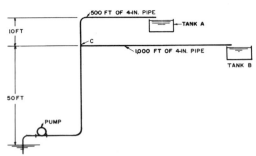

Fig. 18.16 Pumping system involving two tanks at different elevations

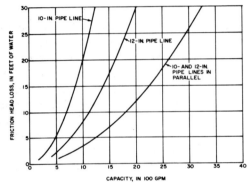

Fig. 18.15 Relationship of capacity to friction head for two pipe lines in parallel

graph. With 20-ft friction head loss, the 10-in. line will have a flow of 1,000 gpm, whereas the 12-in. line will have a flow of 1,615 gpm. Thus, 1,000 plus 1,615 or 2,615 gpm will be the combined flow of the two lines with 20-ft head loss. The capacity for the two lines in parallel for equal head losses can be determined for a number of points and a curve showing combined capacity against head loss can be drawn (Fig. 18.15).

In some systems, such as in a water works distribution system, it is desirable to maintain a nearly constant pressure although the demand varies. To maintain an exactly constant pressure, it would be necessary to vary the speed of the pump or pumps; but, in most systems, it is rarely necessary to

maintain the pressure exactly and some variation can be allowed. Most electric-motor-driven pumps that maintain nearly constant pressure are, therefore, driven by constant-speed motors. Thus, the pressure will depend on the head developed by the pump or pumps operating in parallel at the demand capacity. In order to produce a reasonably constant pressure for the full range of demand, it is desirable to select pumps having head-capacity curves that have a shutoff head of 10 to 20 per cent more than the head at design capacity.

Installations of divided flow or branch lines, in which the flow is controlled only to prevent overflowing of a tank or reser-

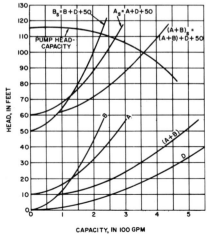

Fig. 18.17 System head-capacity curves for installation in Fig. 18.16

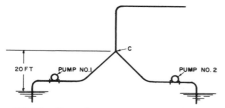

Fig. 18.18 Pumping system involving two pumps that have considerable individual piping discharging into a common line

voir, usually involve two or possibly three branches. In an installation that has two tanks at different elevations (Fig. 18.16), it is obvious that flow will not reach tank A unless the friction head loss in the 4-in. line from point C to tank B exceeds 10 ft, which is the difference in static head for the two inlets. Considering the branch from point C to tank B, the system head has no static component and will be frictional head only. This may be calculated (curve B, Fig. 18.17). The system head for the branch from point C to tank A, has a 10-ft static component and a friction component caused by the friction in 500 ft of 4-in. pipe (curve A, Fig. 18.17). The capacities for the two branches for equal system heads up to point C can be added together to give the system head of the two branches as a unit (curve A + B, Fig. 18.17). As the friction loss in the piping from the supply to point C is the same for both branches, there would be no difficulty in establishing the friction head curve (curve D, Fig. 18.17). Point C is 50 ft above the suction supply, therefore the system head for the common system up to point C will be the friction head plus 50 ft. By adding this value to the head values on curve (A + B), we derive the system-head curve for the entire system as shown in curve (A + B)$_s$ (Fig. 18.17).

If the pump used for this service had the head-capacity curve shown in Fig. 18.17, the resulting total flow would be 358 gpm. This capacity would require a head at C of 31 ft, as shown by projecting a line from the intersection of curve (A + B)$_s$ and the pump head-capacity curve to curve (A + B).

By projecting lines to curve A and to curve B from this point, we see that 193 gpm would be going to tank A and 165 gpm to tank B.

If the two branches to the tanks are each equipped with a valve actuated by a control that closes the valve when the tanks become full, at times there would be flow to only one of the tanks. The flow to tank A only can be determined by constructing its system head curve A$_s$ (by adding head values on curves A and D plus the 50 ft static head) and determining the point at which curve A$_s$ intersects the pump head-capacity curve. In the same way, a curve showing the extent of flow to tank B if the branch to tank A is shut off can be constructed (curve B$_s$).

If a pump that developed 61.5 ft, or less, total head at 90 gpm capacity had been used, the entire flow would have gone to tank B.

In many cases, especially those involving two or more pumps each with its own piping, valves, and fittings discharging into a common discharge line, the solution of a problem can be simplified by determining the head each pump will produce for its range of capacity up to the point in the system at which their pipes join. A system involving two pumps with separate piping discharging into a common line is shown diagrammatically in Fig. 18.18. The usual head capacity of pump No. 1 (Fig. 18.18) is graphed in Fig. 18.19. The friction loss

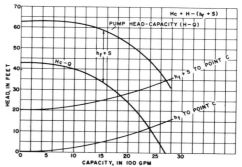

Fig. 18.19 System head-capacity curves for installation in Fig. 18.18

from suction supply to point C is plotted (curve h_f) and the system head (curve $h_f + S$) for pump No. 1 from the suction supply to point C is thus determined. By subtracting the values on curve $h_f + S$ from those on the pump head-capacity curve (H–Q), we derive curve H_c–Q. Curve H_c–Q indicates the head that will be produced by pump No. 1 at point C as a function of the capacity being delivered.

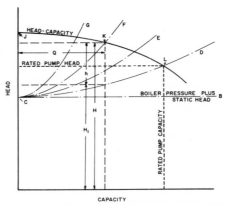

Fig. 18.21   Boiler feed installation system-head curves superimposed on the feed pump head-capacity curve

Types of pumping systems

Pumping systems are of two types—throttled and unthrottled. In a throttled system, the capacity is determined primarily by the demand, and the flow is controlled by throttling the excess head developed by the pump or pumps. In some systems, boiler feed pumps for example, a throttle valve located in the discharge line controls the flow. In others, such as city water-supply systems without a standpipe or reservoir "floating" on the distribution mains, the consumers of the water control pump discharge as they open or close valves. For an unthrottled system in which pumps discharge into a standpipe or a reservoir, the flow depends on the head developed by the pumps and on the characteristics of the system.

Throttled systems

In a throttled system, such as a boiler feed pump installation (Fig. 18.20), the flow is controlled by the throttle valve, which is

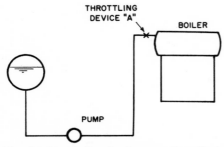

Fig. 18.20   Boiler feed pump installation

A typical throttled system.

usually positioned automatically by the feedwater regulator (valve A, Fig. 18.20). Fig. 18.21 shows the boiler feed system-head curves superimposed on the head-capacity curve of the pump. Curve C–B represents the boiler pressure plus the static elevation. Although slight changes take place in the boiler pressure with changes in load, for the sake of simplicity we shall assume the boiler pressure to be constant.

When water is supplied to the boiler, the pump operates against a pressure that increases with flow because of the friction head losses in the piping, fittings, and valves in the line. With throttle valve A wide open, the system-head curve will be curve C–D (Fig. 18.21). The point at which this curve crosses the pump head-capacity curve (L) is the rated head and capacity of the pump.

If valve A is partially closed, the friction head increases, and the system-head curve may rise to position C–E. Further closing of valve A would produce other system-head curves such as C–F or C–G. If valve A is closed entirely, the pump pressure would go to shutoff (point J). Thus, the system-head curve can be varied by opening and closing the throttle valve so that a family of curves is produced. These curves intersect the head-capacity curve at various

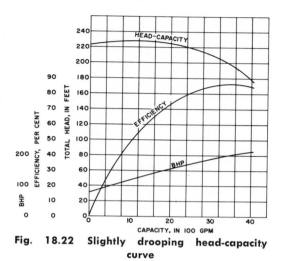

Fig. 18.22 Slightly drooping head-capacity curve

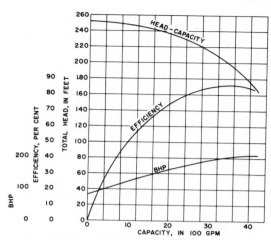

Fig. 18.23 Stable head-capacity curve

points between the fully closed position (J) and the fully open position (D).

In order to supply the boiler with a quantity of water, Q, the throttle valve is adjusted until the system-head curve becomes C–F (Fig. 18.21). This curve crosses the head-capacity curve at K and the head against which the pump operates is represented by the vertical distance H. The actual head required to deliver quantity Q to the boiler on the normal curve C–D is represented by H_1. As the pump develops a head H at capacity Q, valve A will have to throttle an excess head equal to H minus H_1 (distance h, Fig. 18.21).

When a single pump operates on such a system, the shape of the pump head-capacity curve does not matter. The two pumps whose characteristics are shown in Fig. 18.22 and 18.23 could be used alone on a throttled system. If the cost of power were high, the pump shown in Fig. 18.22 would be preferred because the power it requires at part capacities is slightly lower in this particular case. It is not to be inferred that a pump with a flatter head-capacity curve will always have lower power requirements at part capacities. Lower power requirements depend on many factors: Individual impeller and casing designs, ratio of design point to point of maximum efficiency, and

the like. A pump with a steeper head-capacity curve has the advantage in a single-pump throttled system because it is less sensitive—the throttling valve must be moved through a greater distance as more head is throttled off.

Despite having a drooping head-capacity curve, the pump characterized by Fig. 18.22 could be used on a single-pump throttled system with general assurance of satisfactory operation because the operating point will always be determined by the throttled system head. In rare cases, surging has resulted in single-pump throttled systems with pumps having drooping head-capacity curves and, even less frequently, with pumps having stable (constantly rising) head-capacity curves.

Installations of two or more pumps operating in parallel involve piping and fitting losses for each pump as well as for common piping (Fig. 18.24). Instead of the

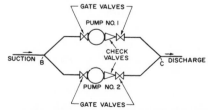

Fig. 18.24 Simplified piping hookup for two centrifugal pumps operating in parallel

true pump head-capacity characteristics, a head-capacity characteristic measured from B to C should be used in the analysis. In throttled systems in general, and particularly in high-head systems such as boiler feed installations, the losses in the individual pump piping are such a small percentage of the total head, that their effect is not noticeable on a curve drawn to reasonable scale.

Two pumps designed for 3,500 gpm, 197 ft total head are shown in Fig. 18.22 and 18.23. They have differently shaped head-capacity curves. The effect of the difference in losses in the individual pump piping in each pump will be ignored in this discussion. The individual head-capacity characteristics for each pump are shown in Fig. 18.25 as A and B. Their combined characteristics are labeled C. Curve C is obtained by adding together the capacities of the individual pumps at the same head. For example, at 210 ft total head, pump A will handle 2,950 gpm and pump B will handle 3,120 gpm; thus they will produce 6,070 gpm together at 210 ft total head.

If these pumps were installed in a water works plant delivering water into a direct distribution system, in which the desired main pressure corresponds to a total head of 197 ft, and the demand was 6,000 gpm, the two pumps in parallel would produce 211 ft total head and the main pressure would be 211 minus 197 or 14 ft (6.06 psi) greater than desired. Of the 6,000 gpm, 2,900 gpm would be delivered by pump A and 3,100 gpm by pump B. If the demand was reduced to 5,000 gpm, the main pressure would be somewhat higher. The pumps would be working against 221 ft total head with pump A delivering 2,270 gpm and pump B delivering 2,730 gpm. If the demand were reduced to 3,000 gpm, the division of pumping capacity would be 500 gpm by pump A and 2,500 gpm by pump B. Finally, should the demand fall to 2,630 gpm, pump B would be delivering all the water while pump A would be delivering none. Thus, at 2,630 gpm or less

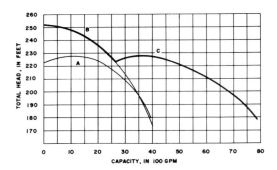

Fig. 18.25 Individual and combined head-capacity characteristics of pumps in Fig. 18.22 and 18.23

demand, pump A would be backed off the line by pump B and would be operating at shutoff—a dangerous situation even for a short duration.

If pump A had been operating alone on a demand of 2,630 gpm or less and pump B were started, pump B would pick up the entire load and back pump A off the line. If pump B had been operating alone at a demand of 2,630 gpm or less, and pump A were started, pump A would be unable to deliver any water to the system. If these pumps were operated in a system in which the change in demand was relatively slow and in which units were cut out when the demand fell to the rated capacity with one less unit in service, they should never be allowed to operate in parallel below 3,500 gpm. For demands less than 3,500 gpm, either pump A or B would be operated alone. In such a carefully supervised situation, the two pumps could be operated successfully in parallel on the throttled type of system. It should be noted that at 5,000 gpm combined flow, pump A with 2,270 gpm flow has 76.8 per cent efficiency, requiring 165 brake horsepower (bhp) and pump B with 2,730 gpm flow has 82.3 per cent efficiency, requiring 185 bhp, a total of 350 bhp for the two pumps. If both pumps were the same as pump A, a flow of 5,000 gpm would have meant 2,500 gpm per pump with 173 bhp per pump, or 346 bhp total. If both pumps were the same as pump B, the power would have been 178 bhp each or 356 bhp total. The use of two pumps that have equal capacities does not necessarily result in power economy.

Difficulties may be encountered on throttled systems with parallel operation of similar pumps if the pumps have even moderately drooping head-capacity characteristics (Fig. 18.22). Figure 18.26 shows, with an exaggerated head scale, the theoretical head-capacity curve of one such pump and of two such pumps in parallel. Let us assume that the friction losses in the individual pump piping are relatively small and that they may be ignored. If the demand was 1,750 gpm with one pump operating alone, the pump would operate against 226 ft total head and exert a discharge pressure corresponding to that against the check valve of the second pump. This discharge pressure is greater than the shutoff head developed by the second pump. If the second pump were started, it would come up to speed against shutoff and would be unable to establish any flow because the pressure on the discharge side of the check valve would have been greater than the pump could develop at shutoff. In some installations with two pumps (Fig. 18.26), various methods are used when it is desired to start the second pump with the first operating on the top of the curve. One, possibly the most common, is to throttle a little on the gate valve of the pump that is running so that the net head (B to C, Fig. 18.24) is less than the shutoff head of the second pump. This, and other tech-

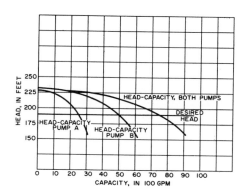

Fig. 18.27   Head produced by two pumps operating in parallel

niques, generally require very experienced manipulation and careful timing.

With the pumps in Fig. 18.26, it is possible to obtain unequal capacities at certain flows even if they are hydraulically duplicates and operating at the same speed. For example, with a demand of 2,230 gpm, one pump could be delivering 500 gpm and the other 1,730 gpm. Actually, it is inadvisable to run two pumps, such as shown in Fig. 18.22 and 18.26, in parallel for capacities at which the developed head exceeds the shutoff head, in this case below 2,100 gpm. First, although the two pumps and their drivers are apparently duplicates, there will be minor differences in the operating characteristics. This will cause unequal distribution of the capacity and, sometimes, backing one off the line. Second, a motor-driven pump operating under apparently stable conditions may have minor speed variations as well as minor variations in hydraulic performance that can result in unequal sharing of the load between the two pumps. This could result in one pump operating at shutoff.

Thus, for pumps to operate satisfactorily in parallel in a throttled system, it is desirable: (1) That they have stable (steadily rising to shutoff) head-capacity curves; and (2) that over the operating head range the pumps have approximately the same percentage reduction in capacity, or at least

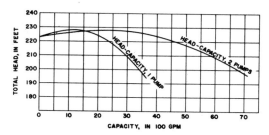

Fig. 18.26   Individual and combined head-capacity characteristics of two pumps with slightly drooping head-capacity curves

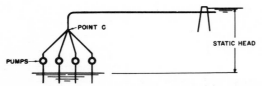

Fig. 18.28 System involving four pumps with individual piping but with a common transmission line

deliver some capacity. As previously mentioned, the increase in head from design capacity to shutoff should not be too high, otherwise excessive pressure is developed at part-capacity flows. In a system in which it is desired to maintain a constant minimum pressure at the pumping plant despite varying demand, the design heads and shutoff heads of all the units are usually the same, or approximately so. Thus, if flows are less than the units in service will produce at rated head, the capacity delivered by each pump will be about the same proportion of the rated capacity. For example, if a 2,500-gpm pump (pump A) and a 5,000-gpm pump (pump B) operate in parallel (Fig. 18.27), and the demand is 5,500 gpm, the head developed by the two pumps would be 211 ft (21 ft above that desired). The capacity delivered by pump A would be 1,800 gpm or 72 per cent of rated capacity, whereas that delivered by pump B would be 3,700 gpm or 74 per cent of rated capacity.

Usually, two or more pumps with stable head-capacity characteristics, and equal or nearly equal shutoff heads when operating in parallel in a throttled system, will share the load about equally down to a system capacity much below the capacity at which one or more pumps would be taken out of service.

When purchasing a new pump or pumps that are to be placed in parallel to existing units, the purchaser should supply the vendor with the head-capacity characteristics of the existing pumps and information on the operating pressure, to enable the vendor to select a new pump with suitable characteristics.

Some throttled systems utilize a long transmission line between the pumping station and the point at which a minimum pressure is to be maintained. In such cases, the reduction in pipe friction in the line when the flow is reduced will cause increased pressure. If this increased pressure is objectionable, possible solutions are: (1) Maintaining constant pressure by throttling the excess head with some form of valve, (2) varying the speed of the pumps so that the required head is developed at the capacity demand, (3) using one or more booster pumps in series with the pumps operating in parallel so that the head developed by the pumps in service can be increased in steps as the capacity demand in-

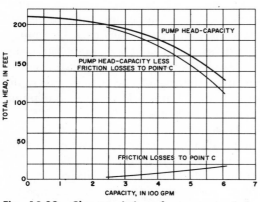

Fig. 18.29 Characteristics of pump no. 1 in Fig. 18.28

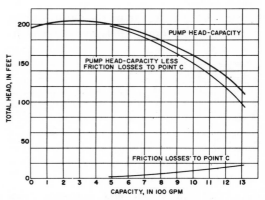

Fig. 18.30 Characteristics of pump no. 2 in Fig. 18.28

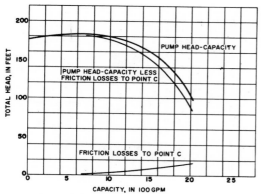

Fig. 18.31 Characteristics of pump no. 3 in Fig. 18.28

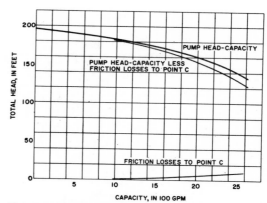

Fig. 18.32 Characteristics of pump no. 4 in Fig. 18.28

creases, or (4) installing a number of pumps so that small increments of capacity can be obtained. The proper solution is usually the one that is economically best and therefore depends, in part, on the cost of power as well as on the cost of personnel necessary for the operation of the system.

Unthrottled systems

For a system in which the flow is not throttled, and in which the capacity is such that the head developed by the pumping system equals the head necessary to deliver the capacity through the system, it is not necessary for pumps to have similar characteristics in order to be operated in parallel. When buying additional pumps to operate in such a system, many purchasers make the mistake of requiring the additional pumps to have characteristics exactly similar to those of their existing units—this is not necessary. A system is shown diagrammatically in Fig. 18.28. The system head beyond point C (Fig. 18.28) is indicated in Fig. 18.33. The desired pumpage rate is from 2,500 to 5,000 gpm. Four pumps that are dissimilar (Fig. 18.29–18.32), may be operated in parallel on this system, since the maximum head against which they will operate is 160 ft. The head-capacity curves plotted in Fig. 18.33 have individual piping and fitting losses deducted.

For unthrottled systems, the most economical pump operation is obtained when there is little variation in the system head as the capacity changes. In many installations, the friction head is so large a part of the total head at maximum capacity that pumps designed for specific capacities and heads are better in the long run than pumps in parallel. Pumps in parallel would operate at poor efficiency at heads other than the designed head. For example, in an installation in which a flow of 6,250 to 10,400 gpm is wanted (Fig. 18.34), three separate pumps designed for 6,250 gpm at 139 ft head, 8,330 gpm at 166 ft head and 10,400 gpm at 200 ft head might be the proper solution. In other cases in which the

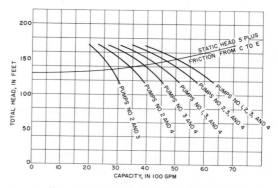

Fig. 18.33 Combined characteristics of pumps in system in Fig. 18.28

Based on head developed at beginning of common line.

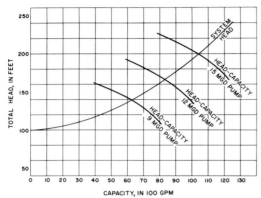

Fig. 18.34 System characteristics for head that has a large frictional component

Different size pumps designed for different heads are required.

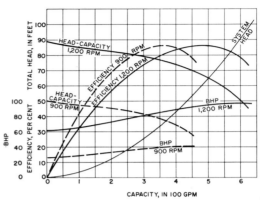

Fig. 18.35 System characteristics of an installation that is all friction

Efficient operation of a pump can be obtained by part-speed operation.

head is practically all friction, the solution might be a full-capacity pump driven by a two- or three-speed motor. The characteristics of a pump driven by a two-speed (1,200 and 900 rpm) motor operating against a system head that is entirely friction is shown in Fig. 18.35. When operated at 900 rpm, the capacity would be approximately three-fourths as much and the pump efficiency would remain almost the same. If pumps are driven by multispeed motors, the capacities that can be obtained at lower speeds depend on the speeds available so that it is not always possible to obtain the exact capacities desired. To obtain exact capacities, a variable speed driver would have to be used.

Frequently, one or more booster pumps are installed, either in the suction line to the main pumps or in the common discharge line to increase the capacity of existing stations. If conditions in the system illustrated in Fig. 18.28–18.33 should change so that the maximum demand at times is 5,900 gpm, one solution would be to have all four pumps discharge into a 5,900 gpm booster pump that has a total head of 30 ft. Booster pumps are particularly practical when an increase in head would cause considerable reduction in capacity of the main pumps.

Although the foregoing examples have described systems handling water, the basic principles apply to systems handling other liquids as well. There are sometimes certain limitations when liquids other than water are used. In systems handling volatile liquids minimum pressures must be maintained at every point. These requirements must be checked when analyzing the system.

Pumps with high specific speed impeller designs have steeper head-capacity curves than pumps with low specific speed designs. Thus, in systems involving low heads for which a high specific speed type of pump will be used, a greater variation in percentage of total head can be met efficiently than in systems involving high heads for which a

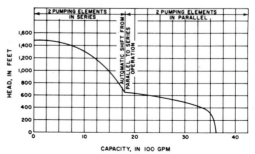

Fig. 18.36 Characteristics of series-parallel type of mine dewatering pump

low specific speed pump must be used. Occasionally, some low-head systems will require the use of a low specific speed type of pump in order to accomplish the desired operation. There are a few systems in which there is a wide variation in head range with no need for a fixed capacity to be delivered at any specific head. Such a system is met in dewatering a flooded mine. This is an unusual application of a centrifugal pump because the total head against which the pump must work varies from approximately zero to a high maximum that occurs when the mine is almost clear of water. A mine-dewatering pump should be designed not for a single point of head and capacity, but for the greatest possible capacity at all heads within the capacity of the motor. For installations involving final heads for which multistage pumps are required, the best possible design is a parallel-series unit (Fig. 18.36). With this unit the dewatering takes place almost twice as fast at the beginning than it would if the various stages were arranged to pump only in series, but the power expenditure is the same.

III
CONTROLS,
DRIVERS,
and
PRIMING

19 *Controls*

Centrifugal pumps are much simpler to control than either reciprocating or rotary pumps. They owe this ease of control to the flexibility of their characteristics, which enables them to adapt well to the varying requirements of the systems in which they can be applied. In order to understand the fundamentals of centrifugal pump control, however, it is necessary to have a clear conception of the relationship between the pump performance characteristics, the characteristics of the system in which the pump operates, and the operating conditions of head and capacity.

Pump and system characteristics

As shown in Chap. 18, a centrifugal pump installation consists of the pump and the system in which the pump is to operate. The characteristics of the system are determined by varying conditions of flow and the resulting heads, whereas those of the pump indicate the ability to produce flow in the system. The superimposition of the head-capacity curve of the pump over the system-head curve will indicate the operating conditions in the system, since a pump will always operate at those conditions of head and capacity that correspond to the intersection of the two curves.

Fundamental control functions

In most installations, pumps are operated continuously in a system for relatively long periods. Often, the only interruption of operation occurs as a result of some mechanical defect in the system or the pump itself, or because of a scheduled examination or overhaul. Nevertheless, a certain amount of flexibility in operating conditions always exists, and sometimes rather extreme variations may be encountered. The fundamental functions of centrifugal pump controls are, therefore, directed towards permitting the pump to meet the required variations in operating conditions, including the complete absence of delivery demand.

For simplification, centrifugal pump controls can be divided into two main groups according to function:

1. Controls that completely interrupt flow or cause it to resume.

2. Controls that vary the operating conditions—either the pump capacity, the total head, or both.

Interruption controls

The reasons for starting or stopping pump delivery will vary markedly; the following list must therefore be considered

251

representative, rather than exhaustive. Pump delivery may be stopped if:

1. The source of supply has been drained either completely or to the desired level, as (generally) with sump pumps, mine-gathering pumps, or tank-car emptying pumps.

2. The vessel into which the pump is delivering has been filled to the desired level. A typical example is the filling of a reservoir tank for service or for drinking water supply.

3. The pressure of the system into which the pump is delivering has reached its required magnitude, typical of hydraulic systems with dead-weight or air-bottle accumulators. However, pump design capacity and accumulator size generally should be selected to assure that maximum and average demand are such that the pump capacity remains within reasonably narrow limits in the neighborhood of its design value.

4. A batch process served by the pump has been completed. A typical example is the use of centrifugal pumps in a steel rolling mill descaling system, in which delivery through the descaling nozzles is interrupted between consecutive steel sheets.

5. A system is served by a battery of several pumps operating in parallel, and the required rate of flow has decreased to a point where the delivery from one or more of the pumps can be reduced to zero, to permit the remaining pumps to operate at a more economical point of their characteristic curve (as described in Chap. 18).

6. Mechanical trouble has occurred in the pump, its driver, or in some part of the system—overloading of a driver, overheating of an electric motor or of the pump, overspeeding of a steam turbine, burst piping in the system, and the like.

Pump delivery must be resumed when the conditions described above are reversed, as in an increase in supply level, reduction of delivery level, reduction of system pressure, or resumption of the batch process.

Controls which interrupt flow or cause it to resume can be subdivided into two separate classes: (1) Controls to start or stop the driver, and (2) controls to open or close a valve in the path of the flow. Each of these two classes of controls can be further divided into source of impulse and controlling medium. These subdivisions are related to the primary function of the controls, as the impulse source is created by the reason which dictates interruption or resumption of flow. Drivers can be started or stopped by various controls, as discussed below.

Pushbutton controls for electric motors

Pushbutton controls are widely used because this method requires only that the operator be informed of the necessity for flow interruption or resumption.

Float switches

Float switches can be applied to maintain certain predetermined maximum or minimum levels in tanks or reservoirs into which liquids are discharged, or from which liquids are removed by electric-motor-driven centrifugal pumps. In such an application, a float switch makes an electrical contact when the reservoir level reaches a predetermined height, and the pump motor starts. When the reservoir level reaches the other predetermined height, the contact is broken, and the pump motor stops. The same function can also be performed by level-sensing electrodes connected through relays that will make or break electrical circuits.

Pressure switches

Pressure switches are similar in principle to float switches but are operated by changes in pressure and are used to maintain pressure or vacuum, within certain selected limits, in a system or in a closed tank. Generally, the pressure to be controlled is applied to a diaphragm actuating the switch. When the system or tank pressure reaches a predetermined value, the diaphragm pressure is changed, an electrical contact is made, and the motor starts. When the pressure reaches the other predetermined value, the contact breaks and the motor stops.

Pressure switches can sometimes be used interchangeably with float switches, as they can be made to maintain reservoir or tank levels. One method of employing pressure switches uses the static head of the fluid in the reservoir as a source of varying pressure. In another method, a quantity of air is trapped in a closed tank; the pressure of this air varies directly with the liquid level.

Power consumption controls

A power consumption control may be used to stop a unit when the driver has been overloaded for any number of causes. This type of control can also be applied to stop the unit when the pump has lost its prime and is operating dry. In this case, an inverse-current relay will stop an electric motor as soon as the input current (that is, the power consumption) has dropped to 50 per cent of the normal operating minimum value.

Thermostat control

Thermostat controls are used in a number of centrifugal pump applications, especially in process industries, in which certain definite temperatures are to be reached or maintained by means of heat transfer, recirculation, or similar method. In such cases, a thermostat control can be applied in the same manner as a float or a pressure switch; however, temperature levels are substituted for either static levels or pressure differences.

Overspeed governors

Overspeed governors are applied mainly to steam turbine drivers, although they can also control internal combustion engines or water turbines used to drive centrifugal pumps. Governors are intended to protect the driver, the pump, and the system in which the pump is operating. The pump

and the system could be damaged if the driver possesses sufficient power to drive the unit at high enough speeds to cause excessive pressures. A typical use is that of the overspeed governor that acts on the inlet valve of a steam turbine and is generally set to trip out the turbine, should the operating speed exceed 5 to 10 per cent more than the design operating speed.

Complete interruption of flow without stopping unit

Many installations require interruption and resumption of flow without stopping the unit. This may be necessary because variations in demand are too frequent to warrant bringing the unit to a stop; because starting the driver is too lengthy and complex a process, once the unit has stopped; or because interruption of flow by stopping the driver does not prevent possible injurious effects, such as reverse flow through the pump. Flow can be interrupted or resumed by positioning a valve without affecting driver operation as follows:

1. Discharge valve manual controls. Some installations require that the unit be operating at all times, as a safety measure, even though no demand exists for flow delivery. A gate, globe, or plug valve located in the pump discharge can be manually adjusted to cut off the flow entirely. This is a frequent arrangement when several units are operating in parallel and when one or more of the units is installed for standby duty. This form of control requires a bypass to prevent the pump from overheating when it is operating at shutoff. In many cases, especially with across-the-line electric motor starters, the unit can be put on the line more quickly by starting the motor than by manipulating a valve in the discharge. Therefore, when the time element is important, it is preferable to start and stop the motor, rather than manipulate the discharge valves, which, if performed too quickly, may cause water hammer.

2. Discharge line check valves. Check valves are almost universally used in centrifugal pump discharge lines. A check valve acts to protect the pump, its driver, and the suction portion of the system against possible damage caused by reverse flow, if the external head in the discharge system exceeds the head generated by the pump, or if the pump comes to a standstill and develops no head whatsoever. Reverse flow through a pump would cause it to operate as a water turbine running opposite to the normal direction of rotation; therefore, if the driver is not suitable for this reverse operation, it could become seriously damaged. Even if the pump or its driver were not harmed, excessive pressures could be imposed on the suction piping and fittings. A check valve is fully automatic in its operation, closing when adverse conditions occur and reopening immediately on the resumption of normal conditions.

3. Float control valves. Variation in suction or discharge levels can be transmitted to a mechanically or electrically operated valve in the pump discharge so as to fulfill basically the same function as a float switch. Generally, this type of valve is used for throttling action, but it can be provided with a mechanism to open or close the valve completely.

4. Pressure control valves. Pressure control valves bear the same relation to pressure switches as float control valves bear to float switches. One specific application, however, differs somewhat in its fundamental function, that is, the use of pressure control valves as relief valves. Many installations require relief valves as protection against either the building up of excessive pressure through operation at very reduced flows, or the overloading of a driver in the same range of operating conditions when the power consumption curve rises sharply with the head developed.

5. Thermostat control valves. Thermostat control valves utilize temperature variations to open or close discharge valves instead of operating on the driver itself.

The last three control applications are relatively rare, except where they perform the added function of modifying the pump operating conditions over a wide range and where they, incidentally, are permitted to travel to the full closed position if externally imposed conditions require it. This phase of their operation will be considered later. It is important that a centrifugal pump should never be permitted to operate against a completely closed discharge and that a bypass recirculation line be provided whenever a valve positioning control fully interrupts flow delivery.

Variation of operating conditions

As many reasons exist which may require a variation in the operating conditions as those which may require interruption or resumption of flow. A few possible reasons are:

1. The need to maintain a constant level, either in the suction or discharge reservoir, regardless of the variation of supply or demand. A typical example of this requirement is the modern boiler feedwater system, where a constant boiler level must be maintained, regardless of the fluctuation of the load imposed upon the boiler.

2. The need to maintain a constant rate of flow with varying differences in static elevation between the source of supply and the ultimate delivery. Such conditions occur most frequently when a centrifugal pump takes its suction from a body of water subject to tidal variations.

3. The need to maintain a constant pressure in a system, or a constant pressure margin over some specific pressure, regardless of the total flow into the system.

4. The need to maintain a constant temperature in a batch process by means of cooling or heating through heat exchangers, regardless of temperature variations in the cooling or heating medium.

Since the operating conditions of a centrifugal pump are established by the intersection of the head-capacity curve at operating speed and the system-head curve, as described previously, it follows that variation of the operating conditions can be obtained by either one of two methods: (1) Modification of the pump head-capacity curve; or (2) modification of the system-head curve.

Modification by speed variation

The pump head-capacity curve is modified by variations in the pump speed except in some very rare applications where the head-capacity curve is modified through mechanical changes of the interrelation of internal pump parts while the pump is in operation. Pump speed can be varied either by modifying the speed of the pump driver itself or by applying a variable speed power transmission mechanism, such as a hydraulic coupling or a magnetic drive between the pump and its driver. The controlling mechanism itself is generally unaffected by the choice between these two forms of speed change. This choice is generally dictated by the choice of the type of driver best suited for the installation and the relative economy of the several possible driver combinations. If electric motors are used, variable speed operation can usually be obtained only with d-c motors or a-c slip-ring-wound rotor motors. In a few isolated cases, squirrel-cage induction motors can be operated at variable speeds through frequency variation, but this method is seldom used at present. Steam turbines and internal combustion engines lend themselves most conveniently to speed variation and, therefore, are seldom connected to the driven pump by means of variable speed transmission mechanisms.

Methods of controlling speed

Pump operating speeds may be controlled by means of:

1. Manual control. Manual control may require a rheostat for motor-driven units, or a throttle valve control for steam turbines and internal combustion engines.

2. Flow control. Flow control is generally used when it is necessary to maintain a constant flow despite variations in the system-head curve. The control speeds up the pump whenever the intersection between the head-capacity and system-head curves falls below the desired value and slows down the pump speed when the intersection rises above the desired value.

3. Float or level control. Float or level control principles of operation are similar to float switches, except that instead of a start-and-stop action, they perform the continuous function of slowing down and speeding up the pump to maintain the desired level.

4. Pressure control. Pressure control is exemplified by constant pressure steam turbine regulators that maintain whatever operating speed is necessary to produce a constant discharge pressure, regardless of flow variations.

5. Temperature control. Temperature control functions to increase or decrease the pump delivery to maintain a constant rate of heat exchange, despite temperature variations between the cooling medium and the liquid to be cooled.

Modification by valve positioning

In the majority of installations, it is either impossible or impractical to change the pump operating speed, and it is necessary to alter the shape of the system-head curve, by varying one of its components, in order to alter the operating conditions. In this case, neither the terminal pressure nor the static level difference can be expected to permit variation, and the only possible solution lies in varying the friction head loss in some part of the system. An artificial source of friction loss, such as a valve, must be introduced. This friction loss is controllable through valve positioning. This positioning, in turn, is controlled by the quantity (capacity, pressure, level, or tem-

TABLE 19.1   FUNDAMENTAL CONTROL FUNCTIONS

Interruption and resumption of flow	Start and stop of driver	Pushbutton on motor Float switch Pressure switch Power consumption Thermostat Overspeed governor	
	Valve positioning	Discharge valve manual control Check valve Float or level control Pressure control Thermostat	
Modification of operating conditions	Modification of pump head-capacity curve by speed variation	Driver speed control	Manual control Flow control Float or level control Pressure control Temperature control
		Power transmission control	Flow control Float or level control Pressure control Temperature control
	Modification of system-head curve by valve positioning	Manual control Flow control Float or level control Pressure control Temperature control	

perature) to be controlled. Figure 19.1 illustrates the operation of a centrifugal pump at constant speed and the effect of a throttling valve used to vary pump capacity. The design capacity is obtained with the throttling valve fully open, while curves A to D represent the system-head curves corresponding to several positions of the valve, as it is closed. Therefore, the pump operating conditions will correspond to the intersection of the head-capacity curve with curves A to D, or any intermediate curve, depending on the required capacity and consequently on the position of the throttling valve.

Methods of valve positioning

The positioning of the throttle valve may be controlled by:

1. Manual control. Manual control is the most widespread application and is used when operating conditions are expected to require infrequent changes. The pump operator progressively closes or opens a gate or globe valve in the pump discharge. Throttling the pump suction is seldom recommended as equally satisfactory or better results can be obtained by controlling the discharge without running the risk of pump cavitation.

2. Flow control. The throttling valve may be automatically operated to maintain the flow at its desired magnitude when the static head is reduced. An uncontrolled system-head curve would intersect the pump head-capacity curve at an increased capacity (Fig. 19.2).

3. Float or level control. Float or level control operation is similar to that directed at varying pump speed, except that the level is maintained by artificially altering the system-head curve. The boiler feed-water regulator is a typical example of this application. Its operation is exactly as shown in Fig. 19.1 with the terminal pres-

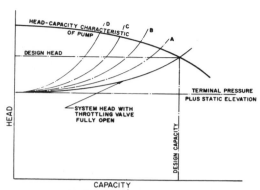

Fig. 19.1   Effect of throttling valve used to vary pump capacity

sure being equivalent to the boiler pressure.

4. Pressure control. The pressure regulator operates in a manner similar to that of the flow or level regulator, except that a constant discharge pressure at the pump is maintained by throttling off the excess pressure as capacity demand varies. Conversely, the valve opens up if the pressure falls below its desired value.

5. Temperature control. The most common application of temperature control is in cooling or refrigerating systems, when the type of driver does not permit speed variation.

Although these controls are intended to provide several dissimilar services, such as starting and stopping the driver, modifying the pump head-capacity curve, or modifying the system-head curve, they are all based on a group of similar impulse sources: Flow, level, pressure, or temperature. This similarity is illustrated in Table 19.1, which shows an analysis of the fundamental control functions.

Analysis of control elements

The description of the fundamental functions of centrifugal pump controls leads to a logical subdivision of controls into two distinct classifications, namely, corrective and protective controls. These two classifications are practically self-defining.

Corrective controls are obviously directed at changing some relationship within the pump system to compensate for changes in the conditions imposed upon the system. For example, when the boiler water level, which must be maintained constant, approaches the desired height, the delivery of the boiler feed pump must be reduced.

Protective controls on the other hand, are directed at protecting the pump or the system against certain harmful combinations of conditions or, in the event such conditions have arisen, at eliminating them in the shortest possible time. Thus, the bypass control in a boiler feed pump discharge, intended to prevent operation of the pump at flows so reduced that overheating will occur, is a protective control. This classification is important because a great many features of pump control will depend on whether corrective or protective controls are involved. Of course, once in a while, the same control can be applied for either corrective or protective purposes.

For example, if a centrifugal pump is to be stopped when sufficient water has been delivered to a reservoir, the float switch stopping the pump is a corrective control, as no specific harm will be done if the pump continues to run a little while longer. If, however, the same float switch is intended to prevent pump operation beyond the point where the water in the suction vessel has been completely drained, so that

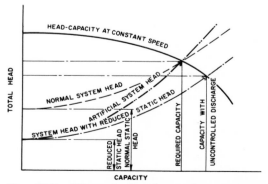

Fig. 19.2   Effect of throttling valve automatically positioned to maintain required flow under reduced static head

the pump will not run dry, the control is a protective one.

A mechanism or a process is easier to understand when it has been separated into its component parts before it is considered as a whole. This holds true for centrifugal pump controls, which will be broken down and discussed in their functional parts, before being combined again. Pump controls can be logically separated into four individual functional parts:

1. Measuring element
2. Impulse element
3. Relay element
4. Power element.

Measuring element

All control problems encountered in the operation of centrifugal pumps can be reduced to the problem of balancing flows, pressures, temperatures, or combinations of two or more of these. As a result, the measuring element must determine some force or forces set up in the pumping system which change in magnitude as the quantity to be controlled varies. If this quantity is fluid flow, the force can be created by a pressure differential across an orifice. If liquid levels are to be controlled, the difference in static head between the level to be controlled and some fixed arbitrary reference level provides this force. If a certain pressure or pressure difference is to be controlled, this pressure in itself provides the force to be measured. The main concern is to select the measuring element and its location in the pumping cycle so that it gives the simplest and most reliable indication of the quantities to be balanced.

Impulse element

The impulse element is that portion of the control which does most of the thinking, deciding when the measured variable has reached a predetermined value, or when it is properly balanced against some other variable with which it must remain in a certain correlation.

If the control function is such that variation in the quantities to be balanced is to be constantly accompanied by the desired change in valve setting, pump speed, or similar changes, the impulse and measuring elements are integrated. For example, if the pump capacity is to be reduced by throttling in some proportion to the reduction in the supply reservoir level at the pump suction, the force set up in the measuring element by the magnitude of the level acts directly as the impulse, which will ultimately be transmitted to the throttling valve control.

If control of operation and change in relationship is to occur only when the measured quantity reaches some predetermined magnitude, the impulse function becomes divorced from the measuring element and becomes operative only when the forces involved are balanced at the desired value. As an illustration, it may be assumed that a valve in the pump discharge is to be closed off whenever the suction level drops to some set value. The measuring element records the suction level at all times and balances a force proportional to this level against a selected force representing the desired minimum. As long as the measured force exceeds the predetermined force, the impulse is inoperative. On reaching a balance between the two forces, the impulse mechanism sets certain controls in motion to close the valve as desired.

Amplification of forces. Sometimes the force set up through quantity variation is very negligible and cannot be used directly to actuate any control mechanism. An amplifying feature must then be incorporated into the control mechanism. This amplifying feature, however, might be more logically considered as part of the relay than of the impulse element.

Relay element

The relay of the impulse may or may not exist as a mechanism, depending on whether

this portion of the pump control is automatic or not. In other words, if the function of the impulse element is to ring an
alarm bell and warn an operator when certain conditions are detected by the measuring element, the function of the relay element will be accomplished when the operator walks over to the valve which the alarm
has indicated must be throttled. However,
in the majority of control mechanisms, the
function of the relay element is to transmit
the decision reached by the impulse element to the power mechanism which actually operates the control. This function
can be discharged in many ways and can
be performed mechanically, hydraulically,
pneumatically, or electrically.

Power element

The power element, the last of the four
functional parts of a complete pump control, is that element which changes some
relationship in the combination of pump
and pumping system, such as a change in
valve setting, a reduction or increase in
pump speed, or stopping or starting the
pump. The source of the power used in this
element may be either mechanical, hydraulic, pneumatic, or electrical.

How four control elements are inter-related in practice

It has been assumed so far, in the interest of simplicity, that the four elements of
a centrifugal pump control can easily be
separated and examined. Actually, the four
elements and their functions are frequently
interlaced, introducing an element of uncertainty into the analysis. For example,

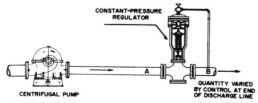

**Fig. 19.3   Constant-pressure regulator installed
in pump discharge**

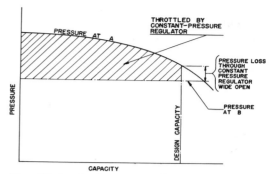

**Fig. 19.4   Performance of constant-pressure
pump in Fig. 19.3**

consider a constant pressure pump regulator installed in the discharge line of a
centrifugal pump, as shown in Fig. 19.3.
The pump is operated at constant speed,
and, therefore, the discharge pressure will
vary with the quantity delivered by the
pump. Assume further that the pump capacity is varied by some other control not
pertinent to the present analysis. In the
valve itself, the valve disk is mounted on
the valve stem, the position of which is
controlled by a diaphragm. The control
chamber above the diaphragm is connected
to the pump discharge line at point B,
where a constant pressure should be maintained. The balancing force, corresponding
to the desired discharge pressure, is provided by a spring. Any increase in the pump
discharge pressure at B will act against the
force exerted by the spring and cause gradual valve throttling. The valve will become
partially closed, increasing the frictional
losses through it, and, therefore, increasing
the difference between the pressures at A
and at B, until the pressure at B corresponds to the desired value and balance
has been re-established. The pump performance is illustrated in Fig. 19.4.

In this particular case, the measuring
element is the diaphragm, which reacts to
variations in the pressure to be controlled.
The impulse is provided by the spring,
which is calibrated to correspond to the
desired pressure at B. The relay does not

exist in this application, while the power element is, in the last analysis, a variable one. When the pressure at point B exceeds the desired value, the force required to close the valve partially and create additional artificial friction is provided by the controlled element itself. Conversely, when the pump delivery is increasing and the pressure at point A decreases, carrying with it a decrease of pressure at point B, the spring intervenes to provide the power required to reopen the valve. The power element could also be described as consisting of the difference between the hydraulic and spring forces, respectively. The actuating force is positive for valve closure when the hydraulic force exceeds the spring force and negative for valve opening, in the reverse case. Regardless of the interpretation given to this phase of the control analysis, it is apparent that measuring, impulse, and power elements are definitely interrelated in this particular application.

Influence of operating medium

The same example may serve to illustrate certain differences which arise from the selection of the operating medium, either for the power or for the impulse element. In the control shown in Fig. 19.3, it was assumed that the desired discharge pressure at point B was represented by a valve spring. The same function, however, could have been fulfilled either by a weight or by some preset hydraulic or pneumatic pressure, to be introduced on the side of the diaphragm opposite the chamber where the pressure to be controlled is applied. While both spring-loading and weight- or pressure-loading have their advantages, it is necessary to visualize the basic differences between the two types to determine the proper type for any given application. The important characteristic of spring-loading is that the spring power is not constant, but varies with compression. As the pump delivery decreases, therefore, the spring is compressed, exerting more pressure in counteraction to the controlled pressure.

Since the discharge pressure at B is balanced by the spring action, the pressure increases with a decrease in pump capacity, although this is not shown on Fig. 19.4 to avoid confusion.

If the maximum movement of the spring is very slight, the valve-loading will be practically constant. If, however, the spring is to be operated in a wide compression range, spring-loading is not well suited to provide constant loading and weight- or pressure-loading is preferred.

Automatic and manual controls

While a great number of centrifugal pump controls are automatic in their operation, and while attention is generally directed to such controls, in many instances a choice must be made between manual and automatic controls. This choice is considerably complicated by factors more psychological than technical. Automatic controls often lead to neglect and indifference on the part of the operators. Should a control failure occur, the operator may be unable to correct any resulting harmful effects, either through lack of knowledge, or simply because he might be so dependent upon the proper operation of the control that he would not be present where his attention becomes necessary. However, human response is never as rapid as mechanical response and requires an excessively attentive attitude, all of which may lead to an equal or greater incidence of failures than in the case of automatic operation. The decreased operating personnel requirements, furthermore, more than overbalance the greater initial financial investment in the automatic control.

No over-all rule or preference can be established in the choice between manual and automatic operation. First, it is very important to realize that the differentiation between the two types is not as sharp as some engineers may be led to believe. Between the two extremes of fully manual and fully automatic control, a large number of intermediate arrangements exists, all of them semiautomatic in some phase of the control mechanism. Referring to the divisions outlined previously, and separating control mechanism into measuring element, impulse, relay, and power element or positioning, it appears that each of these components can be manually or automatically operated. To illustrate this, a typical control operation, involving the throttling of a centrifugal pump discharge in response to level changes in the reservoir at the pump discharge, is described in Table 19.2. In addition to the fully manual and fully automatic controls five other methods of control are shown, in which some of the elements are operated manually and others automatically. Since there are four separate component control steps and two different methods of operation, that is manual or automatic, for each step, there will be sixteen different possible combinations altogether. One of these will be fully manual, a second fully automatic, and fourteen semiautomatic. It is obvious that some of these combinations will be unpractical. For example, it would be illogical to have automatic operation of all steps but the measuring element. However, the combinations described in Table 19.2 indicate that a rather complex chain of reasoning is necessary to determine the most suitable combination and that to base the ultimate decision on a broad preference for manual or automatic controls is decidedly shortsighted.

Factors in analyzing control method

Some of the most important questions indicating the general character of the analysis which must precede a choice between manual and automatic controls are presented below, but not necessarily in order of relative importance.

Frequency of operation

Is the control operation required infrequently or at frequent intervals, possibly constantly? The starting up of a standby pump on failure of the main pump, an operation which is apt to occur very rarely,

is less in need of automatic controls than, for instance, the maintenance of a constant pressure in the discharge header, when variations in the flow delivery require constant repositioning of the regulating valve.

Expectancy of operation

Is the control operation to occur at regularly scheduled intervals or is the requirement sudden and unexpected? For example, starting up additional boiler feed pumps in a power plant before an expected increase in load at certain times of the day can readily be performed manually. On the other hand, the start-and-stop operation of a transfer pump which delivers into a storage tank depends wholly on the varying rate of usage from the tank and, therefore, cannot be accurately predicted as to time of occurrence.

Urgency of operation

Must the controlled operation take place instantly upon the occurrence of certain conditions, or can it take place at leisure? Basically, the difference is between corrective and protective controls. In the case of protective controls, it is almost imperative that the measuring and impulse elements be automatic. Certain corrective controls must operate almost instantly, as, for example, centrifugal boiler feed pump controls intended to maintain a constant level in high pressure boilers. Other corrective controls can rely upon operator judgment and, hence, can be manual.

Difficulty of operation

Is the application of the control suitable for manual operation or are the forces required to set the control in motion excessive? Taking two extremes for illustration, it is obvious that while a 1-in. valve can readily be throttled manually, a 24-in. valve can be opened or closed much more easily if it is provided with automatic electric drive.

TABLE 19.2 COMBINATIONS OF MANUAL AND AUTOMATIC OPERATIONS

	Fully manual	Semiautomatic	Semiautomatic	Semiautomatic	Semiautomatic	Semiautomatic	Fully automatic
Measure	Manual: Operator measures level in reservoir with rod	Manual: Operator measures level in reservoir with rod	Automatic: Direct reading gauge glass indicates level in reservoir	Automatic: Float in reservoir registers level	Automatic: Float in reservoir registers level	Automatic: Direct reading gauge glass indicates level in reservoir	Automatic: Float in reservoir registers level
Relay	Manual: Operator realizes that level has increased	Manual: Operator realizes that level has increased	Manual: Operator sees that level has increased	Automatic: Float switch rings bell	Automatic: Float switch rings bell	Manual: Operator sees that level has increased	Automatic: Float switch makes contact on rise in reservoir
Impulse	Manual: Operator walks over to valve	Manual: Operator walks over to push-button station	Manual: Operator walks over to valve	Manual: Operator hears bell and walks over to valve	Manual: Operator hears bell and walks over to push-button station	Manual: Operator walks over to push-button station	Automatic: Impulse from float switch transmitted electrically
Power	Manual: Operator closes valve partially and throttles pump delivery	Automatic: Motor-operated valve throttles pump delivery	Manual: Operator closes valve partially and throttles pump delivery	Manual: Operator closes valve partially and throttles pump delivery	Automatic: Motor-operated valve throttles pump delivery	Automatic: Motor-operated valve throttles pump delivery	Automatic: Motor-operated valve throttles pump delivery

Facility of detection

Is the quantity to be measured subject to accurate determination by human, senses or is it impossible to obtain sufficiently accurate measurements except by mechanical means?

Economics of personnel attendance

Is an operator always available where needed, or does the manual operation of the control require additional personnel not needed otherwise? The best example of this question is the application of automatic controls to start and stop numerous cellar-draining or flood-protection pumps.

A large number of other questions may arise which apply to individual cases and which must be considered in the ultimate decision between manual, semiautomatic, and fully automatic control operation. It is a popular misconception that once automatic operation is provided, all manual control is precluded. Not only can most automatic controls permit manual operation, but the installation must always be provided with a means to switch from automatic to manual operation, at least as far as the measuring, impulse, and relay elements are concerned.

Pilot devices

A large number of automatic controls are electrically operated, so that while the measuring element may be mechanical or hydraulic, the impulse and relay elements are based on electrical contacts and transmission or interruption of electrical currents. Such automatic controls are principally directed at the interruption or resumption of flow, based on level or pressure fluctuation. Because the electric currents involved in the operation of the pump drivers are generally excessive for the control apparatus, the latter serve as pilot devices. Thus, electric motors may use 2,300-volt current, while control equipment may use 220-volt current to actuate the 2,300-volt starters.

Typical control applications

This chapter is not intended as an exhaustive catalogue of all the forms and types of control equipment which may be encountered in centrifugal pump applications. Instead, its purpose is directed at understanding the various functions of pump controls and the effect the operation of these controls may have on the performance of centrifugal pumps. Therefore, a complete explanation of the functioning of one or two typical centrifugal pump controls will make the application of other types of control more easily understood.

The boiler feedwater level regulator represents one of the most typical applications in centrifugal pump practice, that of maintaining a constant level regardless of demand fluctuation.

To understand the problems involved in the maintenance of a constant boiler water level, it is essential to remember the statement made earlier in this chapter, namely, that a centrifugal pump operates on a system-head curve. In other words, it operates at a capacity and pressure corresponding to the intersection of its head-capacity curve with the system-head curve. Therefore, in a given feedwater system, if the static pressure (boiler drum pressure plus static elevation) and the friction losses remain constant, the boiler feed pump will deliver a constant capacity into the boiler.

This, however, will not permit the maintenance of a constant boiler water level, inasmuch as the steam outflow from the boiler is not constant, but depends on the turbine load. It becomes necessary, therefore, to introduce an artificial source of friction loss, which is controlled by changes in the boiler drum water level. This artificial source of friction loss is provided by the feedwater regulator. While the operation of such a regulator could be handled manually in a few isolated cases, automatic operation of this control is required in all but a few extremely small installations.

Consequently, the feedwater regulator consists basically of two elements: (1) A throttling valve, and (2) a controlling mechanism which determines the setting of this valve, depending on boiler level fluctuations.

Figure 19.1 is typical of the operation of a centrifugal boiler feed pump at constant speed and illustrates the effect of the feedwater regulator throttling valve at several different positions corresponding to various flow requirements of the boiler. It has been assumed in this case that the boiler pressure remains constant at all loads. Depending on the boiler load and, consequently, on the position of the feedwater regulator throttle valve, the pump discharge capacity will correspond to the intersection of the head-capacity curve with curves A to D or with any intermediate curve.

The feedwater regulator represents the simplest but most important of boiler feed pump controls. Without it, boilers would have to be fed manually by an operator who would watch the boiler gauge glass and open or close a valve as he saw the level fall or rise. Of course, if the pump were to run at variable speed, the throttling required from the feedwater regulator would be reduced. But the choice of a control depends on the degree of refinement desired for the solution of the problem at hand. Variable speed operation of feed pumps is one such refinement; it is described later. For direct comparison between the simpler and more complex forms of control, Fig. 19.5 shows a feedwater system where the pumps are operated at constant speed and where the only control is a feedwater regulator.

Many feedwater regulators are available today. The specific manner in which boiler level fluctuations are transmitted to the throttling valve, as well as the design of the valve itself, vary with manufacturer.

Differential or excess pressure regulators

The constant boiler level feedwater regulator can be compared to an orifice in a pipeline through which the feedwater flows. The quantity of flow through this orifice depends upon its area as well as the pressure drop across it. As the water level in the boiler drum rises or falls, the regulator valve moves so that the valve opening will provide the correct rate of feed to the boiler. However, if the pressure drop across this valve is allowed to vary, the flow will vary even though the valve area remains constant. While these variations in flow will eventually result in changing the boiler level, and consequently the area of valve opening, they will have an undesirable effect on the boiler operation since the flow should vary only as required for load changes.

Since the discharge pressure of a centrifugal pump operated at constant speed increases from 10 to 20 per cent between its design operating condition and extremely low flows, while the required system head decreases as flow decreases, the excess pressure generated over and above the system head requirements increases appreciably at light loads. The excess pressure must be absorbed (throttled) in the feedwater regulator valve; this results in a wide variation in pressure drop across the valve (see Fig. 19.5).

Another disadvantage of an increasing pressure drop across the feedwater regulator is that since the valve ports are designed for maximum boiler capacity, the valve will operate through a very small portion of its travel when the boiler is operating at reduced ratings, with a resulting sacrifice of accuracy of control.

To eliminate the disadvantages of a varying pressure drop across the feedwater regulator, differential or excess pressure regulators are used. The function of these regulators is to maintain a more or less constant pressure drop across the feedwater regulator valve either by:

1. Throttling the excess pressure generated by the pump. This is used with constant speed pumps.

2. Decreasing the pressure generated by the pump at various loads by varying the pump speed.

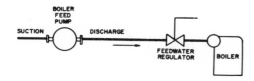

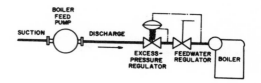

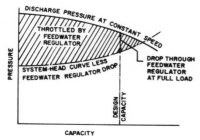

Fig. 19.5 Constant speed boiler feed pump controls

Feedwater regulator only.

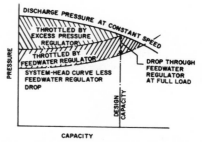

Fig. 19.6 Constant speed boiler feed pump controls

Feedwater regulator and excess-pressure regulator.

A comparison of Fig. 19.5 and 19.6 will show how an excess pressure regulator maintains a uniform pressure drop across the feedwater regulator valve. Figure 19.6 illustrates the hookup of this refinement in pump control and the pressure and capacity relations in a feedwater system so controlled.

Basically, a differential pressure regulator consists of a throttling valve located ahead of the feedwater level regulator valve. The throttling valve is actuated by a metal bellows subject to, and influenced by, the differential pressure across the feedwater valve, as the two sides of the bellows are connected to the upstream and downstream sides of the feedwater valve respectively. As the pressure drop across the feedwater regulator valve increases, the differential pressure regulator closes until the pressure drop is again restored to its initial value.

The differential pressure regulator valve should be located as far from the pump and as close to the feedwater regulator as possible, to minimize the pressure variations caused by the friction losses in the inter-vening piping and closed heaters. However, this is not always possible and the valve is often located almost immediately following the boiler feed pump discharge.

In some cases, instead of maintaining a constant pressure drop across the feedwater regulator, a constant discharge pressure is maintained at the boiler feed pump. Instead of being actuated by a difference of pressures across a bellows or a diaphragm, the regulator is influenced by the pump discharge pressure on one side and a constant pressure set up by a spring or an air-loading mechanism, this constant pressure corresponding to the desired value at full load.

Control of variable speed pumps

When a centrifugal pump is driven by a variable speed motor or, especially, by a steam turbine, it becomes possible to apply the differential pressure control to even greater advantage. Figures 19.7–19.9 show three different arrangements driven by a steam turbine, all employing a boiler feed

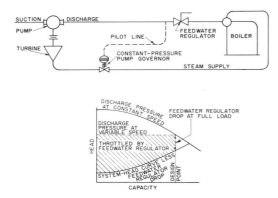

Fig. 19.7 Variable speed boiler feed pump controls

Constant discharge pressure.

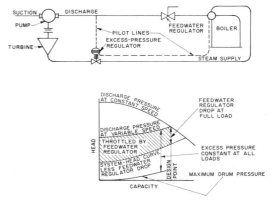

Fig. 19.8 Variable speed boiler feed pump controls

Constant excess over boiler pressure.

pump with a variable speed governor. In all cases it has been assumed that the boiler drum pressure is not constant but increases with the load to maintain a constant steam pressure at the superheater outlet. The increase in drum pressure compensates for increased superheater losses at increasing steam flows.

The three arrangements differ only in the location of the pilot lines leading to the governor. Figure 19.7 illustrates a constant pressure governor applied to the steam turbine driving the boiler feed pump. Only one pilot line is provided from the discharge of the boiler feed pump, and the discharge pressure remains constant at all loads.

In Fig. 19.8, which shows a constant excess pressure governor, the pilot lines lead to the pump discharge and also to the boiler. The pump speed is varied in such a manner that the difference between the pump discharge pressure and the boiler drum pressure remains constant at all loads. The feedwater regulator pressure drop is not constant but, instead, throttles the difference between the pump discharge pressure at variable speed and the pressure representing the system-head curve, less the feedwater regulator drop.

In Fig. 19.9 the pilot lines lead to the two sides of the feedwater regulator valve, and

the pump speed varies to maintain a constant pressure drop across this valve.

All these controls can be arranged to operate separate valves in the steam line ahead of the turbine governor valve or to operate the turbine governor valve itself.

Similar controls can be applied to vary the speed of electric-motor-driven centrifugal pumps, if these are slip-ring-wound rotor motors. Such applications are relatively rare because of the greater cost of such motors. A more frequent solution today is the use of hydraulic couplings or of magnetic drives. The control of these variable speed devices is similar to that of variable speed turbines, in that the same impulse elements are used to transmit the necessary signal to the particular speed-varying mechanisms of the drives.

While the control of centrifugal boiler feed pumps cannot be termed as typical of *every* kind of pump control which may be encountered, certain similarities will occur in most cases. Therefore, an understanding of what may be accomplished in this particular field of application will prove helpful in solving most pump control problems.

The controls described here, however, are basically corrective controls. Protective controls are characterized by certain problems and certain features which make them considerably different.

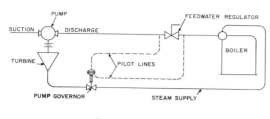

Fig. 19.9 Variable speed boiler feed pump controls

Constant pressure drop across feedwater regulator.

Protective controls

By their very nature, protective controls are automatic in their operation, since their main purpose is to complement the work of the operating personnel and perform functions which would normally be performed manually, if the personnel were on hand and aware of the need. Most protective controls are intended to bring the pump and its driver to a stop whenever certain harmful operating conditions arise. However, a few of these controls cause some alteration in the operating conditions instead.

In general, the impulse for the operation of the protective control is selected from the conditions to be avoided or eliminated. A typical case of a protective control used to bring a unit to a stop is that of a check valve provided with a switch which can control the electric motor driving the pump. Such a check valve is intended to protect the pump against damage caused by loss of water during operation. While the pump is discharging water, the check valve flap is in a raised position and the switch is held closed. If for any reason the flow ceases (due to lack of water at the suction, for instance), the valve flap falls

and opens the switch, stopping the pump automatically before it is damaged by running dry.

Since under certain conditions a centrifugal pump may be permitted to operate against a closed valve in its discharge, the simple check valve is a protective control device by itself, as it prevents reverse flow through the pump whenever the pump discharge pressure falls below the pressure in the header into which the pump is discharging. However, it must be remembered that a pump operating against shutoff will become damaged through overheating in most cases, and, therefore, every check valve installation must be checked from this point of view, to assure that it will not operate at shutoff for any length of time. Wherever danger of such conditions exist, a manually or automatically operated bypass of some sort is provided.

Protection against loss of prime, provided by the check valve and switch described above, could be incorporated into the motor starter itself. In such an arrangement, an inverse current relay is used to interrupt the electric circuit whenever the power consumption, and therefore the current input to the motor, falls to 50 per cent of the minimum value for normal operating conditions. For instance, if a pump has a maximum of 500 bhp and a minimum (at shutoff) of 280, the inverse current relay will stop the pump if the power drops to 140 bhp, indicating that the pump has lost its prime and is in imminent danger of running dry. A time relay must be incorporated in this control, to permit starting the pump.

Driver protective controls

The various forms of motor controls which provide for protection against undervoltage, overload, frequency change, and other motor conditions are not centrifugal pump controls and will be left out of this discussion except for a few remarks. Motor overload can easily be caused by pump operating conditions, therefore, provision must be made for this protection if driver overload is possible in the particular system. The most important consideration in the application of motor protective controls is that the operator should know when one of these controls has stopped a motor. A bell or a signal light, therefore, will allow detection of faulty operating conditions as soon as they occur, and they can be corrected immediately.

In many cases, the driver and its pump should be started again when normal conditions resume; therefore, the interrupting device is provided with features to re-establish current. In other cases, however, such as when a pump must be primed manually before starting, the contactors must be prevented from closing again until the circuit is re-established manually by pressing the starting pushbutton.

Automatic operation of standby pumps

Another example of a protective device is the automatic operation of standby boiler feed pumps in a power plant. The exact

moment of a pump or drive failure is unpredictable and can occur when the operators are at some remote portion of the station, unable to start the standby unit without a dangerous delay. Most protective controls directed towards starting standby pumps are based on pressure conditions in the main feedwater header. This header pressure is transmitted to one side of a spring-loaded diaphragm, the minimum permissible header pressure providing the selected value of spring compression. The diaphragm-operated valve is held in a closed position as long as the header pressure remains in excess of the predetermined minimum. If the header pressure falls below this value for any reason, the valve opens. This valve is generally used as a pilot valve and can be made to operate any suitable mechanism that will start one or more of the standby units.

Protection against operation at shutoff

The automatic operation of a boiler feed pump bypass is a typical example of a pro-

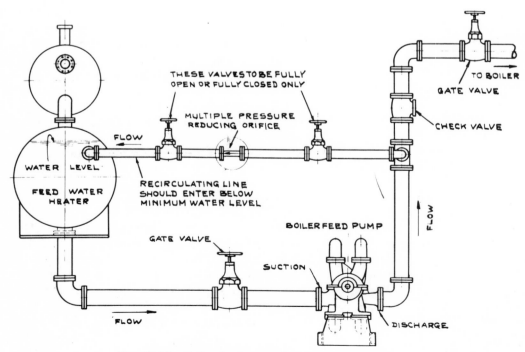

Fig. 19.10   Boiler feed pump bypass arrangement

tective control which eliminates the possibility of shutoff operation.

A high-pressure pump, such as a boiler feed pump, operated at shutoff would soon overheat and seize. Therefore, this type of pump must not operate at abnormally low flows in order to protect it from the damage incidental to an excessive temperature rise.

This is achieved by installing a bypass in the discharge line from the boiler feed pump, between the pump and its discharge gate and check valves, leading to some region of lower pressure in the boiler feed cycle. The capacity of the bypass is such that should the demand at the boiler fall to zero, or should one of the valves become closed while the pump is running at full speed, the flow through the pump will not fall below the predetermined permissible minimum. The minimum is generally based on a maximum temperature rise of 15°F, corresponding to approximately 30 gpm for every 100 bhp at shutoff conditions. The operation of the bypass is then regulated by means of a valve, controlled either manually or automatically.

Generally an arrangement such as shown in Fig. 19.10 is used. Two valves are used, one on each side of the orifice, which controls the actual capacity in the bypass. One of these valves always remains open and is closed only to isolate the orifice during inspection or renewal while the pump is in operation. The second valve is the control valve.

Within the last few years power plant operators have indicated a growing interest in automatic control of the bypass, especially in high-pressure installations where the trend is towards automatic control of all phases of operation.

Automatic control of the bypass involves some sort of actuating mechanism that will open and close the bypass at the proper time. The actuating mechanism itself must be subjected to an impulse governed by the flow through the boiler feed pump; a great number of impulses are available for this purpose:

1. Pressure drop across the feedwater regulator
2. Position of the feedwater regulator
3. Relief valve operated by the discharge pressure of the pump
4. Flow meter control in the suction or discharge line
5. Thermostatic control.

Of these five, flow meter control is the safest, most reliable, and most frequently used at the present time.

Flow meter bypass control

The actual operation of the flow meter bypass control (Fig. 19.11) is quite simple. A flow meter element, located in the pump suction line, operates two mercury switches. The contacts are so adjusted that the first contact completes the circuit on a decreasing flow, but does not permit bypass until the second contact has been made at a still lower flow. This provides an overlap to prevent hunting since the interval between the two contacts exceeds the amount of

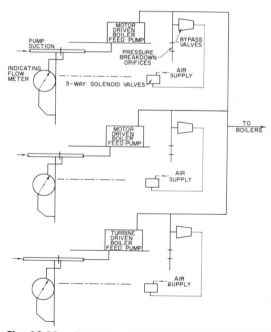

Fig. 19.11 Automatic pump bypass controls on three boiler feed pumps in parallel

(*Courtesy Bailey Meter Co.*)

the bypass flow and since the circuit is not broken until the total flow has increased to a value above the point of the first contact. The circuit operates a solenoid valve which admits air pressure to the control diaphragm of the bypass valve itself.

The circuit can be arranged so that the relay controlling the air loading pressure to the bypass valve is de-energized at the low flow conditions; therefore, the bypass would operate in case of a failure of electric power to the contacts, as well as at reduced flows. Similarly, since the bypass valve is kept in the closed position by air pressure, a failure of air pressure will also operate the bypass and the boiler feed pump is protected against damage under all conditions.

Bypass controls are not necessarily restricted to boiler feed pump applications, since they perform as important a function in any installation of centrifugal pumps where the system cycle may introduce the possibility of operation at extremely reduced flows or against a completely closed discharge. However, when light flow operation occurs at extremely frequent intervals, as in a steel mill descaling installation, it may be more practical to maintain the bypass valve constantly in the open position or to provide a three-way valve so that when the main flow is shut off, the bypass flow is automatically established.

Protection against reverse flow

The use of check valves in the pump discharge to prevent reverse flow in the pumping system has already been mentioned. The application of foot-valves to pumps operating with a suction lift clearly belongs in the same class of protective pump controls, although it serves the additional duty of retaining water in the pump casing, allowing the pump to remain primed at all times.

However, ordinary check valves have the characteristic of closing very quickly, possibly causing excessive pressures in the system through the creation of water hammer.

It is therefore desirable in certain cases to install slow-closing power-operated valves such as cone valves or butterfly valves. These valves are generally hydraulically or pneumatically operated and use either the hydraulic pressure in the pumping system itself or an external source of liquid or air under pressure for their source of power. These protective controls can be applied not only to prevent reverse flow, but also to protect the system against situations arising from the rupture of the pumping line, preventing the loss of pumped fluid. This function is performed by making the valve control responsive to excessive velocities in one direction only, or in either direction, depending on the particular features of the installation.

The automatic siphon breaker is another protective device used to stop reverse flow through a pump under certain conditions. It may be applied where the pump discharges into a siphon system, and where power failure and the resulting pump stoppage would be followed by the loss of water from the discharge reservoir through the siphon unless siphon action is interrupted. The siphon breaker valve is located at the top of the siphon and is maintained in a closed position by means of an energized solenoid. Power failure de-energizes the solenoid and opens the valve, admitting air into the system and breaking the siphon action.

Relief valves

A few pumping systems may become subject to excessive pressures or even to water hammer under certain operating conditions. It is usual, in such cases, to provide protection both for the pumps and for the system in some form, such as air chambers, surge tanks, or relief valves. The exact nature of the protective controls depends upon all the features of the system into which they are incorporated and must be solved individually. As a result, it would be misleading to indicate any over-all rules or suggestions on the application of such controls. However, relief valves may be used to protect certain separate portions of the system against pressures which may be suitable for the remainder of the system but excessive to that portion. One typical example of this application is the use of relief valves in the suction line of a pumping system provided with both check valves in the discharge and foot-valves in the suction. Foot-valves are seldom suitable for the pressures which may exist in the discharge piping and may need protection against possible failure or leakage past the discharge check valve.

The surge suppressor used in some discharge lines is a reverse form of relief valve, that is, a relief valve which opens far enough in advance of the occurrence of excessive pressures to afford full protection against water hammer. This valve is arranged to open on an excessive *drop* in discharge pressures and close *very slowly* as the pressure is again built up. When the check valve in a discharge line closes suddenly, the excessive pressures caused by water hammer are preceded by a sudden drop in pressure immediately beyond the check valve, and by the time the surge or rise in pressure takes place, the surge suppressor valve is fully open and ready to relieve the situation.

Techniques of control application

The knowledge of the fundamental control functions, of the pump characteristics that these controls are intended to modify, of the component elements of controls, and the understanding of some typical applications are not a sufficient guarantee of perfect handling of all control problems which may be encountered in the installation of centrifugal pump systems. In addition, a certain familiarity with the technique of control application is needed, a familiarity best acquired through experience. However, it is practical to acquire some knowledge of this technique analytically. The following, therefore, describes some of the basic principles of the technique of control application. Because the majority of pump controls are based on throttling either the main delivery stream itself, or some part of the stream in a take-off branch, this description will be devoted primarily to the application of manually, automatically, and semiautomatically controlled throttling valves.

Control characteristics

All control apparatus must have the following characteristics:

1. Accuracy
2. Sensitivity
3. Speed
4. Power.

These characteristics are built into the control apparatus by the manufacturer. However, if the engineer applying the controls does not utilize these characteristics, then all the care the designer exercised to insure that the control possesses such characteristics will have been wasted.

Accuracy. Accuracy is possible only if the variable to be controlled is measured at a point in the system where a true reading can be obtained. Such precautions sound obvious, but, unfortunately, pressure or flow measurements are often attempted at locations where the reading is definitely distorted. Measurements can also be distorted if the measuring element or the piping next to it is dirty, clogged, or incrusted. For example, flow measurements by means of venturi meter or orifice should be made at a point where the flow of liquid in the piping system has been properly straightened. Control manufacturers generally recommend the length of straight piping required ahead of the measuring element; these recommendations should be carefully followed.

To be accurate, controls must respond to the slightest deviations from the desired value of the variable being controlled. To obtain such sensitivity, it is necessary to consider the application thoroughly. For instance, if the controlling or measuring element is excessively oversized in relation to the quantity to be measured, all sensitivity is lost. In some cases where the range of operating conditions varies within wide margins and where sensitivity is required under all conditions, it may be practical to split the quantity to be measured into two portions. Two measuring elements would integrate the total quantity for the upper range of its value, and a single element would be used whenever the total quantity is reduced to less than one-half of its maximum value.

It is preferable, although not always practical, that the greater accuracy of regulation occur in that portion of the operating range where sensitive control is most desired.

Flow-meter control of a boiler feed pump bypass presents a typical example of the analysis required to insure sensitivity in the proper range of the quantity being measured. This control is intended to open the bypass valve whenever the feed pump flow falls to dangerously low values. The impulse used to operate the control consists of the differential pressure across a metering orifice. This differential is highest at the maximum flows and lowest in that particular range where it must operate the bypass control. The size of the metering orifice must, therefore, be selected to compromise between two contradictory requirements:

1. The differential at low flows must be sufficient to provide sensitivity in the bypass operating range.

2. The differential at maximum flows must not be excessive, or the resulting head loss may become an adverse factor in the boiler feed pump installation.

This differential varies with the square of the feedwater flow, and the solution lies in locating the orifice so that excessive differentials at high flows will not affect operation. This explains the frequent recommendation to place the orifice in the boiler

feed pump discharge rather than in the suction piping where the resulting head loss would rob the pump of an appreciable portion of the available net positive suction head.

Sensitivity. Sensitivity of corrective pump controls is affected by valve size as much or more as by proper selection of the measuring element. Valve manufacturers frequently point out that the proportions and contours of the fluid passages must be given special attention to avoid excessively turbulent flow and ensuing erosion. Although this phase of the control problem belongs in a treatise on valve design rather than on the application of controls to centrifugal pumps, the relation between the valve lift, the area of the opening, and the resulting flow do play an important part in the successful application of valves. If the valve size selected is too small for the range of flow capacities encountered, friction losses will be higher than necessary and valve wear will be excessive. The penalty for undersized valves, therefore, will be an unjustified increase in power consumption because of the increased total head required in the pumping system and in valve maintenance.

Just as a valve may be too small for its intended service, it may also be too large. Since the area of a valve opening must be made commensurate with the quantity of liquid due to pass through the valve, the valve would be almost closed under most operating conditions. This may lead to wire-drawing and will certainly cause a decrease in sensitivity. Consider, for example, the valve curves in Fig. 19.12. If a selected gate valve is considerably oversize, that is, if the maximum flow in the system corresponds to approximately 50 per cent of the maximum permissible flow through the valve, then the valve stem would have to be held at approximately 23 per cent of its maximum travel. A reduction of 50 per cent in demand (or to 25 per cent of the possible maximum flow through the valve) would call for the movement of the valve

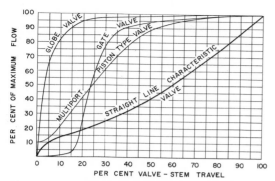

Fig. 19.12 Flow characteristics of different types of valves

(Courtesy Bailey Meter Co.)

stem to a position at 20 per cent of its travel, a reduction of only 3 per cent of the total travel from its former position, corresponding to a negligible angular displacement of the control wheel. Such a situation is even further magnified in the case of a globe valve and explains the reason for the unpopularity of the latter for throttling service.

It is apparent, then, that to fulfill the requirements of accurate and sensitive regulation throughout the full operating range of the controlled equipment, valve design should be chosen to give substantially equal increments of flow for equal increments of valve-stem travel, regardless of valve position. Such a design, usually called a "straight line characteristic valve" is illustrated in Fig. 19.12, and is generally available.

The disadvantages of valve oversizing do not apply to fixed, nonthrottling valves used for on-or-off service. In this case, oversizing will reduce friction losses, at the penalty of increasing initial costs to a point of diminishing returns.

Situations may arise, however, which call for the use of valves much too large for normal operating conditions, yet properly sized when certain emergency increased flows occur. Valves installed in parallel present definite advantages whenever such wide variations in service demands are en-

countered. Only one valve is open at very light flows; additional valves are opened when the demand increases.

Speed. Except for those applications where the impulse is transmitted manually, speed of impulse transmission is generally the problem of the control manufacturer. However, in many instances, the user may handicap the manufacturer of controls in this respect by his insistence on a particular operating medium that precludes suitable speed of transmission under the prevailing conditions and may introduce a time lag that ultimately defeats the entire purpose of the automatic control application.

Power. Electrically operated valve controls are generally available in the semiautomatic or in the fully automatic types. In both cases, an electrically operated valve consists of a reversing torque motor, which drives the valve stem through a train of gears. The motor control is arranged to stop the valve in any position required to obtain the desired control. In semiautomatic operation, the impulse is manual and is controlled by the operator at a pushbutton station. In a large installation, requiring the manipulation of a great number of valves, full control of each valve can be provided by a centrally located battery of pushbuttons. Such control permits the installation of the valves where best plant design and not accessibility considerations dictate. When fully automatic operation is required, the measuring element is made to transmit the necessary impulse to the control panel where, in turn, an electric circuit operates the main valve motor in the desired direction.

Applied techniques

Some suggestions on the techniques of control application may appear self evident, and yet attention must be directed to certain pitfalls. A typical example of an error in application would be the use of the system hydraulic pressure to operate certain protective controls for the very centrifugal pump activating this system. If the function performed by the protective control is not necessary should the pump be suddenly stopped, the use of the system pressure for the power element of the control is acceptable, at least from the point of view of the safety of the supply. If, however, the protective control is required to function either on the stoppage of the pump or during the starting and stopping period, there will be no system pressure to be used as the power element.

Another problem is a tendency toward "over-engineering" in control selection, as in the selection of any other equipment. Over-engineering refers to the selection of equipment that is actually too good for the intended job, and of controls designed to perform a more difficult function than required in the given instance. This leads to unnecessary expense and, at times, to excessive outage and maintenance which could easily be avoided by the application of less complex mechanisms.

In considering automatic control, it should be remembered that in most cases automatic control is directed at supplementing, not supplanting, proper instrumentation and careful observation by the operators. An indicating or recording instrument should always be installed in proximity to the point in the centrifugal pumping system where control action is applied. This instrument, which registers flow, level, pressure, or temperature, depending upon which variable should be controlled, is especially important in the case of fully automatic controls, as it will indicate at a glance whether the control operates properly; it can even be arranged to warn the operator of any failure in the functioning of the control.

Selection of controls

It is necessary, when selecting or ordering centrifugal pump controls, to give complete information to the control manufac-

turer. Some data most essential to the contemplated installation are as follows:

1. Service or control function. Specify what is to be accomplished.

2. Nature of fluid handled. The character of the fluid will have an effect on the selection of the type and the size of the required valving. If the liquid is corrosive, indicate complete analysis. Also, if the viscosity of the liquid differs appreciably from that of water, indicate average or, at least, the range of viscosities. Specify liquid temperature and whether it is constant or variable.

3. Capacities of the flow being controlled and values of the initial and final pressures, if pressures are regulated. If steam flow to a turbine driving a centrifugal pump is to be controlled, full steam conditions must be indicated.

4. Character of system operation. Is it constantly under load or is the operation intermittent?

5. Manual or automatic control. If automatic, which elements of the control must be automatic and which can be manually operated? Is remote control indicated or can it be self contained?

6. Required accuracy of regulation.

7. Required range of adjustment.

8. Type of service. Is dead-end service involved for the regulating valve or is the flow continuous and small leakage, as in double-seated valves, permissible?

9. Control safeguards. Should failure of control place the regulating valve into fully open or fully closed position? This question is especially important in the case of protective controls.

10. Structural details. Position of valve (vertical or horizontal), and type of connections for valve nozzles (if flanged, state desired facing and drilling).

11. Range. For all of the data, indicate the full operating range of all variable factors, such as capacities, pressures, and temperatures, including definite maximum and minimum values.

20 *Drivers*

Most centrifugal pumps are electric-motor driven—direct connected or geared—but steam turbines are also often used. Gasoline engines, diesel engines, and gas engines are less frequently employed, and water turbines drive centrifugal pumps even less often. Steam engines are rarely used now, but were popular in the early days of centrifugal pumps. Probably every form of prime mover and source of power has been used, with some form of intermediate transmission, if necessary, for centrifugal pump drives.

The choice of rotative speed is limited with certain types of drivers such as gasoline engines. The speed of the ideal pump for the service conditions may be much higher or much lower than the speed of the driver. For continuous service units, a speed-increasing or -decreasing device is warranted. For standby units, used occasionally, a compromise pump design can often be employed, saving the cost and complication of a speed changer at some sacrifice in pump performance and, in some cases, with some increase in pump cost.

Size of drivers

Local conditions will determine the load permissible for different drivers. For example, the design of an electric motor, as well as the temperature and density of the surrounding air, and the variations in available electric current will determine the safe load that can be carried by the motor.

The driver should be capable of carrying the loads that will be imposed upon it over the entire range of operating conditions of the pump under the most adverse operating conditions. The meaning of the phrase "most adverse conditions" cannot always be interpreted literally. Sound economics will not justify the selection of a squirrel cage induction motor for a rating under conditions which might last only a few hours per year. For example, if an unusually high ambient temperature might occur for 20 or 50 hours a year, it may have to be disregarded. Thus, if the temperature is 50°C instead of 40°C for a limited number of hours, economics will justify selecting the motor on the basis of the 40°C ambient temperature. Theoretically, this would shorten the thermal life of the motor insulation somewhat, but, practically, it may make little change in the actual life of the motor because so many other conditions also determine the operating life of the motor.

If the driver can operate at variable speeds and the installation requires operation over a wide head range, necessitating a wide power range as well, the unit is sometimes run at reduced speed to keep the maximum power requirement within the available driver size.

The power required to drive a centrifugal pump from shutoff to maximum capacity varies with its specific speed type and its individual design (see Chap. 17). In the lower specific speed types, the power requirement is minimum at shutoff and increases to maximum at, or near, maximum capacity. With this type of pump, operation at heads below normal will result in an increased power requirement and may impose an overload on the driver.

High specific speed types, however, have the opposite power requirements, and a reduction in head causes a reduction in power. In a high specific speed pump, an increase in operating head will cause an increase in power consumption and may overload the driver.

Electric motors

If the brake horsepower of a centrifugal pump exceeds the safe operating load of the motor, which depends on the type used, the motor may be damaged or burned out. The shape of the power characteristic curve and the system-head characteristics will determine if pump operation may exceed the safe loading of its electric motor. Careful attention must also be paid to the shape of the speed-torque curve of the motor and the voltage supply of the power system. A motor with ample speed-torque characteristics at rated voltage may not be capable of handling the load at some reduced voltage.

Steam turbines

A steam turbine driving a centrifugal pump cannot be overloaded. With a fixed throttle position, a steam turbine develops a constant torque; the speed of the unit will always be at a point where the turbine torque equals the pump torque. If the turbine is equipped with a constant-speed governor, the governor will throttle the steam supply until—at the required speed—the developed torque and the torque demand are balanced. With an increase in torque demand, the governor will open up the throttle valve until a point is reached where the torque demand tends to exceed the available torque. Since the governor can no longer provide for the increase in steam required for a torque increase, the turbine slows down until a balance is reestablished between the required torque and the available torque at some speed lower than the design speed.

When the turbine is equipped with a constant- or excess-pressure regulator, the effect is identical to that described for a constant-speed governor; the only difference is the source of the impulse that tends to maintain the required torque balance until external conditions occur to interrupt the functioning of the governor.

A turbine driving a centrifugal pump should be capable of carrying the maximum load that will be imposed on it by the pump under the most adverse steam conditions. These can be determined fairly accurately in large units, but less so in smaller units. As wear causes increased leakage through the wearing rings, the pump capacity will be reduced. It is desirable, therefore, to increase the speed of a turbine-driven pump slightly to restore the desired capacity. Generally, small pumps are not checked as carefully as large pumps, and the leakage may become more pronounced before the leakage joints are restored to their original clearances. In small steam-turbine-driven units, the turbines are generally selected for some excess power (roughly 5 per cent over maximum expected loading) while in large units, turbines are selected for a rating equal to, or just slightly in excess of, the maximum loading.

Internal combustion engines

Power ratings of some internal combustion engines are given for the load they can carry continuously at the rated speed, whereas in other cases, ratings are given for

Fig. 20.1   Gasoline engine

This type is often employed as a fire pump driver and for other standby applications.

the load the engine can develop on test with fully open throttle. In the latter category, the power rating may include the power required by the auxiliaries normally attached to the engine, such as the circulating pump or the fan. Automotive type engines are generally rated for their maximum developed power on dynamometer test while stripped of all auxiliaries (Fig. 20.1). Usually, these engines can safely be operated with a continuous loading of 75 to 80 per cent of their developed power.

Most automotive engines are of the high speed type and have a very limited life when used for continuous operation at full load. Medium speed engines with a longer life and low speed engines, which have a very long life with a minimum of maintenance, are also available.

Although internal combustion engines

rated on the basis of the load they will carry constantly can carry an overload and although engines rated on the basis of maximum developed power can be operated while loaded to a higher percentage of that developed power, prolonged overloaded operation will generally result in mechanical failure.

The engine best suited for any particular application must therefore be selected with due consideration to the service requirements.

Variable speed drives

When varying operating conditions exist in a centrifugal pump installation, variable speed operation may be desirable; drivers that economically permit speed variation are preferred for such installations. A wide

Fig. 20.2   Variable speed transmission for variable pump output

Cover removed from drive to show construction, which uses V-belts and adjustable pitch sheaves.

choice of drivers suitable to variable speed operation is available: Steam turbines, gasoline or diesel engines, wound-rotor motors, or, in certain cases, d-c electric motors. Synchronous or squirrel cage induction motors can be used with a variable speed transmission, such as a hydraulic coupling or magnetic drive. Variable speed devices with variable pitch sheaves and V-belts will fulfill low power requirements (Fig. 20.2).

Advantages of variable speed

Many centrifugal pump applications require operation at varying capacities and total head. Since the operating conditions of a centrifugal pump are determined by the intersection of its head-capacity curve with the system-head curve, the only way to vary operating conditions is to alter the system-head curve by throttling if the pump is operating at constant speed. It may often be more practical and more economical to change the intersection point described above by varying the operating speed and, therefore, the head-capacity curve of the pump.

Many factors enter in the evaluation of the most economical method of operation

under such conditions. Whereas considerable pump horsepower can be saved by varying the speed of a steam turbine or of an engine, the efficiencies of wound-rotor motors and of hydraulic and magnetic transmissions are practically proportional to the ratio of their operating speed and full load speed. Both the actual torque and the ratio of speeds must be taken into consideration. For example, if there is a 20 per cent speed reduction at 100 per cent of full load torque, the slip losses are 20 per cent. But if the speed reduction is 20 per cent and the torque at reduced speed is only 50 per cent of full load torque, the slip losses are only 10 per cent. Therefore, the actual power consumption as well as the increased cost must be considered in the evaluation of a variable drive against a constant speed drive.

A description of both hydraulic couplings and magnetic transmissions appears in a later portion of this chapter.

Dual drives

The pump profession defines a dual-driven pump as one that can be driven by either of two drivers. The two drivers are generally of different types; one is for regular use, such as an electric motor, and the second is a standby type such as a gasoline engine, which is used if the power supply for the regular driver fails (Fig. 20.3). The pump selection may have to be compromised by the available drive speed.

Dual-drive units are generally arranged with a driver on each end of the pump so that the unused driver can be disconnected. It is then necessary to use a double-extended pump shaft; end-suction pumps cannot be arranged in this manner. Some dual drives are arranged with a double extended-shaft motor, with both driving units on one end, in which case end-suction type pumps can be used.

In a few installations the pump is driven by two drivers, each carrying part of the load. The most common installations of this type involve an electric motor and a steam or a water turbine without governors, or with governors set at a higher speed.

When induction motors are used, the speed of the unit varies slightly with changes in operating conditions in order to reach a balance in which the power developed by the two drivers equals the power required by the pump.

The speed of a squirrel cage induction motor changes very little with change in load. In the larger sizes, 200 hp and larger, the change from no-load speed to full-load speed is probably not more than 1½ per cent for most ratings, and in the case of two-pole motors, it is less than 1 per cent.

Belt drives

Before electric drive became almost universal for centrifugal pumps, many isolated pumping plants were steam operated and used steam engines for drivers. In many cases the speed of the pump could not be matched to the slower engine speed,

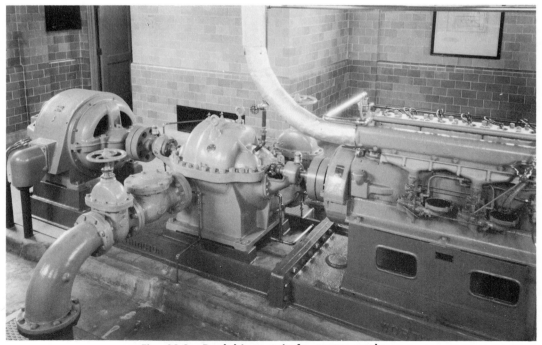

Fig. 20.3 Dual-driven unit for water works

Pump is normally electric-motor driven. The gasoline engine driver is used if power fails.

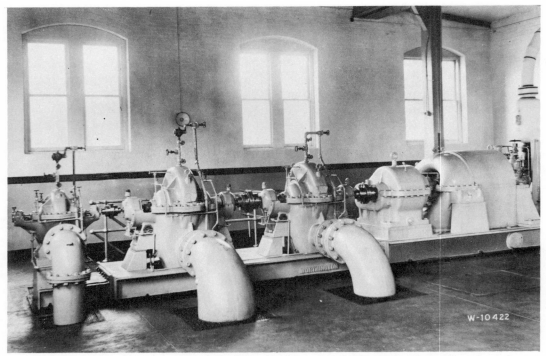

Fig. 20.4   Large capacity series pumping unit driven by condensing steam turbine through reduction gears

Small pump, driven at end of main pump, supplies circulating water to condenser serving the turbine.

Fig. 20.5   High-pressure high-speed boiler feed pump

Step-up gear permits a 1,450-rpm motor to drive pump at 3,800 rpm, through a hydraulic coupling that gives variable speed control.

Fig. 20.6   Pump installation driven by slower speed diesel engine through step-up gears

and the engine drove the pump through a flat belt with different diameter pulleys to give the correct pump speed. Flat-belt-driven pumps are seldom used now except in small sizes and for irrigation pumps that are belted to tractor engines.

Multi-V-belt drives are frequently used for centrifugal pump drives. In some cases the pump is an auxiliary mechanism, such as a cooling water pump for diesel or gas engines. Occasionally a Multi-V-belt drive can be used as a speed-increasing or speed-decreasing transmission to connect a driver to a pump operating at a different speed. Provision has to be made for adjustment of the belt tension and for sufficient slack in order that the belts can be put on the sheaves. The pump location is usually fixed because of its suction and discharge piping. This adjustment must therefore be accomplished by moving the driver or, if that is impractical, by introducing an adjustable idler pulley.

Gear drives

The optimum range of pump speed often does not match the optimum or even practical range of driver speed (Fig. 20.5). These conditions exist in most steam-turbine-driven pumps (Fig. 20.4) other than for low capacity and high heads and in the great majority of units driven by heavy-duty diesel and gas engines (Fig. 20.6). For such installations, gears are used as speed changers and are very satisfactory.

Gears connecting horizontal-shaft drivers to horizontal-shaft pumps are usually double-helical, single-reducing or increasing gears. Right angle gears are often used to connect horizontal-shaft drivers to vertical shaft pumps. These drives were basically developed for engine-driven, vertical turbine pumps and are capable of carrying a large downward thrust load on their vertical output shafts. Most of these gears are the hollow-shaft type favored in the vertical turbine pump field.

THE SELECTION OF ELECTRIC MOTORS FOR CENTRIFUGAL PUMP DRIVE

Drivers for centrifugal pumps must be selected carefully so that the unit will be satisfactory. When the drive is to be a steam turbine or an internal combustion engine, full details of the requirements are usually given to the manufacturer of the driver. Frequently, less care is exercised with electric motors, resulting in the use of motors with unsuitable electrical or mechanical constructions. This misapplication is due mainly to the fact that so-called general purpose motors are offered by the various motor manufacturers for ratings of 200 hp, or less, and 450 rpm, or more, without restriction to a particular driver application. The user must realize the requirements of the application to insure that the proper motor is employed (Fig. 20.7 and 20.8).

Electrical systems

In d-c power supply systems the flow of electricity is in one direction and the power in kilowatts is volts $\times$ amperes/1,000. D-c systems have certain limitations and disadvantages that curtail their use; therefore most power systems are a-c systems.

In a-c systems, the flow of current reverses or alternates periodically. The voltage and the current vary in amount and direction with time. The current and voltage can reach their maximum at the same time, or the peak of the current can be ahead of, or behind, the peak of the voltage. The measure of this lead or lag is called the power factor, and it can range from 0 to 1.0 or 0 to 100 per cent. The power in kilowatts is:

volts $\times$ amperes $\times$ power factor $\times$ K/1,000

where K is 1.0 for single-phase systems, 2.0 for 2-phase, 4-wire systems, and 1.732 for 3-phase systems.

TABLE 20.1   STANDARD VOLTAGES AND CORRESPONDING NORMAL MOTOR SIZE RANGES

Boldface indicates the voltages most commonly used.

Supply	Voltage	Reasonable motor sizes, in hp [1]	
		Min	Max
Direct current	**115**		30
	230		200
	550–600	0.5	
Single-phase alternating current	110–**115**–120		1.5
	220–**230**–240		10
	460–550	5	10
2- or 3-phase alternating current	110–**115**–120		15
	208–**220**–230–240		200
	440–550		500
	2,300	40	
	4,000–**4,160**	75	
	6,000–**6,600**	400	

[1] Where no minimum is given, very small fractional horsepower motors are feasible. Where no maximum is indicated, very large motors exceeding normal commercial sizes are possible.

An a-c cycle is the change occurring from the peak of current flowing in one direction to zero to the peak in the opposite direction to zero and again to the peak in the first direction. The number of cycles occurring per second is the frequency. A number of different frequencies were once used in the United States; 60-cycle current is almost universal now, although some 25-cycle and 50-cycle systems, as well as a few isolated installations of 30-cycle, 40-cycle, and other frequencies, still exist.

A-c systems are classified also as to the number of phases; single-, two-, or three-phase systems are used, with single- or three-phase systems the most common. Two- and three-phase systems are known as polyphase systems.

Table 20.1 lists the standard voltages for a-c systems. The voltages most commonly used are shown in bold type. The table also lists the normal range in motor sizes that can be obtained with these voltages.

The speed of an a-c motor is a function of the frequency of the supply and the number of poles, always an even number as the poles must be in pairs. The synchronous speed of a motor in rotations per minute is 120 × frequency/number of poles. The 25-, 50- and 60-cycle synchronous speeds are shown in Table 20.2.

Direct-current motors

D-c motors are available in three types: Shunt wound, series wound, and compound wound, although the commercial stabilized shunt-wound d-c motor has some compound windings. The general purpose type of d-c motor is not suitable for centrifugal pump drive as the full load speed of such motors may be 5 to 7½ per cent above or below their rated speeds. If the actual speed of the motor is below rated speed, the pump will not produce its rated conditions, while if the actual speed of the motor is above rated speed, the pump will deliver excess head or capacity with increased

Fig. 20.7   Small-size motor-mounted pump

power requirements, possibly imposing a dangerous overload on the motor. D-c motors intended for centrifugal pump drive, therefore, must be specifically ordered, and a design is required that will provide a hot full load speed within plus or minus 3 per cent of the rated full load speed.

This requirement can generally be provided only in the case of larger d-c motors. With small d-c motors, a small fixed resistor is necessary to obtain the speed desired for pump drive.

The speed of a regular shunt-wound, d-c motor can be increased with little change in efficiency by the use of a field rheostat. The motor speed can be reduced by intro-

Fig. 20.8   Motor-driven centrifugal pumps for water works

TABLE 20.2   SYNCHRONOUS SPEEDS FOR 60-, 50-, 40-, 30- AND 25-CYCLE MOTORS

Motor poles	60 cycles	50 cycles	40 cycles	30 cycles	25 cycles
2	3,600 rpm	3,000 rpm	2,400 rpm	1,800 rpm	1,500 rpm
4	1,800	1,500	1,200	900	750
6	1,200	1,000	800	600	500
8	900	750	600	450	375
10	720	600	480	360	300
12	600	500	400	300	250
14	514	428	343	257	214
16	450	375	300	225	187
18	400	333	267	200	
20	360	300	240		
22	327	272	218		
24	300	250	200		
26	277	231			
28	257	214			
30	240	200			
32	225				
36	200				

ducing resistance in the armature circuit. The latter method of speed control causes a considerable loss of efficiency.

Alternating-current motors

Single-phase motors

Single-phase, a-c motors are available in the following types: split-phase, series (universal), repulsion, repulsion-induction, and capacitor. Generally either the repulsion-induction or capacitor-type motor is used for centrifugal pump drive.

Squirrel cage motors

Squirrel cage motors are the simplest polyphase electric motors and are most commonly used for centrifugal pump drives. They have wound primary (stator) windings and squirrel cage secondary (rotor) windings that take power from the primary windings by transformer action, without any separate electrical connection. Mechanical construction permitting, squirrel cage motors can be run in either direction. They are reversed by reversing any two leads of a three-phase, or the two leads of one phase of a two-phase motor.

Common types of squirrel cage motors are listed below:

1. Normal-torque, normal-starting current (NEMA Class A) motors
2. Normal-torque, low-starting current (NEMA Class B) motors
3. High-starting torque, low-starting current (NEMA Class C) motors
4. High-starting torque, high slip (NEMA Class D) motors
5. Multispeed motors.

Two special types of squirrel cage motors are: Low-torque, normal-starting current, and low-torque, low-starting current motors.

Centrifugal pumps do not require motors with high-starting torques; therefore, types (3) and (4) are seldom used for centrifugal pump drives. A few unusual applications, however, where the power supply is very limited and an autotransformer type starter is used with a high-torque motor, require these motors. The starter will use a low tap, perhaps 50 per cent of the available voltage, thereby limiting the current drawn from the line. With this very low voltage, the torque may be insufficient to drive some pumps, such as vertical turbine pumps, which require more than the normal amount of starting torque for a pump. By using a high-starting-torque motor for this application, it becomes possible to start the pump despite the low voltage available from the autotransformer tap. The alternative is to use a wound-rotor motor with secondary control.

Multispeed motors, (5) above, are made in two types. In one type the stator winding is wound so that by external switching the number of poles can be changed. Therefore, the winding of a 6-pole (1,200-rpm, 60-cycle) motor can be reconnected by external switches to make it a 12-pole (600-rpm, 60-cycle) motor. Mechanical and electrical problems limit the combinations of poles and therefore the choice of speeds available. Commercial single-winding motors, for example, are limited to ratings in which the lower speed is half the higher speed, such as 3,600 and 1,800 rpm, 1,800 and 900 rpm, 1,200 and 600 rpm, and similar combinations.

The second type of multispeed motor has two separate stator windings, each wound for a different number of poles. One speed is obtained with one set of windings, whereas the other set yields a different operating speed. The motor can be built so that the number of poles in one or both windings can be changed, thus obtaining a three- or four-speed motor, as for example, a 1,800/1,200/900/600 rpm motor.

Constant horsepower and both constant-torque and variable-torque multispeed motors are available. The power required by a centrifugal pump varies with the cube of the speed, and, consequently, the torque varies with the square of the speed, so that a variable-torque multispeed motor functions well as a centrifugal pump drive. Three- and four-speed motors are seldom used as centrifugal pump drives since the spread is too wide and the resulting capacity and head at reduced speed is very low. They can be employed when the pump is used on a circulating system in which the head is entirely frictional. In systems with

Fig. 20.9   External view of wound-rotor motor

Showing provision for leads from slip rings and plate covering opening, which allows access to slip rings for servicing. (Courtesy General Electric Co.)

a small static head component it is sometimes feasible to utilize a drive with three or four available speeds by coupling two single- or two-speed (two winding) motors together, or on either end of the pump. Constant speed motors are used for most centrifugal pump applications, however.

Normal-torque squirrel cage motors, of both normal- and low-starting-current types, are the most commonly used. Normal-torque, normal-starting-current types may require some type of reduced voltage starting equipment in order to meet the starting current inrush limits set by the public utilities. To combat this problem, the electrical industry offered the low-starting current type, which, with full-voltage starting, drew no more current than the normal-current motor with a reduced voltage starter. (Reduced voltage starters with low-starting-current motors are sometimes necessary to meet power requirements where the power system is very weak.) The low-starting-current motor, complete with a full-voltage starter, became less expensive than a normal-starting-current machine and a reduced-voltage starter.

Normal- and low-starting-current motors sell at the same price, but the reduced-voltage starter costs more than the full-voltage starter. As a result, the normal-torque, low-starting current motor (equivalent to NEMA design B) has become the standard model.

Fig. 20.10 Bracket-bearing synchronous motor with exciter mounted on outboard bracket

(Courtesy Electric Machinery Mfg. Co.)

Wound-rotor motors

Wound-rotor motors (Fig. 20.9) have both wound primary and secondary windings. The primary or stator winding is the same as that of a squirrel cage motor. The secondary or rotor windings are connected to slip rings so that external resistance can be introduced in the secondary winding for starting, or for speed regulation. This reduces the flow of current and also affects the torque-speed characteristics of the motor. The major field of application of wound-rotor motors is for installations where variable speed or low-starting current is necessary. The resistors used for speed control must dissipate the heat resulting from the flow of current through them. Commercial speed controls cut the resistance into the rotor circuit in steps so that the resulting speed change is also in steps. For fine adjustment of speed, a control with a large number of contact points or a liquid slip regulator is required. Where very fine adjustment of speed is required, hydraulic and magnetic drives are finding more applications.

The starting characteristics of the wound-rotor motor offer one of its major advantages. With proper resistance in the rotor circuit, the current demand throughout the entire starting period can be kept very low. The standstill line current specification at rated torque is generally within 130 per cent of rated current. (This current difference is due to the power factor being lower at standstill than at rated load and rated speed.) This type of motor, therefore, is desirable if it must be operated on a limited power supply, or if the size of the supply line is limited.

Synchronous motors

Synchronous motors usually consist of a wound stator connected to the a-c supply and a rotor with wound poles connected to a d-c circuit (Fig. 20.10). The direct current is obtained from a separate source of supply or from a d-c generator, called an exciter, which may be attached to the end of the motor. This motor runs at its synchronous speed regardless of the load, hence its name. Basically, this motor has a very low starting torque characteristic; therefore the modern commercial version has an auxiliary squirrel cage winding built into the rotor for starting duty. The motor is started as a squirrel cage motor, and when the speed has reached approximately 95 per cent of rated capacity, the field (rotor) current can be applied when the position of the rotor properly matches the a-c flow in the stator. This is called "synchronizing." Actually, synchronizing is matching an incoming unit to an existing system with respect to voltage, speed, and phase position; therefore a synchronous motor is not synchronized—it is "pulled into step." The term synchronizing is actually incorrect, although commonly used. With present-day automatic control, the field current can be applied at the proper instant, greatly simplifying the starting of a synchronous motor.

Because a synchronous motor is started by its auxiliary winding as a squirrel cage motor, the starting torque and starting current characteristics depend basically on the squirrel cage element. While the motor manufacturer can provide a squirrel cage element with characteristics like the regular types of squirrel cage motors, space limitations, as well as other factors, curtail the design characteristics of squirrel cage mo-

Fig. 20.11 Open motors can be used to drive pumps installed in clean, dry location
(Courtesy Electric Machinery Mfg. Co.)

tors that can be incorporated in a given synchronous motor design.

Part windings are available for relatively large synchronous motors, in which the a-c stator winding is formed of several parallel windings. In the starting cycle of this motor, the current is applied to one group of windings only; therefore, the initial current drain equals that of a smaller size motor. As the motor reaches part speed, the second group of windings is cut in to carry it up to the pull-in speed.

Mechanical characteristics of motors

The motor used for most pump applications is a horizontal type with an open frame (Fig. 20.11) and with ball or sleeve bearings supported by brackets attached to both sides of the frame. Large synchronous motors are often made with pedestal sleeve bearings. The insulation and elec-

trical design of these are suitable for most pump applications. Where this type of motor is not suitable, however, more applicable designs or modifications should be obtained and used where conditions warrant.

Construction modifications

Both solid- and hollow-shaft vertical motors with various forms of mountings are obtainable (Fig. 20.12). Hollow-shaft motors are used almost exclusively on deep-well turbine pump and vertical propeller pump applications. Both shaft types are made to carry external thrust loads. Small and medium vertical motors require anti-friction thrust bearings; large motors require Kingsbury or similar bearings.

The open frame motor is often not suitable, and a special enclosure is necessary. Drip-proof, splash-proof, weather-protected, and totally-enclosed, nonventilated or fan-cooled frame designs and modifications are commonly needed. A drip-proof motor is designed to prevent liquids or particles falling at an angle of 15 deg to the vertical from getting into the windings (Fig. 20.13). Some manufacturers have made their regular open, general-purpose designs drip-proof (Fig. 20.14). Drip-proof motors have a tem-

Fig. 20.13   Pump driven by drip-proof motor

Installed in damp, dirty location.

perature rise rating of 40°C rise with class A insulation or 60°C rise with class B insulation (maximum temperature rise at rated load with 40°C ambient temperature). A splash-proof motor is one designed to prevent entrance of liquids or particles coming toward the motor at an angle of 100 deg to the vertical (Fig. 20.15). As ventilation is restricted, splash-proof motors are regularly rated at 50°C rise with class A insulation or 70°C with class B insulation. Motors with 40°C rise can be obtained but at higher cost.

The various types of totally enclosed motors manufactured are classified first as to the type of cooling and second as to the design of enclosure for the application. Totally enclosed nonventilated motors are sometimes built in small horsepower ratings; larger horsepower motors, however, must have fans incorporated in the design to provide adequate cooling (Fig. 20.16 and 20.17). Both types are rated at 55°C rise with class A insulation or 75°C rise with class B insulation. Some totally enclosed, separately ventilated motors are arranged to use air from an outside source for cooling. The self- or pipe-ventilated motor has an air-circulating fan built into it, whereas

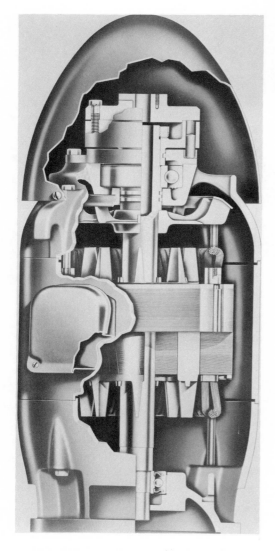

Fig. 20.12   Vertical hollow-shaft motor

Used for vertical turbine pumps and many other vertical wet-pit pumps. (Courtesy West-inghouse Corp.)

the separately or forced-ventilated type depends on a separate, forced air supply for ventilation.

Enclosed or pipe-ventilated motors normally operate at 40°C rise without the ventilating ducts being connected to the frame. Enclosed forced-ventilated motors should be provided with sufficient air from the outside cooling source to operate at 40°C rise.

Totally enclosed motors are used in locations where the surrounding air contains dust, corrosive vapors, or flammable gas or vapor. The design of the enclosure required varies with the application, and the problem presented to the motor designer varies with the character of the hazard. Explosion-proof electrical equipment, for example, is designed so that if an explosion occurs within the equipment, it will not damage

Fig. 20.14 Modern horizontal-shaft, general-purpose squirrel cage motor with open drip-proof frame

(Courtesy Westinghouse Corp.)

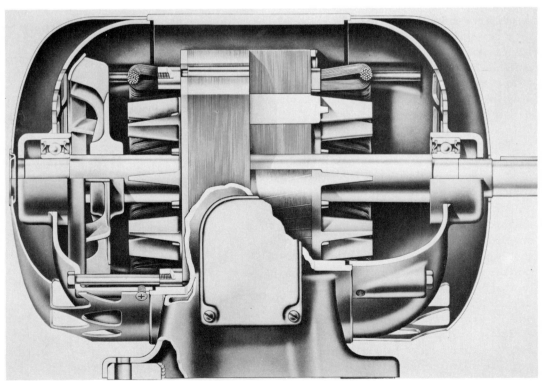

Fig. 20.15 Splash-proof motor

End bells have internally shielded air inlets and outlets at bottom with air openings at top so splashing liquid cannot get into windings. (Courtesy Westinghouse Corp.)

the equipment or permit the explosion to go beyond the interior of the equipment. The National Electric Code has designated hazardous atmospheric conditions as follows:

1. Class I—Gas Hazards
   Group A—acetylene
   Group B—hydrogen, or gases or vapors of equivalent hazard
   Group C—ethyl ether vapor
   Group D—acetone, flammable alcohols, gasoline, naptha, common petroleum distillates, pyroxylin-lacquer solvents, toluene, and natural gas

2. Class II—Dust Hazards
   Group E—metal dust
   Group F—coal or coke dust
   Group G—grain dust.

Motors used in gaseous locations in coal mines have to be constructed and installed to meet the rules of the US Bureau of Mines, not the National Electric Code. It is important when ordering motors for hazardous conditions to specify clearly the hazard involved so that a motor of suitable design is obtained.

In some hot, humid, dusty, or corrosive atmospheres, open, splash-proof, or drip-proof motors with special insulation can be used. The insulation regularly used on standard motors is suitable for a moderate amount of moisture, weak acids, alkalies, and abrasives, but not conducting dust, oil, or other similar materials. For more severe conditions, more resistant types of insulation are available. For example, a combination of conducting or abrasive dust with sulphur fumes is encountered in some applications and requires suitable insulation.

Motors used in the tropics require special protection against fungus, vermin, and excessive moisture and heat (Fig. 20.18). Motor manufacturers recommend tropical protection for motors to be used in the tropics and provide special insulation for such motors, screens over all openings, and

Fig. 20.16 Totally enclosed fan-cooled motor of small horsepower (15 hp or less)

(Courtesy Westinghouse Corp.)

special treatment of insulation and metal surfaces against fungus, insects, and corrosion.

Motors for installation in damp locations on machines that will not be used frequently can become damp, especially in cold weather, with resulting damage to the motors. Under these conditions larger motors are often protected by installing electric space-heater elements in the motor. The heaters are turned on when the unit is not in operation, thereby keeping the motor warm and dry.

The average motor develops some noise in running, mostly due to the fan circulating the cooling air. In locations such as hotels or office buildings, the amount of

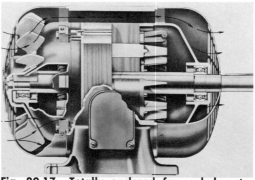

Fig. 20.17 Totally enclosed fan-cooled motor

(Courtesy Westinghouse Corp.)

Fig. 20.18 Motor with screened air inlets and outlets

Prevents small animals and vermin from getting into motor. (Courtesy Westinghouse Corp.)

noise developed with a regular motor design is objectionable. Therefore, special motors, known as quiet operating motors, which produce a minimum of noise, should be used. In locations where noise is very critical, acoustical treatment may be required in the area immediately around the motor, in addition to using a quiet operating motor.

Heat dissipation

The electrical losses in a motor go into heat that must be dissipated to the surrounding atmosphere or to the cooling air. The amount of heat generated is dependent upon the load, and the actual temperature of the motor is dependent upon the load and the surrounding cooling air temperature. The safe operating temperature of the motor depends upon the insulation used in its manufacture. With class A insulation, the allowable observable total temperature is 90°C (194°F); with class B insulation, it is 110°C (230°F). If the ambient temperature is low, a greater temperature rise can be permitted in the motor. Since the temperature rise in the motor is very nearly proportional to the load squared, the load that can be carried by the motor within the safe operating temperature limit is greater. If the surrounding air temperature is high, how-

ever, the maximum load that the motor can safely carry will be lower.

Motors are rated on their maximum temperature rise above an assumed ambient air temperature of 40°C (104°F), when carrying rated load. Most open motors are 40°C rise motors; that is, with an ambient cooling air temperature of 40°C, the temperature rise of the motor operating at full load will not exceed 40°C. Under such conditions the actual temperature of the motor might be 176°F, too hot to touch, but considerably lower than the safe operating temperature of the motor. Regular 40°C rise motors are designed with a service factor of 1.15, meaning that the motors can carry 115 per cent of their rated load continuously without the temperature rise in any part of the motor exceeding a safe operating temperature, providing the air temperature is 40°C or less. It is, therefore, safe to load 40°C rise motors up to 115 per cent of their nameplate rating. This 15 per cent overload safety factor often permits the use of a motor whose rating is less than the required maximum load at a saving in cost. It is desirable to have ample driver power in case a centrifugal pump develops excess head or capacity, or in case adverse operating conditions, such as low voltage or high cooling air temperature, is encountered. Therefore, the margin provided by the service factor is usually utilized only for occasional maximum conditions or on units that will operate infrequently for short periods. The latter motors are cold when started and will be in operation for some time before they reach their maximum operating temperature.

Motors rated at 50°C and 55°C rise are not designed with a service factor and should be selected so that they are not overloaded under any conditions.

Motors are usually cooled by the surrounding air. Rarefied air, found at higher altitudes, is a poorer cooling medium. Standard motors are designed for air densities that will be encountered from 0 to 3,300 ft above sea level. For use above 3,300

TABLE 20.3 APPROXIMATE EFFECTS OF VARIATIONS IN VOLTAGE AND FREQUENCY ON A-C MOTOR CHARACTERISTICS [1]

Characteristic	Voltage		Frequency	
	110 Per cent	90 Per cent	105 Per cent	95 Per cent
Torque				
Starting and maximum running	Increase 21 per cent	Decrease 19 per cent	Decrease 10 per cent	Increase 11 per cent
Speed				
Synchronous	No change	No change	Increase 5 per cent	Decrease 5 per cent
Full load	Increase 1 per cent	Decrease 1.5 per cent	Increase 5 per cent	Decrease 5 per cent
Per cent slip	Decrease 17 per cent	Increase 23 per cent	Little change	Little change
Efficiency				
Full load	Increase 0.5–1 point	Decrease 2 points	Slight increase	Slight decrease
¾ load	Little change	Little change	Slight increase	Slight decrease
½ load	Decrease 1–2 points	Increase 1–2 points	Slight increase	Slight decrease
Power Factor				
Full load	Decrease 3 points	Increase 1 point	Slight increase	Slight decrease
¾ load	Decrease 4 points	Increase 2–3 points	Slight increase	Slight decrease
½ load	Decrease 5–6 points	Increase 4–5 points	Slight increase	Slight decrease
Current				
Starting	Increase 10–12 per cent	Decrease 10–12 per cent	Decrease 5–6 per cent	Increase 5–6 per cent
Full load	Decrease 7 per cent	Increase 11 per cent	Slight decrease	Slight increase
Temperature rise	Decrease 3–4°C	Increase 6–7°C	Slight decrease	Slight increase
Maximum overload capacity	Increase 21 per cent	Decrease 19 per cent	Slight decrease	Slight increase
Magnetic noise	Slight increase	Slight decrease	Slight decrease	Slight increase

[1] Courtesy General Electric Co.

ft, the safe loading of the motor is reduced. Motors especially designed for high altitudes can be obtained at additional cost.

Electrical characteristics of motors

Electric motors are made for standard voltages depending upon their sizes; for example, 30-hp, 1,800-rpm, 3-phase, 60-cycle motors are regularly offered for 208, 220, 440, and 550 volts. A motor can be operated successfully, but not necessarily in accordance with standard guarantees, at voltages 10 per cent above or below its rated voltage (at the same frequency in the case of a-c motors). Motors wound for special voltages can be obtained at additional cost.

The frequency of the power supply varies in some a-c installations, thereby causing a variation in the operating speed and resulting in a variation of the operating characteristics of the pump. In such cases the maximum loading on the motor under the most adverse conditions must be checked carefully to insure safe operation. Standard motors are usually designed to operate satisfactorily at 5 per cent above or below rated frequency. The effects of voltage and frequency variations on motor characteristics are illustrated in Table 20.3. In a number of water works installations where variable pump output is required, the frequency of the current is deliberately varied in order to vary the motor speed and thus the characteristics of the pump.

Some a-c motors are wound for use at either of two voltages, for example, 220 or 440 volts, depending upon how the leads are connected. This construction is standard for some sizes and combinations of voltages but is special for others. The voltage ratios in these dual-voltage motors are necessarily limited to certain fixed values.

By using special designs and special materials, at increased cost, induction motors of more than 200 hp and synchronous motors can be built to operate more efficiently. With units that are to be operated more or less constantly, the saving in power costs

Fig. 20.19   Gear-head motor

The combination of a motor and a reduction gear gives a low output speed when using a light, high-speed motor. (Courtesy General Electric Co.)

will, in many cases, warrant the increased first cost of premium efficiency motors. If power costs are low, however, the increased first cost of premium efficiency motors may not be justified.

Gear-head motors (Fig. 20.19) are regularly made by many motor manufacturers in the smaller horsepower sizes. They are generally four-pole motors (1,800-rpm, 60-cycle) with built-in gear transmissions. This type of motor drive is not used much on centrifugal pumps, except occasionally on low-head units.

In addition to the major electrical and mechanical variations in motor design mentioned above, other variations can be obtained, such as a double-extended shaft for driving from both ends of the motor, split end-shields, and many others. Whenever a regular motor design is not suitable, either electrically or mechanically, the manufacturer should be consulted to obtain a better design.

Starters for electric motors

A starter for an electric motor connects the motor electrically to the source of supply. Manual starters move the starter contactors manually; automatic or magnetic starters move them automatically, by means of magnets, on receiving an impulse from a pilot device.

Direct–current motor starters

A knife switch can serve as a starter for a d-c motor, but this would impose full voltage across the motor terminals with a high current inrush. While this is feasible with small motors, it is generally desirable to have an adjustable resistance, a rheostat, connected in series with the motor armature to limit the initial current. As the motor accelerates, the resistance is cut out until full voltage is applied to the armature and the motor is running at full speed. If this resistance is used solely for starting, it is called a starting rheostat (Fig. 20.20). If the motor is run at reduced speed by leaving some of the resistance in the armature circuit, the rheostat must be designed to dissipate the heat energy used up in the resistance, thereby making the rheostat more expensive.

Alternating–current motor starters

Starters for single-phase, a-c motors are usually some form of an across-the-line starter. Where this type cannot be used because of the limited line capacity, one of the following special starters is used to reduce the voltage initially applied to the motor terminals and to increase it in steps until full voltage is reached.

Squirrel cage motor starters. Starters for polyphase squirrel cage motors are made in three types: Across-the-line, compensator or autotransformer, and resistance. The first two are more common than the last one.

Fig. 20.20 Multistep rheostat used as d-c motor starter

(*Courtesy General Electric Co.*)

An across-the-line or full-voltage starter, as its name implies, applies full voltage to the motor terminals, which results in a high starting current dependent upon the basic design of the motor. With a small line or a limited generator capacity this will result in an objectionable voltage drop. In case a full-voltage starter cannot be used, a reduced voltage starter using a compensator or autotransformer is usually employed. Such a starter is basically a transformer wound so that a reduced voltage (generally 80 per cent of the line voltage) can be initially applied to the motor terminals. When the motor has reached part speed, the transformer is cut out, and full voltage is applied to the motor terminals. This type of starter limits the current drawn from the line approximately in proportion to the square of the applied voltage. For example, a motor with a starting current of $5.5 \times$ its full load current would, with an autotransformer starter giving 80 per cent voltage, have drawn from the line a current of $5.5 \times (0.8)^2$ or $3.52 \times$ full load current.

Compensator starters. Compensator starters are usually made with two taps so that the starting voltage can be either 80 or 65 per cent of rated voltage. If the voltage initially applied is lowered, the initial current inrush will be lowered, but the current inrush when full voltage is applied will be increased. In using a lower reduced voltage in starting, the starting torque is further reduced, and the motor cannot attain as high a speed before it is necessary to apply full voltage. If an extremely low reduced voltage is used, the motor may not start, particularly if it drives vertical turbine pumps where a considerable amount of breakaway torque exists.

Three-step and resistance-type starters. Special three-step starters and resistance-type starters can be obtained for specialized use. These starters are more expensive and are limited to cases where the starting current must be further limited.

Part-winding starters. Part-winding starting is often used and is available from most induction and synchronous motor manufacturers. This starting method provides the advantages of closed transition from one starting step to the next and usually has a higher torque-to-kva ratio than does reactor or resistor-starting, which also provides this type of transition. The auto-transformer starter provides the best torque-to-kva ratio, but unless closed transition starting is used, a large current surge during transfer to full voltage may result.

Wound-rotor motor starters. Starters for wound-rotor motors have a primary switch and resistance for the three leads of the wound rotor. In starting, full voltage is applied to the stator with the maximum resistance cut into the wound-rotor circuit. As the motor speeds up, the resistance in the wound-rotor legs is cut out until, when all resistance is cut out, the motor is running at full speed. With the proper choice of resistance and steps, the wound-rotor motor can be started with very little in-

Fig. 20.21 Magnetic full-voltage starter with magnet to close contacts

Heater elements cause overload relays to open contacts if current becomes excessive. (Courtesy General Electric Co.)

creased current demand and is, therefore, very valuable when the power supply is limited. The resistance used in the wound-rotor circuit can be designed either for starting or for both starting and speed-regulating use.

Synchronous-motor starters

Starters for synchronous motors are basically the same as those used for squirrel cage motors except for the addition of a mechanism to apply the direct current to the motor field circuit when the motor comes up to the pull-in speed.

Manual and automatic starters

Except for the action of protective devices built into some manual starters, motors controlled by these starters are started and stopped manually. With automatic starters, motors are started or stopped either by momentary- or maintained-contact pilot devices, or manually by pushbuttons.

Protective devices

Protective devices in electric starters guard against overload and undervoltage. Heating elements are often incorporated in starters to protect against damage from overload. A sustained excess current will cause the contactors to trip out (Fig. 20.21). The mechanism must then be reset manually before the contacts can be re-engaged.

After being started, a three-phase squirrel cage motor will continue to run on one phase if the other two phases are disconnected. If the same load is carried on the motor, the current at single-phase operation will be greatly increased and the motor may be seriously damaged. Some form of overload protection should be incorporated in any motor installation.

Two methods are employed to protect the motor against low voltage. One is called undervoltage protection and is used with momentary-contact pilot devices. In case of undervoltage, the holding coil in the control releases, and the contactors are opened. The contactors will not re-engage when full voltage is resumed, however, as the opening of the contactors has opened the holding coil circuit. The other method is called undervoltage release and is used with maintained-contact pilot devices. In this form, undervoltage causes the holding coil to release the contactors, which then open. The motor stops, but the control circuit does not open. When the voltage is reestablished, the control will close the motor line contactors and the motor will start. In some applications, undervoltage release is dangerous and undervoltage protection should be used. If the pilot device is a maintained-contact type, undervoltage protection can be used by adding relays to the control circuit.

The electrical installation should provide protection against a possible short circuit as well as providing overload and undervoltage protection for the motor. Most starters (both high- and low-voltage) have an interrupting capacity of ten times the rated motor current. However, most high voltage circuits can produce short-circuit currents far in excess of this value. For both personnel and property safety, equipment should provide adequate protection in case of a short circuit.

STARTING TORQUE CHARACTERISTICS OF CENTRIFUGAL PUMPS

The proper selection of a driver and, in the case of motors, of its control, requires consideration of the starting torque characteristics of the machine to be driven. Centrifugal pumps have favorable starting torque characteristics and rarely cause concern; however, when the pump motor is connected to a line of limited capacity, the starting torque characteristics of the pump may have to be analyzed in detail.

Definition of torque

Torque is defined as, "the moment of a system of forces that causes rotation" and is usually expressed in terms of pounds at a 1-ft radius or, more simply, in foot-pounds. (Small torques are expressed in ounce-inches or in pound-inches to avoid minute fractions.) Figure 20.22 illustrates the moment involved in a 1-lb force acting at a 1-ft radius, using as an illustration a wire wound on a 1-ft radius cylindrical drum with a 1-lb pull on the wire. Torque is independent of rotation; it is the moment tending to cause or causing the rotation. If, in Fig. 20.22, the wire were being pulled off the drum at a rate which would cause 1 revolution of the drum in 1 minute, the distance through which the 1-lb force would have acted in the minute would be equal to the circumference of the drum or $2\pi \times$ 1 ft. The work done, therefore, would have been $2\pi \times 1$ ft $\times 1$ lb or 2π ft-lb. It can be shown that:

$$hp = \frac{2\pi Tn}{33,000} = \frac{Tn}{5,250}$$

or:

$$T = \frac{5,250 \ hp}{n}$$

in which:

$T =$ torque, in pounds-feet

$n =$ speed, in revolutions per minute

$hp =$ horsepower (33,000 foot-pounds per minute).

Accelerating torque

In order to start a load and bring it up to its running speed, the driver must deliver more torque than required by the driven machine at any speed up to its normal running speed. The difference between the torque required by the load and the torque developed by the driver is called the net accelerating torque. Acceleration ceases when the torque required by the driven machine equals the torque developed by the driver. If the required torque exceeds the developed torque, the unit will slow down until the required torque again equals the torque developed by the driver. If the torque of the load at any reduced speed is

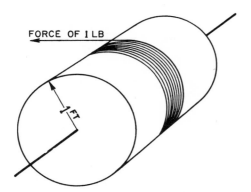

Fig. 20.22 Torque of 1 lb at 1-ft radius—1 ft-lb

greater than the torque developed by the driver, the driver will stall or stop. It is, therefore, a basic requirement that the driver torque must be greater than the torque required by the load at all speeds between rest and full speed if the driver is to start the load and accelerate it to full speed.

The time required to start a load and bring it to full speed depends upon two factors: The margin of torque available, that is, the net accelerating torque, and the mass of the rotating parts to be accelerated. This mass has a flywheel effect; the greater the flywheel effect, the longer the time required to accelerate the rotating parts with the same accelerating torque. The time required to bring a load from one speed to another can be calculated from the formula:

$$t = WK^2 \frac{(n_2 - n_1)}{308 \times T}$$

in which:

t = time, in seconds

WK^2 = flywheel effect of the rotating parts of the driver and the load expressed in foot-pounds-squared

n_1 = original speed, in revolutions per minute

n_2 = final speed, in revolutions per minute

T = average accelerating torque available, in foot-pounds.

The most adverse type of load to be accelerated, therefore, is one that requires a high torque to drive it at all speeds from rest to full speed and has a large flywheel effect, such as a reciprocating compressor. Centrifugal pumps, however, have very low WK^2 (called WR^2 in the past) values and low starting torque characteristics, the latter varying considerably with the type of pump and the method of starting. For this reason the high-starting torque motors necessary for machines with large inertia and high torque values are not required for centrifugal pump applications. There are, however, many cases where the starting current inrush of electric-motor-driven pumps must be kept to a minimum and the pump torques during the starting cycle must be known in order to make the proper motor and control selections.

Definition of terms

Motor manufacturers use the following definitions of torque terms. "Full load torque" is the turning moment exerted by a motor at rated speed and load. "Starting torque," "locked-rotor torque," or "static torque" is the minimum torque developed by the motor for any angular position of the rotor at standstill with rated voltage and frequency applied to the motor terminal. "Pull-up torque" of an a-c motor is the minimum torque it develops during the period of acceleration from rest to full speed with rated voltage and frequency.

"Pull-out torque," "maximum running torque," or "break-down torque" of an a-c motor is the maximum torque an induction motor will develop with rated voltage and frequency, without an abrupt drop in speed, or that torque a synchronous motor can sustain at synchronous speed.

"Pull-in torque" for a synchronous motor is the maximum constant torque under which the motor will pull its connected inertia load into synchronism at rated voltage and frequency, when field excitation is applied.

Electrical manufacturers usually express the various torque values of a motor as percentages of full load torques. Knowing the horsepower and speed, these can be easily converted to foot-pounds, if necessary. Torque values are also given on the basis of rated voltage and, in case of a-c motors, on the basis of rated frequency being supplied to the motor. If reduced voltage is used for starting, the torques are reduced proportionally to the square of the applied voltage. Thus, 0.8 or 80 per cent applied voltage reduces the developed torques to 0.64 or 64 per cent of their full voltage values. There is one exception to torque varying as the square of the applied voltage: The pull-out torque of a synchronous motor varies directly with the applied voltage.

Analysis of torque

Theoretically, any torque, no matter how small, would start a centrifugal pump from rest; in practice this is not quite true because of the friction in the bearings and stuffing boxes. In the case of the former, specifically in the case of babbitt sleeve bearings, the oil between the shaft and the bearing is squeezed out, resulting in metal-to-metal contact. In stuffing boxes, the packing tends to grip the shaft when at rest. The magnitude of this breakaway torque varies

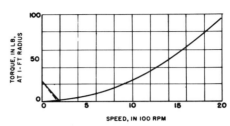

Fig. 20.24   Torque curve of pump of Fig. 20.23, when started against closed gate valve

with the design of the pump bearings, the type and condition of the packing, the tightness of the gland, and the length of time that the pump has been shut down. The torque required to start from rest can be safely taken to be less than 20 per cent of the torque required to drive the pump at design conditions. Usually 15 per cent is used for horizontal sleeve bearing pumps and 10 per cent for ball bearing pumps. It is also impossible to make any general rule as to how quickly the adverse conditions causing this breakaway torque disappear when the pump begins to rotate. It is safe to assume that the breakaway torque decreases as the speed increases, until it reaches practically zero at 10 to 20 per cent speed.

Low specific speed pumps

Most centrifugal pump applications involve relatively low specific speed pumps, which are usually started against a closed

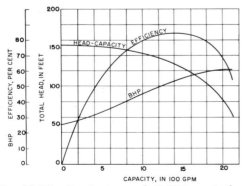

Fig. 20.23   Constant speed (1,770 rpm) characteristics of 6-in. discharge, 8-in. suction (6 x 8) double-suction pump

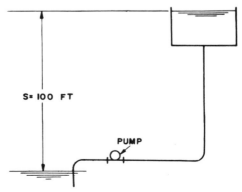

Fig. 20.25   Application for pump characterized by Fig. 20.23

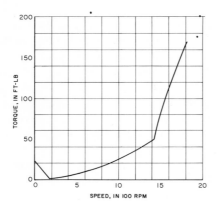

Fig. 20.28 Torque characteristics of 6 x 8 pump when starting against check valve in system shown in Fig. 20.25

the same operating conditions (1,400 gpm, 129 ft total head) as described previously, and the pump is started against a quick-opening check valve with the gate valve open, flow would start immediately. For all practical purposes, again ignoring the acceleration of the water column, the flow would be directly proportional to the speed, the head would be proportional to the square of the speed, the power would be proportional to the cube of the speed, and the torque would be proportional to the square of the speed. On that basis the torque-speed curve would be as shown in Fig. 20.29.

Using this same method of analysis, approximate torque-speed curves can be es-

tablished for any centrifugal pump if the pump characteristics and method of starting are known. Figure 20.30 shows the characteristics of a medium specific speed pump, which requires the same horsepower at shutoff and at the point of maximum efficiency. The speed-torque curve for this pump will be practically the same whether it was started against a closed gate valve or against a check valve. In the latter case it would make little difference whether the system was all static, all friction, or part static and part friction (Fig. 20.31). Curve A shows the torque resulting when starting against a closed gate valve and also when starting against a check valve in an entirely frictional system that would have a head of 25 ft at 6,370-gpm flow. If the pump were operated against a system consisting solely of a 25-ft static head, the torque-speed curve would be like curve B. For any system with combined friction and static heads, the torque-speed curve would be some curve falling between curves A and B.

Propeller-type pumps

The propeller-type pump now used extensively for low-head applications is a high specific speed type and may have a high-speed mixed-flow or true-axial-flow impeller. The typical characteristics of one of these designs with an axial-flow impeller are shown in Fig. 20.26. The shutoff horsepower is approximately 2.5 times that re-

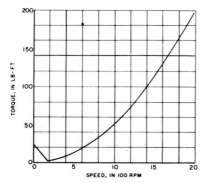

Fig. 20.29 Torque characteristics of 6 x 8 pump when started in wholly frictional system

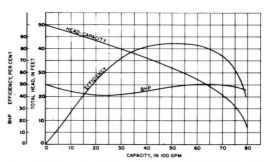

Fig. 20.30 Constant speed characteristics of 16-in. volute pump with mixed-flow type impeller requiring same power at shutoff as at reduced heads

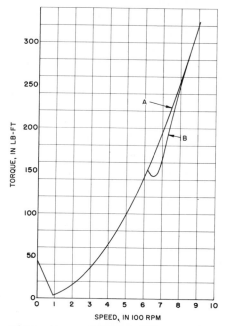

Fig. 20.31 Torque characteristics of 16-in. volute pump shown in Fig. 20.30

Started against closed gate valve or wholly frictional system (A) and started against theoretical 25-ft static-head system with no friction component (B).

tion well. Assuming that one pump was arranged to start when the suction level reached elevation 7.0 ft, that pump, if maximum discharge water level prevailed, would operate against a 14-ft static head.

Two possibilities exist with such an installation: The pump might be empty above elevation 7.0 ft, or, if the pump had been in service only shortly before, it might be completely filled with water. If the pump is filled, the impeller, when it is started, will not deliver any water until after it reaches a speed at which the shutoff head exceeds the static head. The resulting torque curve would be as indicated in Fig. 20.33. If the pump above elevation 7.0 ft is empty, the impeller will immediately begin to fill the column pipe with water. Without a detailed study of the time necessary for the motor to reach various speeds, it would be difficult to predict the torque-speed curve of the pump. The curve would, however, have torque values lower than those shown in Fig. 20.33 until the column pipe, discharge elbow, and discharge pipe were

quired at maximum efficiency. Whereas Fig. 20.26 only shows a flat portion in the head-capacity characteristics, many of these axial-flow designs, especially those over 9,000 specific speed, have an unstable portion in their head-capacity curves. For these two reasons, pumps of this type are applied where the maximum head will be below any head in the unstable range, and the pumps are either installed on a siphon system or started against a check or flap valve. This propeller-type pump might be one of several installed as shown in Fig. 20.32 with maximum discharge level at elevation 21.0 ft and minimum suction level at elevation 5.0 ft. The pumps would probably be controlled by float switches, which would stop all pumps when the suction level was pumped down to elevation 5.0 ft. The various pumps would normally be started at different water levels in the suc-

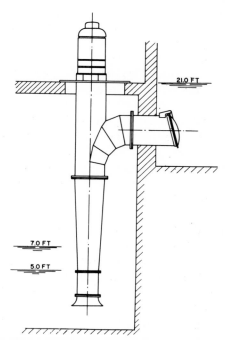

Fig. 20.32 Installation of 30-in. propeller pump of Fig. 20.26 showing water levels

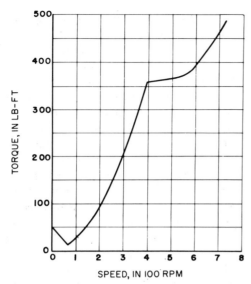

Fig. 20.33 Torque characteristics of 30-in. propeller pump when starting against 14-ft static head with pump full of water

filled, at which time the curves would be identical.

The high-specific-speed, axial-flow pump has the most adverse starting torque characteristics of any centrifugal pump type but offers no serious problem for any normal type of drive when starting against a check or flap valve with a relatively short discharge line. This type of pump, however, can offer a more serious problem when installed in a siphon system in which the establishment of the siphon depends upon the flow washing out the entrapped air. If the pump described in Fig. 20.26 and 20.32 has its discharge higher than elevation 18.0 ft with a siphon discharge, the pump column will fill until water begins flowing over the low point of the siphon as the pump speeds up in the starting period, and, as pumping continues, with a proper design of siphon, the air will be washed out, the siphon established and the head will drop to that of the system. Here again the required pump torque at various speeds would be tied in with the time that the motor would require to reach the various speeds. The pump torque would be affected

by the time, which affects the water level in the column, the operating head, and, therefore, the pump torque. With a relatively high siphon leg, it is possible to have a torque curve rising to a peak at a speed less than rated speed. In an extreme case with a very short starting period, the required pump torque when the unit reaches rated speed could be greater than at rated conditions because flow would not have been established. Such an unusual case would require the combined study of the pump and motor torques in order to establish the approximate starting torque of the pump.

Vertical turbine pumps of the type used in deep-well and open-pit, high-head installations usually have impellers of the mixed-flow type and are built in a number of stages. Some designs have shutoff horsepower lower than required at rated capacity, whereas a few designs have shutoff horsepower greater than required at rated capacity.

Cone-type stop valves

Many installations employ cone-type stop valves, which function as a combined check and gate valve. In practically all installations of motor-driven pumps, the time required for the motor to bring the unit to full speed after the shutoff head reaches a value equal to the head on the discharge side of the valve is very short. The travel of the valve is so small during this interval that the starting of the unit is, in effect, the same as against a closed gate valve. This same effect could result, in whole or in part, with other types of slow-opening check valves. In such installations, however, it is best practice to select and analyze the starting cycle as for an instantaneously opening check valve and for no inertia effect in the hydraulic system. Systems in which propeller pumps are used should employ quick-opening check valves or flap valves or have a siphon discharge.

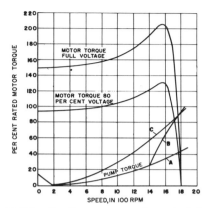

Fig. 20.34 Starting torque characteristics of double-suction pump of approximately 1,700 specific speed

Three different applications compared against torque characteristics of a normal torque squirrel cage induction motor for both 100 per cent and 80 per cent voltage.

Specialized applications

Centrifugal pumps require a very low torque to start from rest, and the torque required to drive them does not reach a maximum value until maximum speed is attained. Situations in which the starting torque characteristics of a centrifugal pump is of any concern, therefore, are rarely encountered. The situations in which starting torque characteristics will be of concern are as follows: (1) When the limited size of the transmission line supplying a motor driving a pump makes a detailed study of the electrical installation necessary to select the proper type of motor and motor control, and (2) when synchronous motors are to be used, and the pull-in torque of the motor has to be compared with the torque that will be required by the pump.

Squirrel cage motors. Normally, a squirrel cage motor develops much greater torque at lower speeds than required by a pump. As the pump has a small WK^2 value (flywheel effect in foot-pounds-squared), the time required for the unit to reach full speed is very short. For the low specific speed pump discussed earlier (but for 1,755-rpm rated speed), Fig. 20.34 shows the start-

ing torque characteristics of the pump: (1) Curve A for starting against a closed gate valve, (2) curve B for starting against a check valve on a system having a static head component, and (3) curve C for starting against a check valve on a system involving all friction head. All the torques are shown in percentages of the rated motor torque. Figure 20.34 also shows the torque developed by a 60-hp, 1,800-rpm squirrel cage motor plotted on the same basis both for 100 per cent and 80 per cent voltage. The excess torque produced by the motor over that required to drive the pump at the lower speeds is very apparent.

Synchronous motors. In the case of synchronous motors, the starting characteristics are dependent upon auxiliary windings that are built so they will bring the motor and its connected load up to 95 per cent rated speed, at which speed the load is transferred to the regular windings.

The pull-in torque that must be developed by a synchronous motor driving a pump is, therefore, the torque required by the pump at 95 per cent speed. Using the same illustration of the low specific speed pump as in Fig. 20.34, but on the basis of 1,800 rpm rated speed instead of 1,755 rpm, the starting torque characteristics from 70 to 100 per cent rated speed would be as shown in Fig. 20.35. At 95 per cent speed, curve A indicates 39 per cent torque, curve

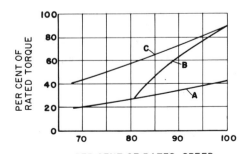

PER CENT OF RATED SPEED

Fig. 20.35 Partial curve of starting torque characteristics of double-suction pump of approximately 1,700 specific speed

Showing torque requirements at 95 per cent speed for three different applications.

B indicates 77 per cent torque, and curve C indicates 81 per cent torque needed to drive the pump. The synchronous motor used to drive this pump must, therefore, be good for at least 39 per cent pull-in torque if the pump is to be started against a closed gate valve (curve A), at least 77 per cent pull-in torque if the pump is to be started against a check valve on a system involving a 100-ft static head (curve B), and at least 81 per cent pull-in torque if the pump is to be started against a check valve on a wholly frictional system (curve C). Some additional leeway would be allowed as the pump might produce a slight excess over the rated conditions.

Check valves. Occasionally the check valve of a centrifugal pump will fail to close when the unit is shut down. When the system on which it operates has a static head component, or when other pumps are delivering into the same discharge header, water will flow back through the pump, causing it to operate as a water turbine. The pump will then run in the reverse direction unless prevented. The speed at which it will run will be the speed at which the torque developed as a water turbine equals the torque required to rotate the driver at that reverse speed. As electric motors require very little torque to drive, the speed reached with motor-driven units will approach the runaway speed of the pump when operating as a water turbine at the net head available under the resulting system conditions. With a net head equal to the design or rated head as a pump, this runaway speed will vary with both the specific speed type and the individual pump design. It might·be expected to be between 100 and 150 per cent of rated pump speed.

In some installations, such as in a water works pumping plant with a number of other units in service, the net head available to drive the pump as a water turbine will be equal to its rated head as a pump. In other installations, such as a single pump delivering through a pipeline to a reservoir, the pumping head is the static head

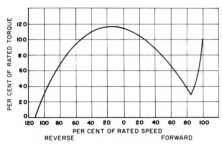

Fig. 20.36 Torque characteristics of double-suction pump of approximately 1,700 specific speed, from runaway speed in reverse direction to rated speed in forward direction

Head measured from suction to discharge being equal at all timcs to rated total head as a pump.

plus friction head loss, whereas under reverse flow, the net head causing the pump to run as a water turbine would be the static head minus the friction loss at the resulting capacity. In cases where the net head is less than the pumping head, the runaway speed will be reduced.

When reverse flow occurs through a motor-driven pump, because of failure of a check valve to close or other reasons, it is generally not feasible to apply power to the motor and again bring it up to speed. While the motor usually has sufficient torque (with full voltage) to bring it up to speed, the reverse torque developed by the pump will prolong the time required to bring it back to full speed in the forward direction. A high current demand will prevail during this period, causing the overload device in the starter to function, thereby kicking out the starter. Figure 20.36 shows the high torque that must be overcome by the motor to carry one specific design of pump from runaway speed in the reverse direction to full speed in the forward direction.

Comparison of pump and motor torques

The torque characteristics of a 60-hp, 1,800-rpm normal-starting-torque, squirrel cage motor for full voltage and reduced voltage (80 per cent tap of the autotransformer) is shown in Fig. 20.37 together with the pump torque curve from Fig. 20.24. With reduced voltage, the initial current inrush is reduced to 64 per cent of the full voltage inrush, but it is still possible for the motor to reach 1,770 rpm before the torque of the motor equals the pump torque. The point at which full voltage is applied would depend upon the timing of the starting cycle. In this case, if full voltage is applied any time after the pump reaches 1,340 rpm, the second inrush as full voltage is applied would not exceed the initial inrush. With automatic reduced voltage starters, the application of full voltage is controlled by timing relays. In most cases the timing relays are set so that the motor reaches the maximum speed possible on reduced volt-

age (1,770 rpm in the example) before full voltage is applied. This is an advantage because in order to apply full voltage, the motor has to be switched from the reduced voltage to full voltage. As a result, the current drops to zero and a second inrush, the magnitude of which depends upon the motor speed, results when full voltage is applied. In the case illustrated in Fig. 20.37, therefore, the second inrush, if full voltage is applied at 1,000 rpm, will not be an increase in demand of 188 per cent (490 per cent less 312 per cent) but the full 490 per cent.

If, instead of starting the pump against a closed gate valve, it is started against a check valve in a wholly frictional system with a resulting torque curve as in Fig. 20.29, the reduced voltage would be able to accelerate the unit until 1,730 rpm is reached. The torque demand of this pump, when started in this manner, is of the same magnitude as the torque demand of a high-speed Francis-vane impeller or of a medium-speed, mixed-flow impeller starting against either a gate valve or check valve (Fig. 20.31), and is close to the same magnitude as a high-speed, mixed-flow or propeller pump (Fig. 20.33). Centrifugal pumps of other specific speed types, therefore, offer starting problems only slightly more difficult than those described above.

It is very unusual to start a squirrel cage motor driving a centrifugal pump on the 65 per cent voltage tap of a reduced voltage autotransformer starter. The use of such a low reduced voltage would be feasible only with a pump having a shutoff horsepower that is considerably less than the horsepower required at rated capacity and which is always to be started against a closed gate valve (Fig. 20.23 and 20.24). With 65 per cent voltage, the 60-hp motor illustrated above would accelerate this pump to approximately 1,700 rpm, resulting in an inrush of 150 per cent current when full voltage was applied. In the case shown in Fig. 20.29, starting the motor on 65 per cent voltage would only allow it to

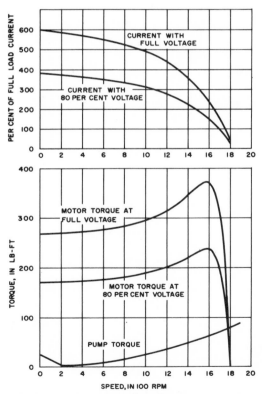

Fig. 20.37 Torque-speed and per cent current-speed curves of 60-hp, 1,800-rpm, 3-phase, 60-cycle squirrel cage motor both for full and 80 per cent reduced voltage

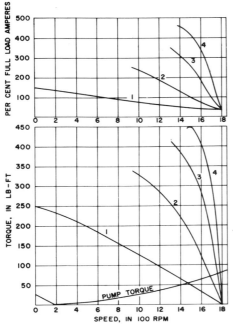

Fig. 20.38 Torque-speed and per cent current-speed curves of 60-hp, 1,800-rpm, 3-phase, 60-cycle wound-rotor motor with four-step starter

accelerate the pump to about 1,200 rpm with a resulting second current inrush of 440 per cent when full voltage was applied.

As previously described, the starting of a wound-rotor motor involves applying current to the stator winding with resistance connected into the rotor windings. The amount of resistance cut into the rotor winding and the amounts of resistance cut out in each successive step in the starting cycle determine the amount of torque available at various speeds as well as the current demand. The number of steps used will depend upon the limit of current allowed. Figure 20.38 illustrates the torque and current characteristics of a 60-hp, 1,800-rpm slip-ring motor with a four-step starter proportioned for the pump torque requirements shown on the same figure. The first step results in a 249-ft-lb torque at zero speed with 150 per cent of rated current. This first speed allows the motor to accelerate the pump to 1,465 rpm. If the second step is cut in when that speed is reached, the motor torque would become 194 ft-lb with the current stepping up to 106 per cent. This step would result in accelerating the pump to 1,685 rpm, when the third step would be cut in with a 193-ft-lb torque and 108 per cent current. The third step would accelerate the pump to 1,775 rpm, whereupon the fourth step (no resistance in the rotor circuit) would be cut in, resulting in a 180-ft-lb torque and 100 per cent current. An ideal starter is not always obtainable and it may be necessary to purchase a starter with more steps than theoretically required to meet a limitation.

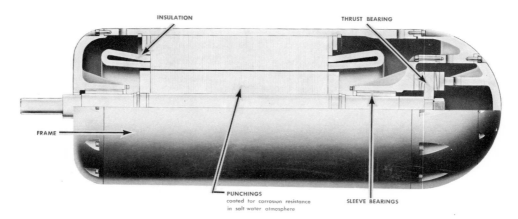

Fig. 20.39 Wet-stator motor

(*Courtesy General Electric Co.*)

The modern synchronous motor has an auxiliary squirrel cage motor built into it to start the motor and bring it and its connected load up to about 95 per cent of the synchronous speed. To pull the motor into step, the load has to be transferred from the squirrel cage element to the regular synchronous motor element. With present-day controls, which incorporate proper timing devices, this no longer requires expert handling. Except for some differences in relative values, the starting torque and starting characteristics of a synchronous motor are similar to those illustrated in Fig. 20.37 for a regular squirrel cage motor.

SUBMERSIBLE MOTORS

The submersible motor was first developed a number of years ago to eliminate some of the disadvantages of the vertical turbine pump when installed at great depths. The main advantages of a submersible pump and motor unit are the following:

1. Elimination of long line shafts and numerous bearings
2. Elimination of a pump house for the protection of motors located at the surface
3. Freedom from motor noise in applications where noise from a surface unit might be objectionable
4. Ease of installation
5. Freedom from maintenance, periodic lubrication, and adjustment.

The first attempts to provide a motor that could be close-coupled to the pump and installed at the pumping level involved the use of air bells, in which the motor housing was kept filled with air under pressure, while the bottom of the housing was open to the water. This construction was only partially successful; eventually, submersible motors were developed that required connection only to an electric conduit line, an air supply being unnecessary.

So many different submersible motor designs exist that it is possible to describe only several of the major variations.

Wet-stator types

In the wet-stator design, no attempt is made to prevent the pumped liquid from contacting the motor windings. The conductors of the stator windings are coated

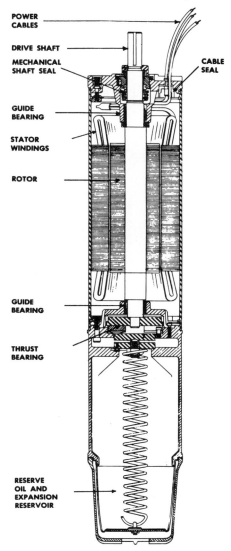

Fig. 20.40 Oil-filled submersible motor

(*Courtesy General Electric Co.*)

Fig. 20.41   Installation of vertical turbine pump with submersible motor

Figure 20.39 shows a submerged drive motor application. For pumps, the motor is manufactured integrally with the pump and both are encased in a shell. Wet-stator motors are available in sizes up to 1,000 hp, and special designs can be obtained in larger sizes. Because the majority of these motors are intended for well or borehole service, submersible motors are very long and thin, often with a 5 to 1 ratio of rotor length to diameter. As a result, submersible motors have lower efficiencies and power factors than normal motors. On the other hand, their small diameter results in lower inertia and a much higher rate of acceleration to full speed.

Some wet-stator submersible motors exclude the liquid pumped from the motor housing by filling the housing with a refined nonhygroscopic dielectric mineral oil, which protects the insulation against breakdown and the internal ferrous parts against corrosion. This oil also provides a superior lubrication to the motor bearings and reduces the wear and maintenance of these bearings in comparison with constructions using water-lubricated bearings. This type of motor (Fig. 20.40) is provided with a mechanical shaft seal that prevents the oil from leaking out and the water and foreign matter from entering.

The installation of a vertical turbine pump with a submersible motor is illustrated in Fig. 20.41. The motor is located at the bottom of the pumping unit. The complete unit is suspended at the upper pump flange and lowered into the well.

with plastic insulation impervious to water, such as polyvinylchloride or polyethylene. Figure 20.39 shows the cross section of a wet-stator motor. The fluid in which the motor operates lubricates the motor bearings and cools the motor.

The motor parts in contact with water, with the exception of the rotor and stator laminations, are generally made of noncorrosive materials. The laminations are made of carbon steel, since they must have magnetic properties and are usually coated for corrosion resistance.

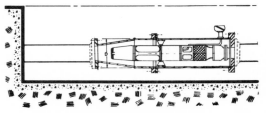

Fig. 20.42   Horizontal pipeline booster pump with submersible motor

(Courtesy Layne & Bowler.)

Consecutive lengths of piping are added until the unit is lowered to the required depth. As each length of pipe is added, the power cable is paid out and clipped to the piping.

One variation of the submersible pump is illustrated in Fig. 20.42. The pump and motor are capable of operating in a horizontal position. In pipeline booster service, therefore, the unit can be suspended horizontally within the pipeline (Fig. 20.43). The entire unit is manufactured with flanges at each end and becomes an integral part of the line.

Dry-stator types

In the dry-stator design, the stator windings are hermetically sealed against the pumped liquid by means of a thin stainless steel tube along the stator bore, sealed at each end. The rotor windings are enclosed within a corrosion-resistant can, hence their name, canned-rotor motors (Fig. 20.44).

The canned-rotor motor was developed primarily for boiler circulating service or as a driver for nuclear reactor coolant pumps. The operating pressure for the pump may reach 2,400 psi and the temperature as much as 650°F. The unit requires neither external seals, which would

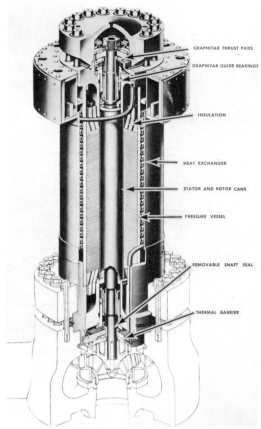

Fig. 20.44 Canned-rotor motor pump

(Courtesy General Electric Co.)

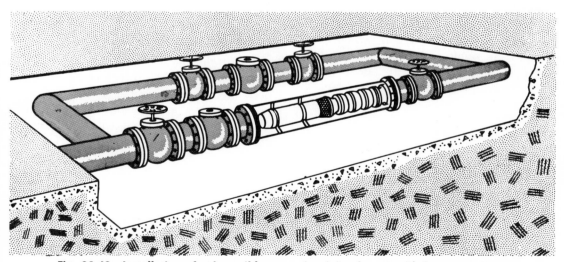

Fig. 20.43 Installation of submersible pump in piping layout with bypass system

(Courtesy Layne & Bowler.)

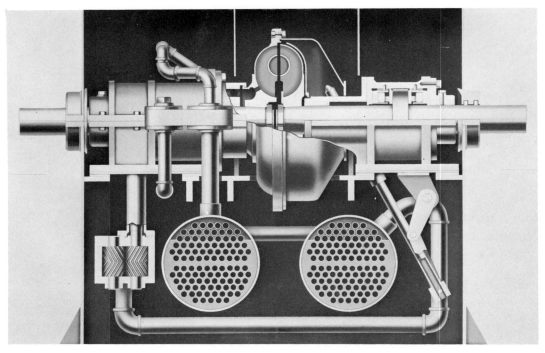

Fig. 20.45   Hydraulic coupling

(Courtesy American-Standard, Industrial Division.)

be difficult to maintain under these high pressures and temperatures, nor cold water injection at a pressure in excess of the internal pump pressure. A thermal barrier limits heat flow from the pump interior to the motor portion of the pump. In addition, a labyrinth shaft seal is incorporated between the pump and the motor to reduce the heat flow from the high-temperature pumped liquid. External cooling water is pumped to an integral heat exchanger, which cools the internal motor water. A small integral pump circulates this motor water for cooling and bearing lubrication.

VARIABLE SPEED DEVICES

As mentioned earlier in this chapter, many installations could benefit from the variable speed operation of the centrifugal pump, but their drivers (squirrel cage elec-

tric motors, for example) can only operate at constant speeds. One device frequently used in such cases to allow variable speed operation is the hydraulic coupling or fluid drive.

Hydraulic couplings

All of the many mechanical designs of hydraulic couplings are composed of four basic elements: An impeller, a runner, a casing, and a means for changing the amount of oil contained within the unit during operation so as to obtain an adjustable speed output while the speed input remains constant.

The impeller and the runner are mounted facing each other on two separate shafts with no mechanical connection between the two. The power is transmitted from the impeller to the runner entirely by a vortex of oil circulating between the two. The impeller, mounted on the driven shaft

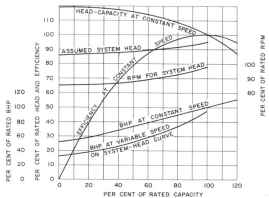

Fig. 20.46 Performance curve of variable speed centrifugal boiler feed pump

that is connected to the driver, acts as a centrifugal pump and imparts energy to the oil, which is accelerated radially outward. The runner is mounted on a shaft connected to the driven equipment and acts as a reaction turbine, absorbing the kinetic energy in the oil as it passes from the periphery inwardly towards the center of rotation. The circulation of oil between the impeller and runner is called the working circuit. A casing is attached to the impeller to confine most of the oil to the working circuit. An adjustable scoop tube can withdraw a variable amount of oil from the working circuit to provide speed variation.

The construction of a hydraulic coupling is illustrated in Fig. 20.45. The entire rotating assembly is enclosed in a welded steel

Fig. 20.47   Variable speed magnetic drive

(Courtesy Electric Machinery Mfg. Co.)

Fig. 20.48 Variable output pump with variable speed magnetic drive

housing, which is split on the horizontal centerline. The bottom portion of this housing serves as a support for the bearings in which both the driving and driven shafts rotate. It also contains an oil reservoir from which the lubricating and operating oil is supplied to the unit. A centrifugal pump, gear-driven from the input shaft, picks up oil from this reservoir and sends it through an oil cooler, from which it returns to the tank as well as to the bearings.

The oil in the working circuit rotates in an annular ring by virtue of the centrifugal force imparted to it by the impeller. Connections are provided between the working circuit and the scoop tube chamber. Therefore, the amount of the oil in the annular

ring can be varied by varying the oil level in the scoop tube chamber. When the tip of the oil scoop tube is extended towards the periphery, the amount of oil in the working circuit is reduced and the output speed decreases. Withdrawing the scoop tube towards the center permits the working circuit to fill with oil and the slip decreases, increasing the output speed.

The hydraulic coupling is ideally suited for variable speed operation of variable torque machines such as centrifugal pumps, because the pump horsepower varies with the cube of the speed, whereas the hydraulic coupling efficiency is almost inversely proportional to the speed. The power input to the hydraulic coupling can

TABLE 20.4 CALCULATION OF INPUT BRAKE HORSE-POWER

Pump capacity, in per cent	Pump brake horsepower at variable speed, in per cent of rated power	Pump speed, in per cent of rated speed	Input brake horsepower, in per cent of rated power
0	32	85	40
20	38	85.2	47
40	46	86.7	55.7
60	58	89	68
80	74	93	83
100	95	98	101

be calculated from the relation: Input brake horsepower = (pump bhp at variable speed)(input speed/pump speed) + fixed losses.

With the working circuit of oil there is a slight slip which may vary from 1 to 3 per cent, depending on the power output and the size of the hydraulic coupling. The pump speed, therefore, will be from 99 to 97 per cent of the driver speed. The fixed losses are the bearing and windage losses and run to approximately 1 per cent.

The effect of using a hydraulic coupling with a centrifugal pump is illustrated in Fig. 20.46, which shows a typical performance of a boiler feed pump with the capacity, head, efficiency, power, and speed expressed in percentages of their values at the design-rated condition. The system-head curve has been superimposed on this figure, and the required rpm for the system-head curve and the pump brake horsepower

Fig. 20.49 Vertical motor and magnetic drive with vertical centrifugal pump

(Courtesy Electric Machinery Mfg. Co.)

when operated at variable speed are also shown. The system-head curve at 100 per cent rated flow is only 95 per cent of the rated head, reflecting the margin of safety that might be added to the pump head in the calculations.

If the hydraulic coupling has a slip of 3 per cent at rated conditions and the fixed losses are 1 per cent, the motor speed will, therefore, be approximately 103 per cent of the pump rated speed. The input brake horsepower, when operating at variable speed is calculated as shown in Table 20.4.

A comparison of the power input listed above with the pump brake horsepower at the same flows, when operated at constant speed, shows the advantages of using hydraulic coupling. For example, at 60 per cent flow, the power consumption is 79 per cent at constant speed. It is only 58 per cent at variable speed, and, including the hydraulic coupling losses, the input horsepower is only 68 per cent.

Magnetic drives

One of the different designs of variable speed magnetic drives is illustrated in Fig. 20.47. It consists essentially of two rotating members: One, called the ring, is driven by the driving motor, the other, called the magnet, is coupled to the driven machine.

The ring revolves at the same speed as the driving motor; the magnet is free to revolve within the ring, and is separated from the ring by an air gap. The magnet poles are energized (or excited) by direct current through collector rings in the shaft.

The difference in speed between the ring and the magnet causes the ring to cut the magnetic flux produced by the magnet, which induces currents in the ring. The induced currents produce poles in the ring, which react with and drag around the poles of the magnet, thus causing the magnet to revolve.

The torque is transmitted magnetically through the air gap between the ring and the magnet and is varied gradually by varying the excitation of the magnet, thereby varying the speed of rotation of the magnet. A magnetic amplifier rectifier supplies the excitation, providing automatic speed control.

Magnetic drives are available in several forms of mechanical construction to fit various methods of mounting and for different conditions of alignment. For centrifugal pump drive, the most common forms are the self-contained unit, with either pedestal-type or bracket-type bearings; they are connected to the motor and to the pump by flexible couplings at both ends as in Fig. 20.48. Other forms are available for vertical pump drive, as in Fig. 20.49, which shows a vertical squirrel cage induction motor and an adjustable speed magnetic drive operating a vertical centrifugal pump. The motor is rated 100 hp, 1,180 rpm. The magnetic drive provides adjustable speed from 1,180 to 350 rpm.

The losses in a magnetic drive are essentially similar to the losses of a hydraulic coupling, namely, slip losses and fixed power losses. The input to the drive, therefore, is basically the same as if the variable speed was obtained by means of a hydraulic coupling.

21 *Priming*

A centrifugal pump is primed when the waterways of the pump are filled with the liquid to be pumped. The liquid replaces the air, gas, or vapor in the waterways. Removal of the air, gas, or vapor may be done manually or automatically, depending upon the type of equipment and controls used. To one familiar with positive displacement pumps of the reciprocating and rotary types only, it might seem strange that a centrifugal pump cannot prime itself. Positive displacement pumps, if properly sealed, will pump air as well as liquid and will, therefore, exhaust any air in the suction line. Centrifugal pumps will also pump air (contrary to commonly held opinion), but because of the low density of air the actual pressure developed when pumping it is very small.

For example, a pump that develops a 200-ft head with water would likewise develop a 200-ft head with air, but a head of 200 ft of air is equivalent to a vacuum of approximately 3 in., measured in terms of a water column. Such a vacuum is obviously insufficient for normal priming.

When first put in service, the waterways of a centrifugal pump are filled with air. If the suction supply is above atmospheric pressure, this air will be trapped in the pump and compressed somewhat when the suction valve is opened. Unless the suction pressure is sufficiently high, however, the air will not be compressed enough to permit the suction waterways and the eye of the impeller to be filled with water, and the pump will not be primed; therefore, the air must be removed. With a positive suction head on the pump, priming is accomplished by venting the entrapped air out of the pump through a valve provided for the purpose.

If the pump takes its suction from a supply located below the pump itself, the air in the pump must be evacuated, either by some vacuum-producing device, or by providing a foot valve in the suction line so that the pump and suction piping can be filled with water, or by providing a priming chamber in the suction line.

Foot valves

Foot valves were very commonly used in early installations of centrifugal pumps. Except for certain applications, their use is now much less common. A foot valve (Fig. 21.1) is a form of check valve installed at the bottom or foot of a suction line. Like an ordinary check valve, it allows flow in

315

one direction only—towards the pump. When the pump is stopped and the ports of the foot valve close, if the valve seats tightly, the water cannot drain back to the suction well.

Unfortunately a foot valve does not always seat tightly, and the pump occasionally loses its prime. The rate of leakage is generally small, however, and it is feasible to restore the pump to service by filling and starting it promptly. This tendency to malfunction is increased if the water contains small particles of foreign matter like sand, and foot valves should not be used for such service. Another disadvantage of foot valves is their unusually high friction loss. To keep friction loss within reasonable limits, a foot valve with a large port area is necessary. Unfortunately, the size of the pipe connection is not a true indication of the area of the valve ports.

A typical installation of a pump with foot valve is shown in Fig. 21.2. The pump can be filled with water through a funnel attached to the priming connection or from an overhead tank or any other source of water. If a check valve is used on the pump and the discharge line remains full of water, a small bypass around the valve permits the water in the discharge line to be used for repriming the pump when the foot valve has leaked. When a pump equipped with a foot valve is primed, provision must be made for filling all the waterways and for venting out the air. The actual point at which water is introduced into the waterways is of no significance but is usually at the priming connection provided on the top of the pump casing (Fig. 21.2).

When the discharge head is low, the discharge check valve is often eliminated and the foot valve acts as a check valve to prevent reverse flow through the pump. As most foot valves are designed for relatively low pressures, such use is not generally practicable with systems having a high discharge pressure. In such an installation, a discharge check valve is necessary, but there

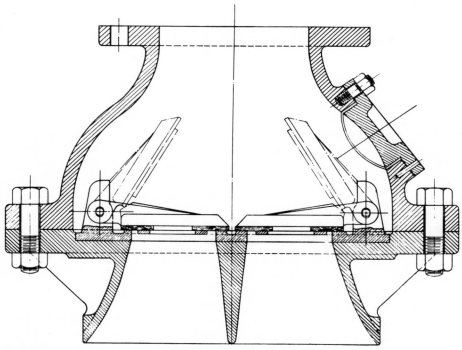

Fig. 21.1   Foot valve

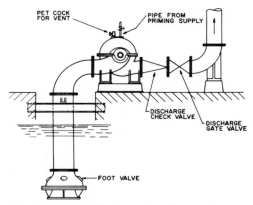

Fig. 21.2   Installation using foot valve

is always the danger that the check valve will not seat as tightly as the foot valve. The foot valve would then be subject to the static discharge pressure. In a system in which the water can drain out of the pump through the discharge line when the pump is shut down, it is necessary to have a discharge gate valve. This valve must be closed before the pump is stopped and opened after it is started.

Priming chambers

A priming chamber in its simplest form is a tank with an outlet at the bottom that is level with the pump suction nozzle and directly connected to it. An inlet located at the top of the tank connects with the suction line (Fig. 21.3). The size of the tank must be such that the volume contained between the top of the outlet and the bottom of the inlet is approximately three times the volume of the suction pipe. Leakage of air when the pump is shut down may cause the liquid in the suction line to leak out, but the liquid in the tank below the suction inlet cannot run back to the supply. When the centrifugal pump is started, it will pump this entrapped liquid out of the priming chamber, creating a vacuum in the tank. The atmospheric pressure on the supply will force the liquid up the suction line into the priming chamber.

Precautions must be taken to prevent vortexing at the outlet of the chamber and to prevent siphoning when the unit is shut down. It is usually more advantageous to buy a commercial priming chamber with proper automatic vents and other features than to attempt to make one's own design. The use of priming chambers is restricted because of their size to relatively small pumps. A well-known design of a priming chamber with automatic features is shown in Fig. 21.4. This type of priming chamber is more satisfactory for most installations than a simple tank (Fig. 21.3).

Types of vacuum devices

Almost every commercially made vacuum-producing device can be used with systems in which pumps are primed by evacuating the air. Formerly, water- and steam-jet primers had wide application, but with the increase in the use of electricity as a power source motor-driven vacuum pumps have become popular. The wet-type vacuum

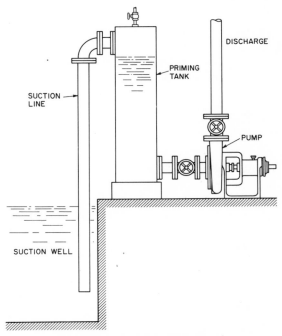

Fig. 21.3   Simple priming tank

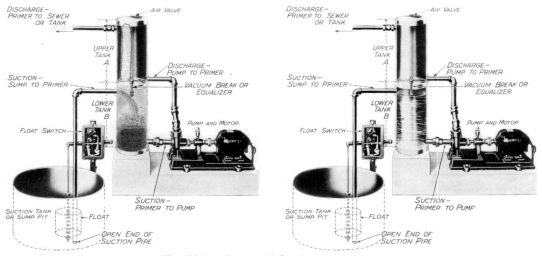

Fig. 21.4   Commercial priming tank

When the pump starts (left), it draws liquid from the lower tank and discharges it through the upper tank. Withdrawal of liquid causes a partial vacuum in the lower tank, and liquid flows from the well into the tank and then to the pump. When pump stops (right), liquid from upper tank runs back into pump and lower tank, by gravity, keeping the pump ready for starting. (Courtesy Valve and Primer Corp.)

pumps are best for manually controlled, electrically driven units because no damage will result if slugs of water are carried over into the vacuum pump. When dry vacuum pumps are employed, some protective device must be interposed between the centrifugal pump and the vacuum pump to prevent water entering the vacuum pump. The dry vacuum pump is used extensively for central priming systems.

Ejectors

Priming ejectors work on the jet principle, using steam, compressed air, or water as the operating medium. In steam or compressed-air ejectors, the steam or air is discharged at high velocity into the throat of a venturi tube carrying with it some of the surrounding air and forcing the combined mixture out of the tail piece of the venturi (Fig. 21.5). A high vacuum can be produced by both steam and water ejectors. Water ejectors are similar to steam ejectors except that multijet nozzles are used when the actuating water pressure is low.

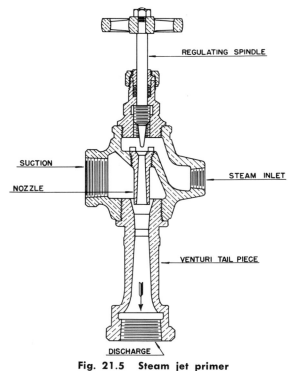

Fig. 21.5   Steam jet primer

(Courtesy Schutte & Koerting Co.)

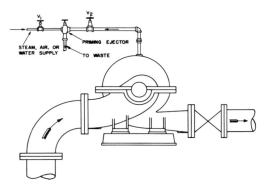

Fig. 21.6 Arrangement for priming with an ejector

A typical installation for priming with an ejector is shown in Fig. 21.6. Valve V_1 is opened to start the ejector and then valve V_2 is opened. When all the air has been exhausted, water will be drawn into, and discharged from the ejector. When this occurs, the pump is primed, and valve V_2 is closed. Valve V_1 is then closed.

An ejector may be used for a number of pumps if it is connected to a header through which the individual pumps are vented through isolating valves.

Dry vacuum pumps

Dry vacuum pumps, which may be of either reciprocating or rotary type, cannot accommodate mixtures of air and water. When they are used in priming systems, liquid must be prevented from entering them.

Wet-vacuum pumps

Any rotary, rotative, or reciprocating pump that can handle air or a mixture of air and water is classified as a wet-vacuum pump. The most common type used in priming systems is the Nash Hytor pump (Fig. 21.7 and 21.8).

This is a centrifugal displacement pump consisting of a round, multiblade rotor revolving freely in an elliptical casing that is partially filled with liquid. The curved rotor blades project radially from the hub, and form, with the side shrouds, a series of pockets and buckets around the periphery. The rotor revolves at a speed sufficient to throw the liquid out from the center by centrifugal force, resulting in a ring of liquid revolving in the casing at the same speed as the rotor, but following the elliptical shape of the casing. This condition alternately forces the liquid to enter and recede from the buckets in the rotor at high velocity.

The complete cycle of operation in a given chamber is illustrated in Fig. 21.8. At Point A the chamber (1) is filled with liquid. Because of centrifugal force, the liquid follows the casing, withdraws from the rotor, and pulls air in through the inlet port, which is connected with the pump inlet. At (2) the liquid has been thrown outward from the chamber in the rotor and has

Fig. 21.7 Partially disassembled Nash pump

(Courtesy Nash Engineering Co.)

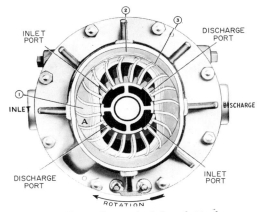

Fig. 21.8 Operating principle of Nash pump

(Courtesy Nash Engineering Co.)

Fig. 21.9 Nash priming pump with sealing liquid reservoir base

(Courtesy Nash Engineering Co.)

been replaced with air or gas. As rotation continues, the converging wall (3) of the casing forces the liquid back into the rotor chamber, compressing the air trapped in the chamber and forcing it out through the discharge port, which is connected with the pump discharge. The rotor chamber is now filled with liquid and the cycle is repeated. Each revolution includes two cycles.

If a solid stream of water circulates in this pump in place of air or an air and water mixture, the pump will not be damaged, but will require more power in order to drive it. For this reason, in automatic priming systems using this type of vacuum pump, a separating chamber or trap is provided so that water will not reach the pump.

Water needed for sealing a wet vacuum pump can be supplied from a source under pressure—a city water supply or some adjacent tank. When water is taken from a pressure supply, the amount of the flow is regulated by a throttle valve. The flow is controlled by a separate shutoff valve so

that the adjustment of the throttle valve need not be disturbed. The shutoff valve can be manually or solenoid operated. If solenoid operated, its operation is connected with the motor control.

If no water under pressure is available, or if it is desired to maintain an independent system, the vacuum pump can be mounted, as shown in Fig. 21.9, on a base containing a reservoir. The sealing water is taken from this reservoir and the pump discharges its mixture of air and water into it. This type of installation is very desirable in locations where freezing may occur for the sealing can be done with a solution of antifreeze.

Hand primers

Hand primers were at one time quite commonly used to prime centrifugal pumps but are seldom used in modern practice. A hand primer is a hand-operated displacement pump (Fig. 21.10). A common type was the old-fashioned pitcher pump, which was often mounted on the priming connection on top of the casing with a shutoff valve interposed.

Tightness of suction lines

Sometimes a pump that has been operating on a lift and that does not have a foot valve will hold its prime when shut down

Fig. 21.10 Portable pumping unit with hand primer

for a considerable period. This is the result of a tight suction line and a stuffing box that is in good condition and properly adjusted. Such units can be restarted because they are fully primed. The duration of the time interval in which a pump loses its prime is a measure of the tightness of the suction line and the condition of the stuffing box. If a pump loses its prime quickly, the suction line should be checked for leaks and the stuffing boxes should be inspected.

Central priming systems

If there is more than one centrifugal pump to be primed in a station, one priming device can serve all the pumps. Such an arrangement is called a central priming system. Figure 21.11 illustrates the method of piping for a central priming system. If the priming device and the venting of the pumps is automatically controlled, the system is called a central automatic priming system.

Vacuum-controlled automatic priming system

A vacuum-controlled automatic priming system consists of a vacuum pump exhausting a tank. The pump is controlled by a vacuum switch and maintains a vacuum in the tank of 2 to 6 in. of mercury, above the amount needed to prime the pumps with the greatest suction lift.

The priming connections on each pump served by the system are connected to the vacuum tank by automatic vent valves and piping. The vacuum tank is usually provided with a gage glass and a drain in order to detect water in the tank as the result of leakage in a vent valve. In such instances, the tank may then be drained.

This type of system (Fig. 21.12) is the most commonly used type of central automatic priming system and has been made with both wet and dry motor-driven vacuum pumps. In the vacuum-controlled automatic priming system, manually con-

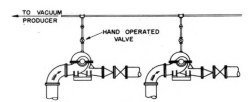

Fig. 21.11 Connections for a central priming system

trolled valves for priming have been replaced by automatic vent valves and operation of the vacuum pump is automatically controlled by the vacuum switch.

The vacuum tank, pump, and controls of a well-known make of this type priming system are shown in Fig. 21.13. The vacuum pump is a belt-driven dry type mounted with its motor on top of the vacuum tank. A vacuum switch on the side of the tank controls the operation of the vacuum pump. The circuit in this switch is closed if the vacuum falls below a certain value (for example, 16 in. of mercury) and is opened if the vacuum reaches a higher value (for example, 20 in. of mercury). Contained in this switch is an unloader to assure that the vacuum pump starts with atmospheric suction pressure and atmospheric exhaust pressure. As the unloader closes, the vacuum pump exhausts through a check valve (shown next to the vacuum switch in Fig. 21.13) from the vacuum tank and builds up the vacuum until the circuit in the switch is broken, thus stopping the motor. The pipe connection for the vacuum line running to the various automatic vent valves is not shown in the illustration.

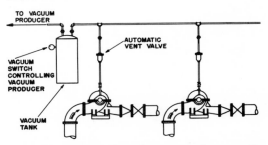

Fig. 21.12 Connections for a central automatic priming system

Fig. 21.13 Vacuum tank, vacuum pump, and control for automatic priming system

(Courtesy Valve and Primer Corp.)

Fig. 21.14 Installation of centrifugal pump served by automatic water jet priming device

A gage glass on the tank shows the presence of water in the tank in case of leakage of any of the vent valves. Above the gage glass is a vacuum relief valve, set to admit air into the tank if the vacuum in the tank reaches a value above the upper limit of vacuum normally maintained. This protects the vacuum pump against operation on a very high vacuum if the vacuum switch fails to open when the upper limit of vacuum is reached.

Automatic priming systems using ejectors

Several automatic priming systems use water ejectors. Figure 21.14 shows an installation of one type that uses water from the discharge line of the pump. The flow of water to the ejector (Fig. 21.15) is controlled by a piston valve that is spring- and water-pressure actuated. This valve is governed by a float chamber (Fig. 21.16) mounted on the suction line.

Air is drawn from the top of the priming chamber until the water rises in it. The rising water lifts the float which, by mechanical linkages, closes the connection to the suction of the jet. A second valve, which when open relieves the pressure on the top of the piston of the water supply valve and allows it to open, is also closed. The closing of the pressure relief valve in the priming or float chamber equalizes the pressure on both sides of the piston and the spring closes the valve, thus shutting off the water supply to the jet.

The same manufacturer has a unit for use in installations where water under sufficient pressure is not always available (Fig. 21.17). It consists of a jet supplied with water under pressure by a small electric-motor-driven pump mounted on a reservoir tank and arranged so that the water is recirculated. The float chamber (Fig. 21.18) uses a switch that is actuated by the float to control the motor of the supply pump. The installation of the float chamber is illustrated in Fig. 21.19.

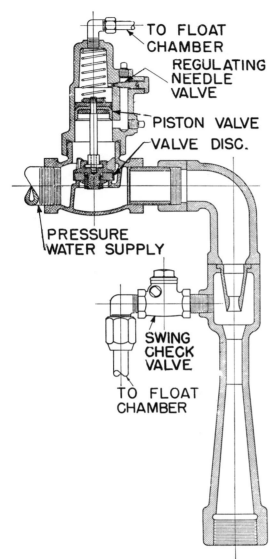

Fig. 21.15 Water jet primer with piston valve to control water supply

(Courtesy Skidmore Corp.)

A third type, Fig. 21.20 and 21.21, made by another manufacturer, is basically similar to the first type except that the water supply to the jet is controlled by a solenoid-operated valve. This is, in turn, controlled by electrodes in the priming chamber, so that the jet operates if the level of the water in the priming chamber falls and the longer electrode is uncovered. The jet

continues to operate until the water level rises and the shorter electrode comes into contact with the water. The priming chamber is mounted directly on the pump to be primed (Fig. 21.21) and is connected by a special venting arrangement to the top or tops of the suction waterways of the pump.

Central priming systems using two vacuum pumps

Most central priming systems are provided with two vacuum pumps (Fig. 21.22). This is desirable for two reasons: First, a breakdown of one does not put the system out of operation; and second, the capacity

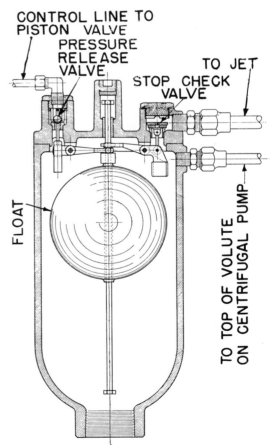

Fig. 21.16 Float chamber to control primer in Fig. 21.15

(Courtesy Skidmore Corp.)

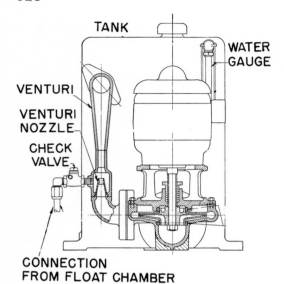

TANK

WATER GAUGE

VENTURI

VENTURI NOZZLE

CHECK VALVE

CONNECTION FROM FLOAT CHAMBER

Fig. 21.17 Vacuum producer using water jet with motor-driven centrifugal pump to supply power water

(Courtesy Skidmore Corp.)

of a vacuum pump is selected for the normal leakage to be encountered after the pumps served by the system have been primed. When one of the units served has been out of service and it is desired to re-establish its prime, the amount of air to be removed is relatively large and the priming time would be quite long if a single vacuum pump were used. Furthermore, there would be a reduction in the vacuum in the priming system particularly when another pump is being primed. This would be objectionable if any one of the units in service had accumulated air and the vent valve had opened to vent this air.

It is usual practice to have the control of one vacuum pump switched on at some predetermined vacuum and the control of the second switched on at a slightly higher vacuum. Thus if the capacity of one pump

FLOAT SWITCH

TO VACUUM PUMP

STOP CHECK VALVE

FLOAT

TO TOP OF VOLUTE ON CENTRIFUGAL PUMP

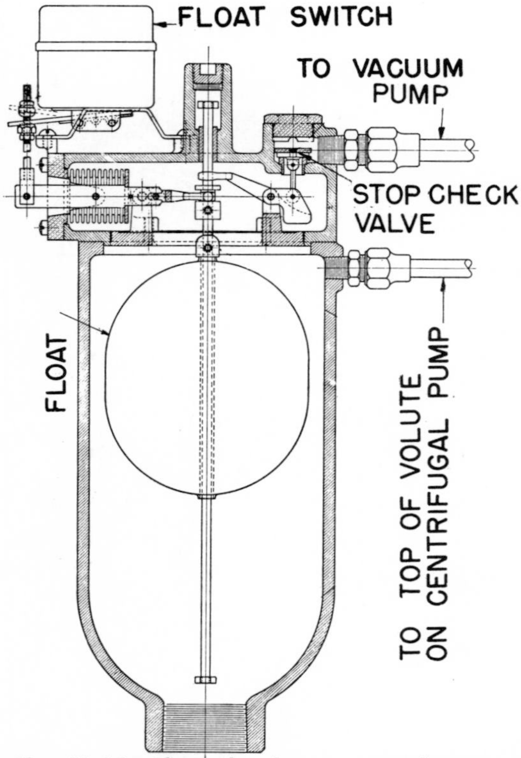

Fig. 21.18 Float chamber to control vacuum producer shown in Fig. 21.17

(Courtesy Skidmore Corp.)

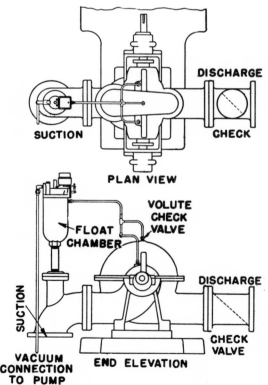

DISCHARGE

SUCTION

CHECK

PLAN VIEW

VOLUTE CHECK VALVE

FLOAT CHAMBER

DISCHARGE

SUCTION

CHECK VALVE

VACUUM CONNECTION TO PUMP

END ELEVATION

Fig. 21.19 Connections for float chamber shown in Fig. 21.18

(Courtesy Skidmore Corp.)

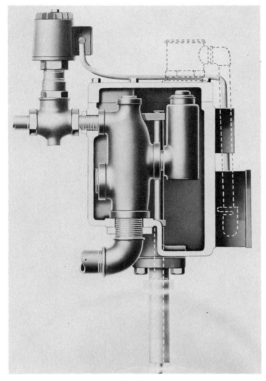

Fig. 21.20 Priming device using water jet with electrode and solenoid valve control

(Courtesy DeLaval Steam Turbine Co.)

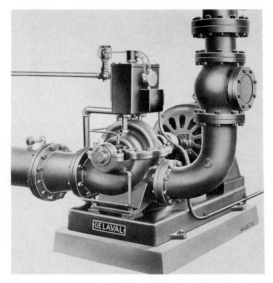

Fig. 21.21 Centrifugal pump equipped with priming device in Fig. 21.20

(Courtesy DeLaval Steam Turbine Co.)

is sufficient to prevent the vacuum in the system from falling, the second pump need not start. Generally, the controls are equipped with switches so that the sequence of starting can be changed and the use of the two pumps can be equalized.

Standby units

A station having a central automatic priming system, using an electric-motor-driven vacuum pump, may have standby units with auxiliary drive in case of electrical power failure. These standby units are usually kept fully primed by the central priming system. If these units are manually started and controlled, a standby, manually controlled vacuum producer is generally installed. Thus in a vacuum-tank type of priming system, with gasoline-engine-driven standby pumps, a gasoline-engine-driven vacuum pump would also be installed and the operator would manually start and stop it as required to maintain the required range of vacuum in the vacuum tank.

Fig. 21.22 Vacuum tank type of priming unit with two vacuum pumps

(Courtesy Valve and Primer Corp.)

Standby priming equipment that is automatically controlled presents a more complicated problem. One solution used in units that are driven by internal combustion engines is to connect a continuously running vacuum pump directly to a relief valve on its suction line or to an unloader in the form of diaphragm valve controlled by the discharge pressure of the centrifugal pump. In either case, proper vent valves and check valves have to be incorporated into the priming piping. For units driven by steam turbines, automatically controlled steam jet primers can be used. Most systems of this nature are especially designed for the requirements of the installation.

Point of air removal

Air is usually exhausted from a centrifugal pump that is being primed through the high point of the volute. Generally, it is possible to start a centrifugal pump if the liquid covers the eye of the impeller. The remaining air entrapped in the casing will be driven out by the flow of liquid through the pump. On some units, the priming is accomplished solely from the high point in the suction waterways. On other units, notably those of large size, air is exhausted both from the high point of the volute and the high point or points of the suction waterways.

Automatic vent valves

Automatic vent valves consist of a body containing a float that actuates a valve located in the upper part of the body. The bottom of the body is connected to the space to be vented. As air or gas is vented out of the valve, water rises in the body until the float is lifted and the valve is closed. A typical vent designed basically for vacuum priming systems is illustrated in Fig. 21.23. The valve is provided with auxiliary openings on the lower part of the body for connection to the auxiliary vent points in the system. When one or more of

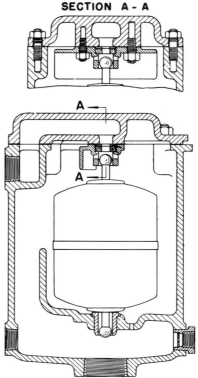

Fig. 21.23 Automatic vent valve

(Courtesy Nash Engineering Co.)

these vent points is at higher pressure, such as the top of the volute of the pump, an orifice is used to limit the flow of liquid. This constant flow of liquid from the discharge back to the suction means a constant loss of effective capacity. When a unit is used more or less constantly, a separate valve is desirable.

In some systems it is necessary to prevent reverse flow of air through the valve. This can be done by adding a check valve between the vent and the vacuum source in the vent piping. One make of automatic vent valve is designed so that a ball can be inserted to act as a check valve.

In addition to their use with automatic priming systems, automatic vent valves are employed alone, without connection to a vacuum source, to .vent air from the passages of pumps and pipes that are under greater than atmospheric pressures. They are used frequently to vent air from pumps that have a positive suction head, thus insuring that the pump is always primed.

Self-contained units

Centrifugal pumps are today available with various designs of priming equipment that render them self-contained units. Some have automatic priming devices, which are basically attachments to the pump, and become inactive after the priming is accomplished. Other units, which are self-priming pumps, incorporate a hydraulic device that can function as a wet-vacuum pump during the priming period. For stationary use, the automatically primed type is more efficient. The self-priming designs are usually more compact and are better for portable or semiportable use.

Self-priming units

In self-priming units the pump design incorporates a chamber or other device that recirculates entrapped air. The air can be removed from the suction waterways and discharged into the discharge line by means of an ejector. When the presence of air in the discharge line is objectionable, an automatically primed unit would be preferred. Since the self-priming type was discussed in Chap. 15, this discussion will be restricted to priming methods and systems that may be used in conventionally designed centrifugal pumps.

Figure 21.24 illustrates an interesting combination of a priming tank and an ejector. In effect, the regular centrifugal pump is made into a self-priming unit. It consists of a regular centrifugal pump with a priming tank that is smaller than would be required if a priming tank alone was used. For proper functioning there must be a check valve in the suction line, but none in the discharge line. The discharge line must be vertical, and the volume of the vertical discharge pipe must equal the volume of the priming tank plus the suction piping to the check valve. When first installed, the pump, the priming tank, the suction line back to the check valve, and the discharge pipe are filled with water level with the top of the priming tank. Two electrodes are installed in the tank and incorporated in the control of the motor. Thus, the unit cannot start unless the water covers the top electrode and will stop if the water falls below the lower electrode.

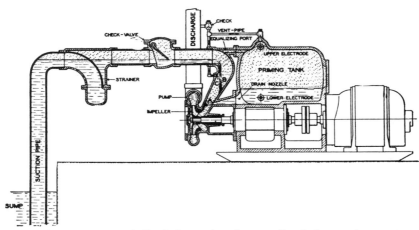

Fig. 21.24 Self-priming unit using small priming tank

(*Courtesy Barret-Haentjens Co.*)

When the pump is started, the water is drawn from the priming tank and the suction line between the pump and the check valve. Thus a vacuum is established, and air is drawn through the check valve from the balance of the suction line (Fig. 21.24). At the same time the stream of water leaving the drain nozzle traps the air in A and carries it into the pump, which discharges it with the water. When the water in the priming tank is pumped out, the lower electrode is uncovered and the control stops the pump. The suction line check valve closes and water from the discharge flows back through the pump, forcing the air in the priming tank (and in the suction line up to the check valve) through the vent pipe and its check valve into the discharge pipe, until the upper electrode is again submerged. The pump starts again, and more air is drawn from the suction line beyond the check valve. When the suction line has been evacuated sufficiently and water partly fills the horizontal run, the water forms a wall at B that aids in carrying air into the pump. As the horizontal run of suction pipe fills, the flow of water through the pump increases until normal capacity is reached. The flow of water past C draws the air from the priming tank until it is again filled with water. After the prime has been established and all air has been exhausted from the priming tank, the priming tank becomes inactive. This device permits priming of a system in which the suction line has a large volume and the priming tank a small volume. The number of cycles needed to prime the system depends on the volume to be evacuated.

The manufacturer of this unit also makes a separate primer that operates on the same principle for use with regular centrifugal pumps.

Automatically primed units

Many automatically controlled priming systems for attachment to centrifugal pumps have been developed in recent years. A pump equipped with such a device is called an automatically primed pump.

A rotary wet vacuum pump, either connected directly or driven by a separate motor, is used in most automatically primed, motor-driven pumps. With the direct-connected vacuum pump, the controls that function when the centrifugal pump is primed open the vacuum pump suction line to the atmosphere, enabling it to operate without a load. With the separately driven vacuum pump, the controls that function when the centrifugal pump is primed stop the vacuum pump.

Direct-connected vacuum pump unit

A typical unit with a direct-connected vacuum pump is shown in Fig. 21.25. The centrifugal pump is mounted on one end of the motor and the vacuum pump, which runs all the time, is mounted on the opposite end. The suction of the vacuum pump is connected to the vent on the suction of the centrifugal pump and to a pressure-operated valve that is closed when the centrifugal pump is not generating pressure.

When the unit is started, the vacuum pump exhausts the air from the centrifugal pump and its suction line until the centrifugal pump is primed. When the centrifugal pump becomes primed, it generates pressure and the pressure-operated valve opens, allowing outside air to flow into the suction of the vacuum pump. The branch of the vacuum pump that is connected to the centrifugal pump suction vent has a check valve to prevent back flow of air when the pressure-operated valve opens, as well as a strainer to prevent dirt or foreign material from being carried over to the vacuum pump. The connection from the discharge waterways of the centrifugal pump to the pressure valve is also connected to the vacuum pump suction line through an orifice. This provides a small flow of sealing water for the vacuum pump. A check valve is required in this line in front of the branch.

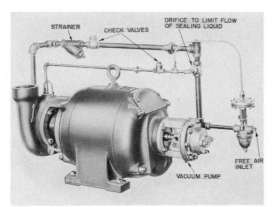
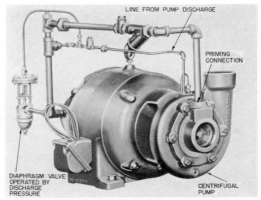

Fig. 21.25   Two views of automatically primed pump with direct-connected vacuum pump

Another type of direct-connected vacuum pump unit is illustrated in Fig. 21.26. This type utilizes a float valve on the vent on the suction of the centrifugal pump. The vacuum line has a relief valve set to open on high vacuum, so that the vacuum pump is not required to pull against a closed suction. This unit incorporates a sealing liquid reservoir also. Other types of direct-connected vacuum pump units might include: A float-actuated valve that shuts off the vent of the pump from the vacuum pump and opens the suction line of the vacuum pump to the atmosphere, or a pressure switch controlling a solenoid valve in the vent pipe either by a vacuum relief valve or three-way solenoid valve.

Separately driven vacuum pump unit

An automatically primed pump using a separate motor-driven vacuum pump is usually electrically connected to enable the main pump motor and the vacuum pump motor to start at the same time. The suction of the vacuum pump is connected to vents at the high point of the suction waterways of the main pump and evacuates the air in the waterways until the pump is primed. When the pump is primed, it actuates a pressure switch in the circuit of the vacuum pump motor. This switch opens to stop the pump. A check valve must be located in the line from the vacuum pump to the suction vent connection to prevent back flow of air when the vacuum pump is

idle. Sealing water is usually provided from a reservoir in the base.

With this and some other automatic priming systems, it is necessary for the centrifugal pump to run empty during the time required for priming. This can be done if the clearance between the rings is ample and the size of the vacuum pump is sufficient to prime the centrifugal pump in less than two minutes. In normal service there is generally some water trapped in the volute of the centrifugal pump—this aids in lubricating the rings during the priming period. A unit with this type of priming is illustrated in Fig. 21.27.

Systems without vent valves

When a pump is primed from the suction side, it is possible to have the vent piping from the pump to the vacuum pro-

Fig. 21.26   Automatically primed pump with direct-connected vacuum pump

Fig. 21.27 Automatically primed pump with separately driven vacuum pump

ducer in the form of an inverted loop of
sufficient height to prevent the water or
liquid being pumped from being carried
over into the vacuum producer. Theoretically, a loop 34 ft above the water level
is correct for water. However, if there is
leakage, air bubbles in the water will lower
the net density of the column and a loop
extending 50 ft or more above the water
level would be necessary to prevent liquid
from being carried over when the vacuum
produced is high. Generally, a system incorporating such a high loop is difficult to
accommodate in a station, and automatic
vent valves and other devices may be used
more advantageously.

Some systems use risers without vent
valves; the vacuum producer is controlled
by a vacuum device that keeps the water
level in the risers within a range corresponding to the vacuum range. If there is considerable variation in the suction level, the
height of the riser has to be high enough
to match the water level in the riser resulting from the maximum suction water
level and the maximum vacuum. Naturally,
the minimum vacuum must produce a

water level in the riser slightly above the
pump with the minimum water level in
the suction supply.

Another system that does not involve the
use of a vent valve is illustrated in Fig.
21.28. This system can also be used as a
central priming system to serve pumps
taking suction from the same source.

Systems for sewage pumps

A pump handling sewage or similar
liquids that contain stringy material can
be equipped with automatic priming. However, a special type must be used. One sys-

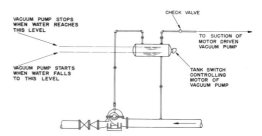

Fig. 21.28 Priming system without vent valves using elevated tank with maintained water level

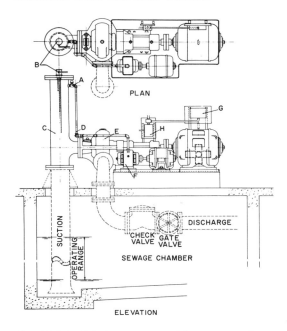

Fig. 21.29 Priming system without vent valves using riser with maintained water level

KEY:

A = *solenoid-operated air valve*
B = *electrode unit*
C = *suction submergence chamber*
D = *suction to vacuum pump*
E = *sewage pump*
F = *vacuum pump—motor controlled by electrode relay*
G = *main pump motor starter*
H = *induction relay.*

(*Courtesy DeLaval Steam Turbine Co.*)

level of liquid falls to the point at which this third electrode is uncovered.

The automatic priming system described previously (see Fig. 21.27) that uses a separate motor-driven vacuum pump controlled by a pressure switch actuated by discharge pressure has also been successfully used on sewage pumps by incorporating an inverted vertical loop in the vacuum pump suction line. This prevents the sewage from being carried over into the vacuum pump because the pump shuts down before the liquid reaches the top of the loop.

tem (Fig. 21.29) uses a tee on the suction line immediately adjacent to the pump suction nozzle, with a vertical riser mounted on the top outlet of the tee. This riser is blanked at the top, thus forming a small tank. Two electrodes are located in the tank at different levels. The top of the tank is vented to a vacuum system through a solenoid valve. This solenoid valve is controlled electrically, closing if the liquid level reaches the top electrode and opening if the liquid level falls below the level of the lower electrode. A third electrode, installed at a still lower level but slightly above the pump suction, is used to control the pump motor, stopping the motor if the

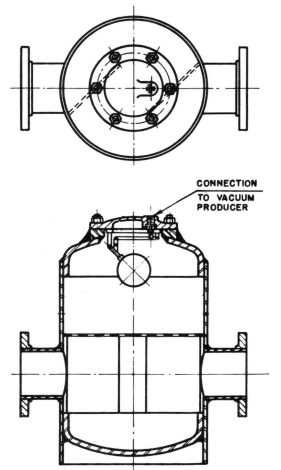

Fig. 21.30 Air-separating chamber for 4- to 12-in. suction lines

(*Courtesy DeLaval Steam Turbine Co.*)

Fig. 21.31 Installation with automatic priming equipment utilizing combined air-separating and sand chamber

(Courtesy DeLaval Steam Turbine Co.)

Systems for air-charged waters

Some types of water, particularly from wells, have considerable dissolved gas that is liberated when the water is pumped with suction lift. In such installations an air-separating tank (also called a priming tank or an air eliminator) should be used in the suction line. One type using a float-operated vent valve to permit the withdrawal of the air or gas is shown in Fig. 21.30. Another arrangement that is often used utilizes a float valve mounted on the side of the tank which controls directly the starting and stopping of the vacuum pump. With this arrangement there is danger of frequent, repeated starting and stopping of the vacuum pump unless the air-separating tank is relatively large and the capacity of the vacuum pump is just a little larger than the amount of air to be evacuated.

When sand is present as an impurity in the water, the air-separating tank can be made to function as a sand trap (Fig. 21.31 and 21.32).

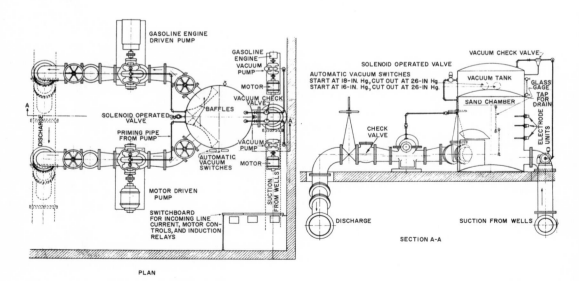

Fig. 21.32 Drawing of installation in Fig. 21.31

(Courtesy DeLaval Steam Turbine Co.)

Fig. 21.33   Series pump with attached vacuum pump and controls to keep unit primed when in operation

Systems for units driven by gasoline or diesel engines

Automatic priming is desirable for any centrifugal pump installation in which air leakage in the suction line might occur. In a unit driven by a diesel engine, an automatic priming system using motor-driven pumps can be used if a reliable source of electric power is available in the station.

An auxiliary gasoline-engine-driven vacuum pump for emergency use might be desirable in case of electric power failure. If a reliable source of electric power is not available, a direct-connected wet vacuum pump with controls similar to those used in motor-driven automatically primed units is very satisfactory (Fig. 21.33).

The choice of the priming device for a gasoline-engine-driven centrifugal pump depends on the size of the pump, the required frequency of priming and the porta-bility of the unit. Most portable gasoline-engine-driven pumps are used for relatively low heads and small capacities, for use in pumping out excavations and ditches, for example. Self-priming pumps of various types are most satisfactory for this service and are preferable to regular centrifugal pumps.

Some relatively small-capacity high-head portable gasoline-engine-driven pumps (notably units made for auxiliary fire pump service), utilize the vacuum in the intake manifold of a gasoline engine as a means of priming or keeping the pump primed. The rate at which the air can be drawn from the pump is relatively low, so that many of these units use foot valves. They are initially primed by filling the pump through a funnel or by means of a hand primer. When the vacuum in the manifold is utilized, water must be prevented from being drawn over into the manifold.

Fig. 21.34 Pump driven by a gasoline engine with belt-driven priming pump using tight and loose pulley

For larger volume low-head portable units, a good priming unit is a wet-vacuum pump belted to the main shaft by a tight and a loose pulley (Fig. 21.34). The wet-vacuum pump can then be stopped when the pump is primed.

For permanent installations, a separate wet-vacuum pump driven by a small gasoline engine is generally preferred.

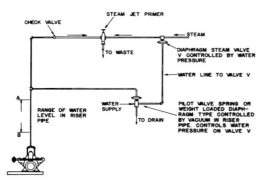

Fig. 21.35 Automatically controlled steam jet primer controlled by vacuum in riser

Systems for steam-turbine-driven units

The most logical means of priming a steam-turbine-driven pump is with a steam air ejector. If the turbine is a condensing unit, the vacuum existing in the condenser can be used with an automatic vent valve to keep the pump primed automatically. However, if there is considerable leakage, the regular steam air ejector serving the condenser may not be able to handle this additional volume of air and the condenser performance will be affected. A regular steam condenser should not be used as a source of vacuum for priming unless the suction lines are tight and the condenser is served by an oversize ejector. With this arrangement care must be taken to prevent leakage of water from the pump into the condenser in order to avoid contamination of the condensate. This requires a constant check of the automatic vent valve to insure that it seats tightly.

With a noncondensing steam-turbine-driven centrifugal pump, a semiautomatic priming system (Fig. 21.35) can be used if there is danger of air leakage into the suction line.

Time required for priming

The time required to prime a pump with a vacuum-producing device depends on: (1) the total volume to be exhausted, (2) the initial and final vacuum, and (3) the capacity of the vacuum producer over the range of vacuums that will result in the priming cycle. The actual calculations for determining the time necessary to prime a pump are complicated. To permit close approximations, jet primers are usually rated in net capacity for various lifts. It is necessary to divide the volume to be exhausted by the rating to obtain the approximate priming time. Unless such a simplified method of calculating the required time is available, the selection of the size of primer should be left to the vendor of the equipment.

Central automatic priming systems are usually rated for the total volume to be kept primed. The time initially required to prime each unit served by the system is not usually considered, as the basic function of the system is to keep the pumps primed and in operating condition at all times.

Use of relief valves

The power required by a vacuum pump increases with the magnitude of the vacuum. If the vacuum pump is allowed to operate uncontrolled after the vent valves have closed, it would continue to exhaust air from the piping and vacuum tank (if they are used) until it had established its maximum vacuum. This would result in an increased load on the driver. To avoid this difficulty, most priming systems of an automatic or semiautomatic type using power-driven vacuum pumps incorporate a relief valve in the suction line of the vacuum pump. This valve is adjusted to admit air to the line if the vacuum exceeds a value above that needed for priming.

Prevention of unprimed operation

Various controls may be used to prevent the operation of a pump when it is unprimed. These controls depend upon the type of priming system used. For most installations, a form of float switch in a chamber connected with the suction line is used. If the level in the chamber or tank is above the impeller eye of the pump, the float switch control allows the pump to operate. If the liquid falls below a safe level, the float switch acts through the control to stop the pump or to prevent its being started. In the system illustrated in Fig. 21.29, three electrodes are placed in the liquid. When the level drops below the longest electrode, the pump stops or, if stopped, will not start.

Since a great number of automatic priming devices and systems are available, care should be taken to use the type or variation best suited to the application. This discussion and the accompanying illustrations do not, of course, cover all makes and modifications for every specific application.

An automatic priming system will often allow units to be operated with excessive air leakage into the suction lines. This is a poor practice because it requires the operation of the vacuum producer for a greater part of the time.

IV
SERVICES
and
SELECTION
of PUMPS

22 *Services*

Centrifugal pumps are used wherever any quantity of liquid must be moved from one place to another. Centrifugal pumps are found in such services as steam power plants; water supply plants; sewage, drainage, or irrigation; oil refineries, chemical plants, and steel mills; food processing factories, and mines; dredging or jetting operations; hydraulic power service; and almost every ship, whether driven by steam or diesel engine. While these pumps have much in common, they are varied to meet the special requirements and particular needs of each service. Although it would be a monumental task to present all the requirements of each different type of application and every individual particularity that distinguishes pumps on different services, this book would not be complete without a general discussion of these requirements as they apply to some of the most common applications.

STEAM POWER PLANTS

Power is produced in a steam power plant by supplying heat energy to the feedwater, changing it into steam under pressure, and then transforming part of this energy into mechanical energy in a steam engine or steam turbine to do useful work. The feedwater, therefore, acts merely as a conveyor of energy. The basic elements of a steam power plant include the heat engine, the boiler, and a means of getting water into the boiler. Modern power plants use steam turbines as heat engines and centrifugal boiler feed pumps for feeding water to the boiler, except in very small installations.

The basic cycle is improved by connecting a condenser to the steam turbine exhaust, thereby increasing the pressure drop through the turbine and using more of the energy in the steam. The condenser also recovers the feedwater, almost entirely eliminating makeup. The cycle is further refined by heating the feedwater with steam extracted from an intermediate stage of the main turbine. This results in an improvement of the cycle efficiency, provides deaeration of the feedwater, and eliminates the introduction of cold water into the boiler and the resulting temperature strains on the latter. The combination of the condensing and feed heating cycle (Fig. 22.1) requires a minimum of three pumps: The condensate pump, which transfers the condensate from the condenser hotwell into the direct contact heater; the boiler feed pump; and a circulating pump, which forces cold water through the condenser tubes to condense the exhaust steam. The cycle illustrated in Fig. 22.1 is very common and is used in most small steam power plants. A number of auxiliary services not illustrated in Fig. 22.1 are normally used, such as heater drain pumps, service water pumps, cooling pumps, ash sluicing pumps, oil circulating pumps, and the like.

As the size and number of central steam-electric generating stations grew, the desire for improvements in operating economy dictated many refinements in the steam cycle. These refinements have created new or altered services for power plant centrifugal pumping equipment. Some of these refinements involved a steady increase in operating pressures until 2,400 psi steam turbines have become quite common. Several plants operating at supercritical steam pressures of 3,500 to 5,500 psi are already in service or under construction.

Other refinements were directed towards a greater utilization of heat through increased feed heating, introducing a need for heater drain pumps—equipment with definite problems of its own. The introduction of forced or controlled circulation as opposed to natural circulation in boilers created a demand for pumping equipment of an entirely special character.

The problem presented by the introduction of multiple-stage heaters between the condensate pump and the boiler centered in the choice between direct-contact heaters and closed heat exchangers. From the thermodynamic point of view, the direct-contact heaters have certain advantages. A separate pump would be required after each direct-contact heater, however, whereas the use of a group of closed heaters permits a single boiler feed pump to discharge through these heaters and into the boiler. The average power plant is based on a compromise system: One direct-contact heater is used for feedwater deaeration, and several

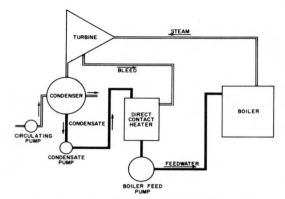

Fig. 22.1 Simple steam power plant cycle

closed heaters are located upstream and downstream of the direct-contact heater and the boiler feed pump (Fig. 22.2). Such a cycle is known as an "open cycle." The major variation of this method is the "closed cycle," where the deaeration is accomplished in the condenser hotwell and all heaters are of the closed type (Fig. 22.3).

Boiler feed pumps

A listing of conditions of service of boiler feed pumps should include not only the pump capacity, suction conditions and feedwater temperature, and discharge pressure, but also such data as the chemical analysis of the feedwater, the pH at pumping temperature, the variation in suction pressure and temperature if any, and all other pertinent information that may reflect upon the hydraulic and mechanical design of the boiler feed pumps. Whenever the power plant designer has any doubts, he should submit a complete layout of the feedwater system and of the heat balance diagram to the boiler feed pump manufacturer. After studying this layout, the manufacturer will often be able to suggest an alternate arrangement of the equipment that will result in more economical operation, in decreasing the first cost of the installation, or even in improving the life of the equipment, thereby reducing the eventual maintenance expense.

Boiler feed pump capacity

Once the maximum flow to the boiler has been determined, the boiler feed pump total capacity can be established by adding a margin to this maximum flow to cover boiler swings and the eventual reduction in effective capacity from wear. This margin varies from 20 per cent in small plants to 8 per cent in the larger central stations.

Heat balance calculations as well as turbine and boiler guarantees are always expressed in pounds per hour, whereas it is customary to define the service to be performed by centrifugal pumps in terms of gallons per minute against a head of so many feet. To convert pounds per hour into gallons per minute, use the formula: Gallons per minute = pounds per hour/(500 × specific gravity). Values of the specific gravity of water for different feedwater temperatures are given in Fig. 22.4.

The total required capacity must then be handled either by a single pump or subdivided between several duplicate pumps operating in parallel. Industrial power plants generally use several pumps. Central stations normally use single full-capacity pumps to serve turbogenerators up to 80,000-kw rating and two pumps in parallel for larger installations. Exceptions to this practice occur as follows: Some engineers prefer the use of multiple pumps even for small installations; and steam-driven turbine boiler feed pumps designed for full capacity are applied for units as large as 450,000 kw.

In most cases, a spare boiler feed pump is included. In central stations today, however, it is becoming common to eliminate spare pumps when two half-capacity pumps are used and, in a few cases, even if a single full-capacity boiler feed pump is installed.

Suction conditions

As discussed in Chap. 17, the net positive suction head or NPSH represents the net suction head at the pump suction, corrected

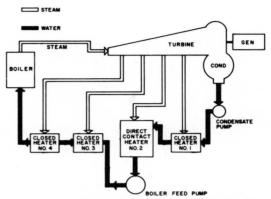

Fig. 22.2   Open feedwater cycle with one deaerator and several closed heaters

to the pump centerline, over and above the vapor pressure of the feedwater. If the pump takes its suction from a deaerating heater as in Fig. 22.2, the feedwater in the storage space is under a pressure equivalent to the vapor pressure corresponding to its temperature. Therefore, the only energy available at the first-stage impeller over and above the vapor pressure is the static submergence between the water level in the storage space and the pump centerline, less the friction losses in the intervening piping. Table 22.1 gives the relationship between water vapor pressure and temperature.

The required NPSH is independent of the operating temperature by virtue of its definition (Chap. 17). Practically, this temperature must be taken into account when

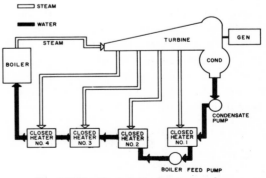

Fig. 22.3   Closed feedwater cycle

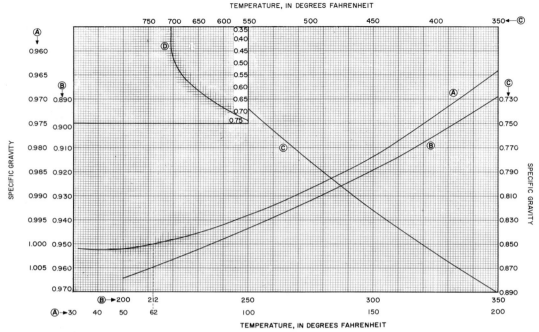

Fig. 22.4 Specific gravity of water at temperatures from 30 to 750°F

Curve A (from 30 to 200°F) is from US Bureau of Standards' circular no. 19; curve B (from 200 to 350°F), curve C (from 350 to 550°F), and curve D (from 550 to 750°F) are from Goodenough Steam Tables. All curves are based on specific gravity of water at 62°F = 1.000.

establishing the recommended submergence from the deaerator to the boiler feed pump. A margin of safety must be added to the theoretical required NPSH to protect the boiler feed pumps against the transient conditions that follow a sudden reduction in load for the main turbogenerator.[1]

Whereas the previous discussion applies primarily to the majority of installations where the boiler feed pump takes its suction from a deaerating heater, it holds as well in the closed feed cycle (Fig. 22.3). The discharge pressure of the condensate pump must be carefully established so that the suction pressure of the boiler feed pump cannot fall below the sum of the vapor

pressure at pumping temperature and the required NPSH.

Normally this is not difficult; however, some installations break up the total pressure to be generated by the boiler feed pump between two separate pumps and locate some of the closed heaters between these two pumps. The first of these is called a "primary feed pump" or a "booster condensate pump," and the second is called the "secondary feed pump" or the "main feed pump." In the event the primary pump is operated at variable speed, great care must be taken to meet the requirement given above. The controlling factor for this speed variation is the flow to the boiler and the controlling mechanism, generally an excess-pressure regulator (Chap. 19), which varies the speed of the pump in question. This control is unaffected by the relationship between the temperature and

[1] For a complete treatment of this problem, see "Sudden Load Drop in Open Feed Cycles of Steam Turbine Power Plants," Karassik, Bosworth, and Elston, *The International Operating Engineer.* July 1957 to January 1958. Worthington Corp. Reprint RP-961.

the pressure at the suction of the main feed pump, and conditions can arise where the speed regulation lowers the suction pressure at the feed pump dangerously close to, or even below, the required minimum. In such installations, therefore, it is absolutely necessary to provide some form of lower limit control to the speed of the pump,

subject to variation. This lower limit is determined on the basis of always assuring sufficient NPSH to the main feed pump.

Discharge pressure and total head

In order to determine the discharge pressure for which the boiler feed pumps are to be designed, it is necessary to obtain the

TABLE 22.1   TEMPERATURE–VAPOR PRESSURE RELATIONSHIP OF WATER

At sea level, the saturation pressure or vapor pressure (PSIG) = vapor pressure (PSIA − 14.7)

Temperature, deg F	Vapor pressure, psia	Temperature, deg F	Vapor pressure, psia	Temperature, deg F	Vapor pressure, psia
32	0.088	232	21.58	400	247.3
35	0.100	236	23.22	405	261.7
40	0.122	240	24.97	410	276.8
45	0.148	244	26.83	415	292.4
50	0.178	248	28.80	420	308.8
55	0.214	252	30.88	425	325.9
60	0.256	256	33.09	430	343.7
65	0.306	260	35.43	435	362.3
70	0.363	264	37.90	440	381.6
75	0.430	268	40.50	445	401.7
80	0.507	272	43.25	450	422.6
85	0.596	276	46.15	455	444.3
90	0.698	280	49.20	460	466.9
95	0.815	284	52.42	465	490.3
100	0.949	288	55.80	470	514.7
105	1.102	292	59.36	475	539.9
110	1.275	296	63.09	480	566.1
115	1.471	300	67.01	485	593.3
120	1.692	304	71.13	490	621.4
125	1.942	308	75.44	495	650.6
130	2.222	312	79.96	500	680.8
135	2.537	316	84.70	505	712.0
140	2.889	320	89.66	510	744.3
145	3.281	324	94.84	515	777.8
150	3.718	328	100.3	520	812.4
155	4.203	332	105.9	525	848.1
160	4.741	336	111.8	530	885.0
165	5.335	340	118.0	535	923.2
170	5.992	344	124.4	540	962.5
175	6.715	348	131.2	545	1003.
180	7.510	352	138.2	550	1045.
185	8.383	356	145.4	555	1088.
190	9.339	360	153.0	560	1133.
195	10.385	364	160.9	565	1179.
200	11.526	368	169.2	570	1226.
204	12.512	372	177.7	575	1275.
208	13.568	376	186.6	580	1326.
212	14.70	380	195.8	585	1378.
216	15.90	384	205.3	590	1431.
220	17.19	388	215.3	595	1486.
224	18.56	392	225.6	600	1543.
228	20.02	396	236.2		

sum of the maximum boiler drum pressure and of all the frictional and control losses that occur between the boiler feed pump discharge and the boiler drum inlet. The calculations of the frictional losses must, of course, be based upon the maximum capacity previously determined. The required discharge pressure will generally vary from 115 to 125 per cent of the boiler drum pressure.

The net pressure to be generated by the boiler feed pump is the difference between the required discharge pressure and the available suction pressure. This must be converted into a total head, using the formula: Total head, in feet = (net pressure, in pounds per square inch $\times$ 2.31)/specific gravity.

Head-capacity curve slope

In the range of specific speeds normally encountered in multistage centrifugal boiler feed pumps, the rise of head from the point of best efficiency will vary from 10 to 25 per cent. Further, the shape of the head-capacity curve for these pumps is such that the drop in head is very slow at low capacities, accelerating as the capacity is increased (see Fig. 20.46 and 25.2).

If the pump is operated at constant speed, the difference in pressure between the pump head-capacity curve and the system-head curve must be throttled by the feedwater regulator. Thus, the higher the rise of head towards shutoff, the more pressure must be throttled off and, theoretically, wasted. Also, the higher the rise, the greater the pressure to which the discharge piping and the closed heaters will be subjected. It is not advisable, however, to select too low a rise to shutoff because too flat a curve is not conducive to stable control, as a small change in pressure corresponds to a relatively great change in capacity, and a design that gives a very low rise to shutoff may result in an unstable head-capacity curve, difficult to use for parallel operation.

When several boiler feed pumps are to be operated in parallel, they must have stable curves and equal shutoff heads. Otherwise, the total flow will be divided unevenly and one of the pumps may actually be backed off the line after a change in required capacity occurs at light flows.

Driver horsepower

A boiler feed pump is selected for a given capacity and pressure and generally it is not expected to operate at any capacity beyond the design condition. In other words, in contrast to a standard general-service pump for which the power consumption may increase with a decrease in head and, consequently, an increase in capacity, a boiler feed pump has a very definite maximum capacity because it operates on a system-head curve made up of the boiler drum pressure plus the friction losses in the discharge. If, as it should be, the design capacity of the pump is chosen as the maximum capacity that can be expected under emergency conditions, there can be no further increase under any operating conditions since the pressure requirement corresponding to an increased capacity would exceed the design pressure of the pump. Even when the design pressure includes a safety margin, the boiler demand does not exceed the design capacity and the feedwater regulator will impart additional artificial friction losses to increase the required pressure up to the pressure available at the pump.

When two pumps are operated in parallel to feed a single boiler, the situation is somewhat different. If one of the pumps is taken off the line at part load, the remaining pump could easily operate at capacities in excess of its design, since its head-capacity curve would intersect the system-head curve at a point lower than the design head (Fig. 22.5). In such a case, it is necessary to determine the pump capacity at the intersection point; the horsepower corresponding to this capacity will be the maximum expected.

It is not necessary to select a driver so large that it will not be overloaded at any

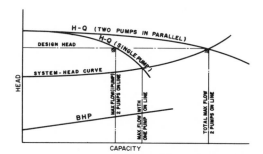

Fig. 22.5 Method of determining maximum pump horsepower for two boiler feed pumps operating in parallel

point of the boiler feed pump operating curve. Electric motors used on boiler feed service, however, generally have an overload capacity of 15 per cent, and it is usually the practice to reserve this overload capacity as a safety margin and to select a motor that will not be overloaded at the design capacity. Exceptions occur in the case of very large motor sizes. For example, if the pump brake horsepower is 3,100, it is logical to use a 3,000-hp motor that will be overloaded by about 3 per cent rather than a considerably more expensive 3,500-hp motor.

Because steam turbines are not built in definite standard sizes but can be designed for any intermediate rating, they are generally selected with about 5 per cent excess power over the maximum expected pump horsepower.

General structural features

As described in Chap. 3, boiler feed pumps used for discharge pressures under 1,250 psi are generally designed with axially split casings (Fig. 3.4 through 3.6). Pumps with radially split double casings (Fig. 3.14, 3.15, and 3.17) are used for higher discharge pressures.

The merits of single-suction over double-suction impellers are discussed in Chap. 4, and the means used to balance the axial thrust of multistage pumps with single-suction impellers are described in Chap. 5 and 6. The single-suction impeller design

has been almost universally adopted for the reasons presented in those chapters. The exception is the double-suction first-stage impeller used in some very large capacity boiler feed pumps. The selection of materials for the boiler feed pump casings and internal parts is described in Chap. 16.

High-speed, high-pressure boiler feed pumps

Boiler feed pump design practices and the present trend to operating speeds in excess of 3,600 rpm are best understood after examining the effect that the use of increasing operating steam pressures has had on boiler feed pumps. As these pressures were increased from 1,250 to 1,800, then to 2,400 and even to 3,500 psi, the total head that had to be developed by the pump rose from somewhere around 4,000 ft to as high as 7,000 and 12,000 ft. The only way to achieve these higher heads at 3,600 rpm was to increase the number of stages. The pumps, therefore, had to have longer and longer shafts, which threatened to interfere with the long uninterrupted life between pump overhauls to which steam power plant operators were beginning to become accustomed.

In order to decrease the frequency of maintenance requirements, the rate of wear at the internal running joints must be decreased. To evaluate the probable rate of wear of several different designs, however, the causes of wear must be considered. Wear occurs not only because of the erosive action of the leakage past internal running joints, but also—to a greater measure—through contact and the rubbing action of two metal parts, even if this contact is infrequent and too slight to cause galling and immediate seizure. In other words, the two prime factors affecting the life of a given high-pressure boiler feed pump are the shaft deflection and the internal clearances. The greater the margin between these two, the less will be the chance of accidental contact and the longer the life of the equipment.

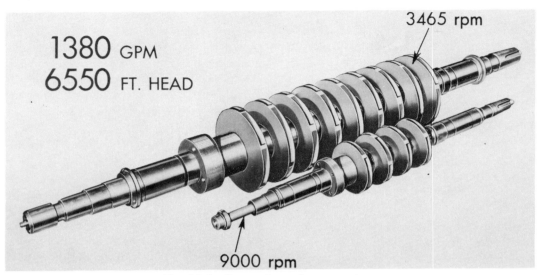

Fig. 22.6   Comparison of pump rotors

The deflection of a pump shaft varies approximately as the fourth power of the shaft span and inversely as the square of the shaft diameter, as shown in Chap. 7. The logical solution, then, is to reduce the shaft span by reducing the number of stages. By 1953, experience had indicated that stage pressures could be increased from the 250 to 350 psi range commonly used at 3,600 rpm to as high as 800 or 1,000 psi.

In turn, these higher heads per stage could better be achieved by increasing the speed of rotation than by .increasing impeller diameters. The beneficial effect of

high-speed operation on the shaft deflection can be seen from a comparison of two pumps, each designed to handle 1,380 gpm against a total head of 6,550 ft (Table 22.2). This comparison indicates that the high-speed pump will have less than one-third the deflection of the low-speed pump. A more dramatic comparison is afforded by Fig. 22.6, which shows the rotors of the two pumps side by side. If the diametral running clearances are kept the same for both designs, the frequency of momentary contact will be materially reduced, if not eliminated altogether. As a result, the wear at the running clearances of the high-speed pump will be due solely to the erosive action of the leakage. These are the reasons for the increasing popularity of high-speed boiler feed pumps in large central stations operating at high pressures.

Boiler feed pump drives

The majority of boiler feed pumps in small and medium size steam plants are electric motor driven (Fig. 22.7). In the past, steam-turbine-driven standby pumps were installed as a protection against the interruption of electric power supply, but this practice has disappeared in central

TABLE 22.2   COMPARISON OF DESIGN DATA FOR HIGH- AND LOW-SPEED, HIGH-PRESSURE BOILER FEED PUMPS

Data	Low-speed pump	High-speed pump
Number of stages	9	4
Pump speed, rpm	3,465	9,000
Impeller diameter, in.	13.8	8.4
Shaft span, in.	87⅜	51.5
Average shaft diameter at impellers, in.	4¾	3¹⁄₁₆
$(L^4/d^2)/10^6$	2.575	0.75

Fig. 22.7   Electric-motor-driven boiler feed pumps in a central steam station

Fig. 22.8   Steam-turbine-driven boiler feed pump in a central steam station

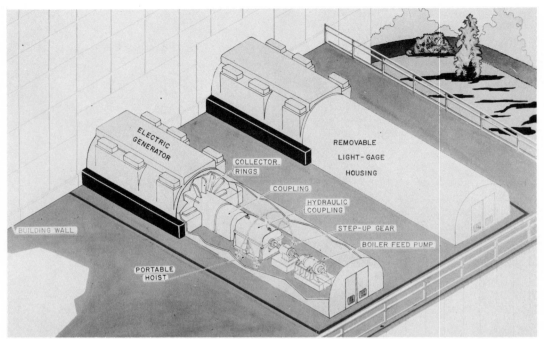

Fig. 22.9   Installation of main turbine-generator-driven boiler feed pumps

Hydraulic coupling and step-up gear are used between generators of cross-compound unit and two half-capacity boiler feed pumps.

steam stations and is encountered rarely even in industrial plants.

A trend away from electric motor **drive** in large central steam stations for units in excess of 150,000 kw has occurred in recent years. The first departure was to steam turbine drive (Fig. 22.8) because:

1. An independent steam turbine increases plant capability by eliminating the auxiliary power required for boiler feeding.

2. Proper utilization of the exhaust steam in the feedwater heaters can improve cycle efficiency.

3. In many cases, the elimination of the boiler feed pump motors may permit a reduction in the station auxiliary voltage, with an appreciable reduction in cost.

4. Driver speed can be matched exactly to the pump optimum speed.

5. A steam turbine provides variable speed operation without an additional component such as a hydraulic coupling.

The most dramatic development in the selection of the boiler feed pump drive, however, has been the use of the main turbine-generator to drive the boiler feed pumps (Fig. 22.9). Speed is varied by interposing a hydraulic coupling or a magnetic drive between the generator and the boiler feed pump unit.

Condensate pumps

Condensate pumps take their suction from the condenser hotwell and discharge to either the deaerating heater in open feedwater systems (Fig. 22.2), or the suction of the boiler feed pumps in closed systems (Fig. 22.3). These pumps, therefore, operate with a very low pressure at their suction, that is, from 1 to 3 in. Hg absolute. The available NPSH is obtained by the submergence between the water level in the condenser hotwell and the centerline of the condensate pump first-stage impeller. Because of the desire to locate the condenser hotwell at as low an elevation in the plant as possible and to avoid the use of a condensate pump pit, the available NPSH is generally extremely low, of the order of 2 to 4 ft. The only exception occurs when vertical-can condensate pumps are used, because they can be installed below ground elevation and, therefore, higher values of submergence can be obtained. Friction losses on the suction side must be kept to an absolute minimum. The piping connection from the hotwell to the pump, therefore, should be as direct as possible, of ample size, and have a minimum of fittings.

Because of the low available NPSH, condensate pumps operate at relatively low speeds, ranging from 1,750 rpm in the low range of capacities to 880 rpm or even less for larger flows.

It is customary to provide a liberal excess capacity margin above the full load steam condensing flow to take care of the heater drains that may be dumped into the condenser hotwell if the heater drain pumps are taken out of service for any reason.

Types of condensate pumps

Both horizontal and vertical condensate pumps are used. Depending on the total head required, horizontal pumps may be either single or multistage.

Figure 22.10 shows a single-suction, single-stage pump with an axially split casing that is used for heads up to approximately 100 ft. It is designed to have discharge pressure on the stuffing box. The suction opening in the lower half of the casing keeps the suction line at floor level. An oversize vent at the highest point of the suction chamber permits the escape of all entrained vapors that will be vented back to the condenser and removed by the air-removal apparatus.

Multistage pumps are used for higher heads. A two-stage pump is shown in Fig. 22.11, with the impellers facing in opposite directions for axial balance. By turning the impeller suctions towards the center, both boxes are kept under positive pressure to prevent leakage of air into the pump. For higher heads and larger capacities, a three-stage pump as shown in Fig. 22.12 may be used. The first-stage impeller is a double-suction type and is located centrally in the

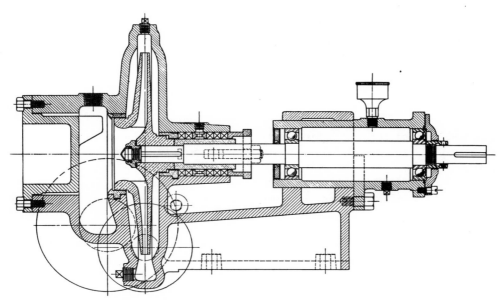

Fig. 22.10 Single-stage, horizontal condensate pump with axially split casing

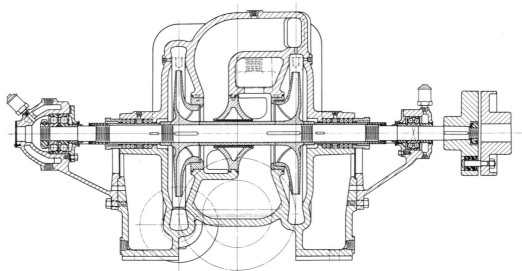

Fig. 22.11 Two-stage, horizontal condensate pump with axially split casing

pump. The remaining impellers are single suction types and are also arranged so that both stuffing boxes are under pressure. Two liberal vents connecting with the suction volute on each side of the first-stage double-suction impeller permit the escape of vapor.

Recently a trend has developed to the use of vertical-can condensate pumps (Fig. 14.17). The chief advantage of these pumps is that ample submergence can be provided without building a dry pit. The first stage of this pump is located at the bottom of the pumping element, and the available NPSH is the distance between the water level in the hotwell and the centerline of the first-stage impeller.

Although condensate pumps are designed so that their stuffing boxes are under pressure, air leakage can still take place into a particular pump, either when partial cavitation of a first-stage impeller reduces its discharge pressure to below atmospheric or when a pump is idle but the suction valve is left open. To prevent this air leakage, condensate pump stuffing boxes are always provided with seal cages. The water used for gland sealing must be taken from the condensate pump discharge manifold beyond all check valves.

Materials

Whereas condensate pumps handle essentially the same water as the boiler feed pumps—very pure and unbuffered—its low temperature makes it considerably less corrosive. In most cases, therefore, standard fitted pumps are used with cast-iron casings and bronze internal parts.

Condensate pump regulation

When a condensate pump operates in a closed cycle, its capacity is controlled by the same means that are used to control the delivery of the boiler feed pump. In other words, the two pumps can be considered as a combined unit insofar as their head-capacity curve is concerned. This head-capacity curve intersects the system-head curve, and flow is varied either by throttling in the boiler feed pump discharge and altering the system-head curve or by varying the speed of the boiler feed pump and altering the combined head-capacity curve of the two pumps.

In an open feedwater system, several means can be used to vary the condensate pump capacity with the load:

1. The condensate pump head-capacity curve can be changed by varying the pump speed (used very infrequently).

2. The condensate pump head-capacity curve can be altered by allowing the pump to operate in the "break" (Fig. 17.52 and 17.53).

3. The system-head curve can be artificially changed by throttling the pump discharge by means of a float control.

4. The pump can operate at the intersection of its head-capacity curve and the normal system-head curve. The net discharge is controlled by bypassing all excess condensate back to the condenser hotwell.

5. Methods (3) and (4) can be combined so that the discharge is throttled back to a predetermined minimum, but if the load, and consequently the flow of condensate to the hotwell, is reduced below this minimum, the excess of condensate handled by the pump is bypassed back to the hotwell.

Operation in the "break," or "submergence control" as it has often been called, has been applied very successfully in a great many installations. Condensate pumps designed for submergence control require specialized hydraulic design, correct selection of operating speeds, and limitation of stage pressures. The pump is operating in the break (cavitates) at all capacities. However.

this cavitation is not severely destructive in nature because the energy level of the fluid at the point where the vapor bubbles collapse is insufficient to create a shock wave of a high enough intensity to inflict physical damage on the pump parts. If, however, higher values of NPSH were required—as with vertical-can condensate pumps—operation in the break would result in a rapid deterioration of the impellers. It is for this reason that submergence control is not applicable to can-type condensate pumps.

The main advantages of submergence control are its simplicity and the fact that the power required for all operating conditions is less than with any other system. Disadvantages occur when the pump is operating at very light loads, however, as the system head may require as little as one-half of the total head produced in the normal head-capacity curve. In this case, the first stage of a two-stage pump produces no head whatsoever, and, if the axial balance was achieved by opposing the two impellers, a definite thrust is imposed on the thrust bearing, which must be selected with sufficient capacity to withstand this condition. In addition, no control is available to provide the minimum flow that may be required through the auxiliaries such as the ejector condenser.

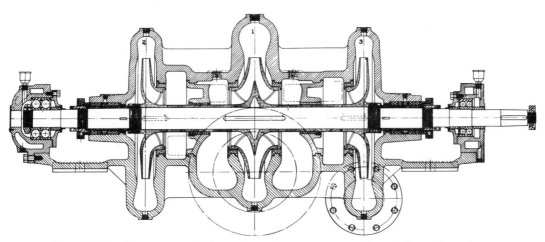

Fig. 22.12 Three-stage, horizontal condensate pump with axially split casing

The condensate pump discharge can be throttled by a float control arranged to position a valve that increases the system-head curve as the level in the hotwell is drawn down. It eliminates the cavitation in the condensate pump, but at the cost of a slight power increase. Furthermore, the float necessarily operates over a narrow range, and the mechanism tends to be somewhat sluggish in following rapid load changes, often resulting in capacity and pressure surges.

When condensate delivery is controlled through bypassing, the hotwell float controls a valve in a bypass line connecting the pump discharge back to the hotwell. At maximum condensate flow, the float is at its upper limit with the bypass closed and all the condensate is delivered to the system. As the condensate flow to the hotwell decreases, the hotwell level falls, carrying the float down and opening the bypass. Sluggish float action can create the same problems of system instability in bypass control as in throttling control, however, and the power consumption is excessive because the pump always operates at full capacity.

A combination of the throttling and bypassing control methods eliminates the shortcoming of excessive power consump-

tion. The minimum flow at which bypassing begins is selected to provide sufficient flow through the ejector condenser.

A modification of the bypassing control for minimum flow is illustrated in Fig. 22.13, which shows a thermostatic control for condensate recirculation. With practically constant steam flow through the ejector, the rise in temperature of the condensate between the inlet and outlet of the ejector condenser is a close indication of the rate of flow of condensate through the ejector condenser tubes. Therefore, an automatic device to regulate the rate of flow of the condensate can be controlled by this temperature differential. A small pipe is connected from the condensate outlet on the ejector condenser back into the main condenser shell. An automatic valve is installed in this line and is actuated and controlled by the temperature rise of the condensate. Whenever the temperature rise reaches a certain predetermined figure, indicating a low flow of condensate, the automatic valve begins to open, allowing some of the condensate to return to the condenser and then to the condensate pump, which supplies it to the ejector condenser at the increased rate. When the temperature rise through the ejector condenser is less than the limiting amount,

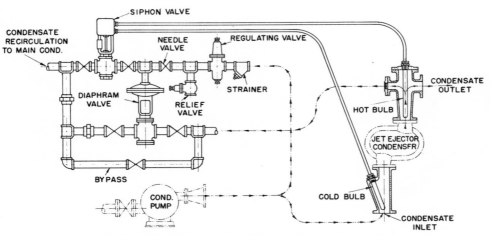

Fig. 22.13   Thermostatic control for condensate recirculation

indicating that ample condensate is flowing through the ejector condenser, the automatic valve remains closed.

Heater drain pumps

Condensate drains from the closed heaters must be returned to the feedwater cycle to avoid wasting water and the heat content of these drains. Two basic methods are available for this: (1) Drains can be flashed to the steam space of a lower pressure heater, or (2) drains can be pumped into the feedwater cycle at some higher pressure point.

When it leaves the hotwell of a closed heater, condensate is at saturation temperature; if it is passed to a region of lower pressure, part will flash into steam and part will remain as cooler liquid. Piping each heater drain to the next lower pressure heater is the simpler mechanical arrangement and requires no power-driven equipment. This "cascading" is accomplished by an appropriate trap in each heater drain. A series of heaters can thus be drained by cascading from heater to heater in the order of descending pressure, the lowest being drained directly to the condenser.

This arrangement, however, introduces a loss of heat since the heat content of the drains from the lowest pressure heater is dissipated in the condenser by transfer to the circulating water. It is generally the practice, therefore, to cascade only down to the lowest pressure heater and pump the drains from that heater back into the feedwater cycle, as shown in Fig. 22.14. Because the pressure in that heater hotwell is low (frequently below atmospheric even at full load), heater drain pumps on that service are commonly described as on "low pressure heater drain service."

In an open cycle, drains from heaters located beyond the deaerator are cascaded to the deaerator. Although the deaerator is generally located above the closed heaters, the difference in pressure is sufficient to overcome both the static and the friction

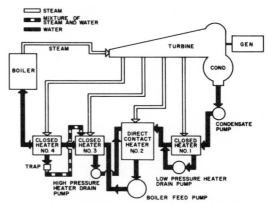

Fig. 22.14 Typical arrangement for heater drain pumps

losses. This difference in pressure decreases with a reduction in load, however, and at some partial main turbine load it becomes insufficient to evacuate the heater drains. Unless a pump is used to "boost" the drains back up to the deaerator, they have to be switched to either a lower pressure heater, or even to the condenser, with a subsequent loss of heat. To avoid these complications, a "high-pressure heater drain pump" is generally used to transfer these drains to the deaerator. Actually this pump has a "reverse" system head to work against—at full load, the required total head may be negative, whereas at light loads the required head is at its maximum.

Heater drain pumps, especially those in high-pressure service, are subject to more severe conditions than boiler feed pumps encounter:

1. Suction pressure and temperature are higher.
2. Available NPSH is generally extremely limited.
3. Transient conditions during sudden load fluctuations are the same as for boiler feed pumps, but are more severe.

Types of heater drain pumps

In the past, heater drain pumps were always designed as single-stage or multistage horizontal pumps, depending upon total head requirements. In the single-stage type, end suction pumps of the heavier "process pump" type were preferred for both low- and high-pressure service (see Fig. 22.44). Recently, however, the vertical-can pump (Fig. 14.17) has been frequently applied on heater drain service, as well as condensate service.

As previously described, the advantages of the vertical-can pump are lower first cost and an additional NPSH obtained because the first-stage impeller is lowered below floor level in the can. The vertical-can pump has several disadvantages, however. A horizontal heater drain pump is more easily inspected than a can pump. The external grease- or oil-lubricated bearings of the horizontal pump are less vulnerable to the severe operating conditions during swinging loads than the water-lubricated internal bearings of the can pump.

Although these disadvantages should not exclude the vertical-can pump from consideration, it is wise to limit its application to low-pressure service and give preference to the horizontal pump for high-pressure applications.

Materials

The same considerations that dictate the choice of materials of the boiler feed pumps should be used for heater drain pumps. In other words, low-pressure heater drain pumps can use cast-iron casings and bronze fittings if no evidence of corrosion-erosion has been uncovered. On high-pressure service, stainless steel fittings are generally mandatory, and 5 per cent chrome stainless steel casings should be used if the feedwater is unbuffered.

Regulation

Heater drain pumps should never be designed to operate on submergence control, because the required NPSH is considerably higher than that needed for condensate pumps and because cavitation at the higher temperature levels is very unfavorable to pump life. Capacity is generally controlled

by a throttle valve in the discharge of the heater drain pump, actuated by the heater hotwell level. Unlike a condenser hotwell, the heater hotwell contains very little storage. The control must be carefully arranged to prevent pumping the hotwell dry and steam from binding the pump. Just as in the case of boiler feed pumps, some minimum flow must be made to prevent an abnormal temperature rise. However, automatic bypass recirculation would not be economical, and it is best to allow a continuous bypass equal to the minimum recommended flow.

Installation

Heater drain pumps handle water at or near saturation pressure and temperature. Like condensate pumps, heater drain pumps should be adequately vented to the steam space of the heater. Because heater drain pumps and especially those on low-pressure service may operate with suction pressures below atmospheric, it is necessary to provide a liquid supply to the seal cages in the stuffing boxes.

Condenser circulating pumps

Condenser circulating pumps may be of either horizontal or vertical construction. For many years, reliable and efficient, low-speed, horizontal, double-suction, volute centrifugal pumps (Fig. 22.15) were preferred. This type of pump was usually located on the condenser room floor above a water intake tunnel. It usually required priming on starting, unless local conditions permitted the installation of the pump in a dry pit, with the impeller centerline located at the lowest prevailing water level. This pump design is simple but rugged and allows ready access to the interior parts for examination and rapid dismantling if repairs are required.

Although some installations of horizontal circulating pumps are still encountered, the larger central stations have generally switched over to vertical pumps. These

Fig. 22.15   Installation of double-suction, horizontal circulating pumps

fall into two separate classifications: (1) The dry-pit type, which operates surrounded by air, and (2) the wet-pit type, which is either fully or partially submerged in the water pumped.

The choice between these two types is somewhat controversial; very strong personal preferences exist in favor of one or the other. A great many installations of both types exist, however, giving equally satisfactory use in the field, both in performance and service.

A number of factors must be considered when choosing between dry- and wet-pit pumps for condenser circulating service. Some of the factors involved lend themselves to a straight economic evaluation, their advantages and disadvantages readily expressed in dollars. Other factors, no less important, are intangible, and only experience and sound judgment can give them their proper and deserved weight.

Mechanical considerations

The most popular type of pump during the past thirty years for condenser circu-

lating service in dry-pit installations has been the single-suction, medium specific speed, mixflo pump (Fig. 14.11). This design combines the high efficiency and low maintenance of the horizontal, double-suction, radial-flow centrifugal pump with lower cost and slightly higher rotative speeds.

Because of their suction and discharge nozzle arrangements, these pumps are ideally suited for vertical mounting in a dry pit, preferably at the lowest water level, so that they are self-priming on starting. They are directly connected to solid shaft induction or synchronous motors, either close-coupled (Fig. 22.16), or with intermediate shafting between the pump and the motor, which is then mounted well above the pump pit floor (Fig. 22.17).

Like the horizontal double-suction pump, the vertical dry-pit mixflo pump is a compact and sturdy piece of equipment. Its rotor is supported by external oil-lubricated bearings of optimum design. This construction requires the least attention, as the oil level can easily be inspected by the operator by means of an oil sight glass mounted at the side of the bearing or oil reservoir.

The pump construction further facilitates maintenance and replacements as the rotor is readily removed through the top of the casing. Therefore, the pump does not have to be removed from its mounting, nor the suction and discharge connections broken in order to make periodic inspections or repairs.

In recent years, power plant designers have shown a preference for the wet-pit, column-type, condenser circulating pump, although most recently this trend has appreciably slowed down, and several major utilities are reverting to the dry-pit type. One of the advantages of the wet-pit pump is that it is possible to locate the impeller below the surface of the water in the suction pit or well. With proper submergence, these pumps can operate at higher rotative speeds than the conventional volute or dry-pit mixflo pumps.

Unless qualified, the term "wet-pit pump" normally implies a diffuser pump. A conventional volute pump has been submerged into a pit and operated as a wet-pit pump, but this is the exception to the rule. The volute pump is essentially a dry-pit pump.

The wet-pit pump (Fig. 14.18–14.20) usually employs a long column pipe that supports the submerged pumping element. It is available with open main shaft bearings, lubricated by the water handled, or, preferably, with enclosed shafting and bearings, lubricated by clean, fresh, filtered water from an external source. Even in the latter case there is some danger of contamination of the lubricating water from seepage into the shaft enclosure tube during shutdowns. A water-lubricated bearing is never the equal of an oil-lubricated bearing even with the best design and care. Consequently, higher maintenance costs may be expected with the use of a wet-pit pump.

Larger power plants are generally located near centers of population and, as a result, often have to use badly contaminated water—either fresh or salt—as a condenser

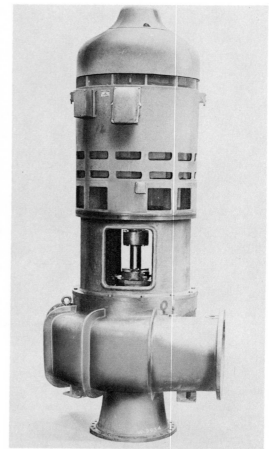

Fig. 22.16 Vertical, dry-pit, mixflo condenser circulating pump, with close-coupled motor drive

cooling medium. With such water, a fabricated steel column pipe and elbow would give short life, and cast iron, bronze, or even a more corrosion-resistant cast metal must be used. This results in a very heavy pump when large capacities are involved. Pulling up the column in a long pump requires special and expensive facilities, and, in addition, the discharge flange must be disconnected when withdrawing the pump and column from the pit.

To avoid the necessity of lifting the entire pump when the internal parts require maintenance, some designs (Fig. 14.20) are built so that the impeller, diffuser, and shaft assembly can be removed from the top without disturbing the column pipe assembly. (The driving motor must still be removed.) These designs are commonly designated as the pull-out type.

The selection of materials for either dry- or wet-pit pumps can vary considerably, depending on the character of the circulating water. A change of water in the dry-pit pump would require few changes in material, since the impeller and rings are the parts that would be primarily affected. The wet-pit pump, however, would require a careful selection of all submerged parts, particularly when sea water is to be pumped. Special alloys are usually required for the impellers and rings. Because of the danger of electrolytic attack, material changes may also often be required for such parts as the shaft, diffuser column pipe, and shaft enclosure tube.

Performance characteristics

Condenser circulating pumps are normally required to work against low or moderate heads. Consequently, it is important to keep friction losses in the system at a minimum by the proper selection of pipe size, conduit size, and types of fittings. Extreme care should be exercised in calculating the system friction losses, which include the condenser friction. If more total head is specified than is actually required, the resulting driver size may be unnecessarily increased by an appreciable percentage. For example, an excess of 1 or 2 ft in an installation requiring only 20 ft of head represents an increase of 5 to 10 per cent in excess power costs.

If a dry-pit pump is to be installed, the range or variation of suction lift must be determined very accurately and checked with the manufacturer to insure that the pump installation will avoid cavitation. Special precautions must be used to insure priming of the pump. Thus, either priming facilities must be provided, or the pump must be installed in a dry pit at such an elevation that the water in the suction channel leading to the pump will be main-

Fig. 22.17 Vertical dry-pit, mixflo condenser circulating pump with extended shafting to motor at higher elevation

tained at the level recommended by the manufacturer. A wet-pit installation does not present priming problems under any operating conditions, because the pump column can be made long enough to provide adequate submergence, even with minimum water levels in the suction well or pit.

The dry-pit pump has a performance advantage over the wet-pit pump. The former will generally have 3 to 4 per cent higher efficiency, resulting in 3 to 4 per cent lower power consumption.

The two types of pumps will also have widely different performance characteristics as far as the shape of the head-capacity and power-capacity curves are concerned. The specific speed of a given pump will very definitely affect the shape of these performance characteristics, and, although the shape of these curves can be varied by changing the impeller and casing waterway designs, the variation which can be obtained without affecting the pump efficiency very adversely is relatively small.

TABLE 22.3 COMPARISON OF WET-PIT AND DRY-PIT PUMPS FOR CONDENSER CIRCULATORS HANDLING SALT WATER

Typical for a 200 mw installation of a single-pass condenser using two half-capacity pumps

	Vertical, wet-pit column type with diffuser (removable rotating element, pull-out design)	Vertical, dry-pit, mixed-flow, volute (removable rotating element)
Parts		
Motor base	Cast iron or fabricated steel	Cast iron or fabricated steel
Impeller, nut, and rings	Zincless bronze	Zincless bronze
Diffuser	Zincless bronze	None
Pump barrel (bowl)	Zincless bronze	None
Suction bell	Zincless bronze	None
Discharge elbow (tee)	Nickel cast iron	None
Column pipe or casing	Nickel cast iron	Nickel cast iron
Shaft	Manganese bronze	Steel
Shaft sleeve(s)	"S" monel	Bronze or monel (in stuffing boxes)
Shaft cover pipe (tube)	Bronze	None
Line bearings	Leaded bronze or cutless rubber	Babbitted sleeve (one required)
Bearing lubrication	Clean, fresh water 5 to 10 gpm at 15 psi	Oil
Bolts, nuts, and screws under water	Monel	Monel
Shaft coupling(s)	Rigid—fabricated steel flanged	Rigid—fabricated steel flanged
Thrust bearing	In motor	In motor
Performance and size		
Capacity, each, in parallel	78,000 gpm	78,000 gpm
Total head, ft	22	22
Full load speed, rpm	290	220
Efficiency, per cent	84	88
Bhp, basis specific gravity 1.0	515	492
Motor hp	600	500
Motor rpm (synchronized)	300	225
Specific speed, N_s	7,900	6,000
Vertical thrust load, lb		
At design conditions	17,000	10,000
At shut-off	37,000	12,000
Minimum submergence required above suction bell		
For two-pump operation	Approx 5½ ft	Will self-prime if casing centerline is at low-water level
For one-pump operation (approx 100,000 gpm)	Approx 7 ft	
Diameter pump discharge and suction	60-in. × 72-in. bell	54 in. × 54 in.
Assumed setting and height	20-ft max—motor base to suction bellmouth	Close-coupled or motor mounted independently 20-ft max above pump casing centerline
Weight, each pump	60,000 lb	50,000 lb
Approximate prices		
Two pumps [1]	100 per cent (base)	72 per cent
Two motors [2]	100 per cent (base)	120 per cent
Total, pumps and motors	100 per cent (base)	86 per cent

[1] Selling price, based on materials listed, includes curb rings or sole plates but no suction piping or elbows. These prices will increase for service on highly aerated or polluted waters approximately as follows: For wet-pit type, add 50 per cent for Ni-Resist discharge elbow, column pipe, and bowl assembly, with stainless steel trim, stainless steel shaft and cover pipe; for dry-pit type, add 20 per cent for stainless steel impeller and rings.

[2] Basis is 40°C rise, 3-phase, 60-cycle, 4,000-volt, squirrel cage induction motors, solid-shaft, high-thrust type, weather-protected for outdoor installation.

The low specific speed, double-suction pump (Fig. 17.33) has a very moderate rise in head with reducing capacities and a non-overloading power curve with a reduction in head. The mixed-flow impeller (Fig. 17.34) with a specific speed of 4,000 has a steeper head-capacity curve and a reasonably flat power curve that is also non-overloading. As the specific speed increases, the head-capacity curve increases in steepness, and the curvature of the power curve reverses itself, hitting a maximum at the lowest flow. Finally, the curve of a high specific speed propeller pump (Fig. 17.35) has the highest rise in both head-capacity and power-capacity curves towards zero flow.

Depending on a multitude of factors, such as the amount of static and friction head, and the relative frequency of operation of two pumps in parallel, and of a single pump, there may be some advantages in a flatter or a steeper curve. But the steeply rising power curve is always a disadvantage, since it may introduce problems from the point of view of inrush current limitations. In general, it can be stated that the head range developed by the mixflo pump is ideal for condenser service; this pump is usually furnished with an enclosed impeller, which produces a relatively flat head-capacity curve and a flat power characteristic.

System hydraulics

The dry-pit pump is not overly affected by the suction well design, since the inlet piping and the formed design of the suction passages into the pump normally insure a uniform flow into the eye of the impeller. However, it is always advisable that the pump manufacturer be requested to review the installation drawings and suggest any changes that would improve the pump operation.

On the other hand, the higher speed wet-pit pumps are more sensitive to departures

from ideal flow conditions than the slow speed centrifugal volute pump or the medium speed mixflo pump. This sensitivity is especially pronounced with regard to irregularities occurring at the inlet side of the pump. Thus, special emphasis must be placed on the design of the suction intake. The design is particularly important because the small floor space occupied by each unit offers the temptation to reduce the size of the installation still further by placing the units closer together. If this is done, the suction arrangement may not permit the proper flow of water to the pump suction intake.

A discussion on the arrangements recommended for wet-pit pump installation is presented in Chap. 14.

Drivers

Whether a dry-pit or a wet-pit pump is used, the axial thrust and weight of the pump rotor are normally carried by a thrust bearing in the motor, and the driver and driven shafts are connected through a rigid coupling. The higher rotative speeds of the wet-pit pumps act to reduce the cost of the electric motors somewhat. This difference may be offset considerably, however, by the fact that the thrust load of the wet-pit pump is considerably higher than that of the dry-pit pump.

General considerations

The vertical dry-pit pump is generally cheaper in first cost than the wet-pit pump, especially when long column lengths (more than 20 ft from motor base to suction bell) are required, and special bronze or stainless steel construction must be used to avoid trouble from salt, brackish, highly aerated or polluted waters. Table 22.3 shows a typical comparison of the performance, construction, and relative costs for a two-pump installation. Each pump is designed to handle 78,000 gpm against a total head of 22 ft with a 20-ft setting. The 60-in. wet-pit hiflo pumps operate at 290 rpm and

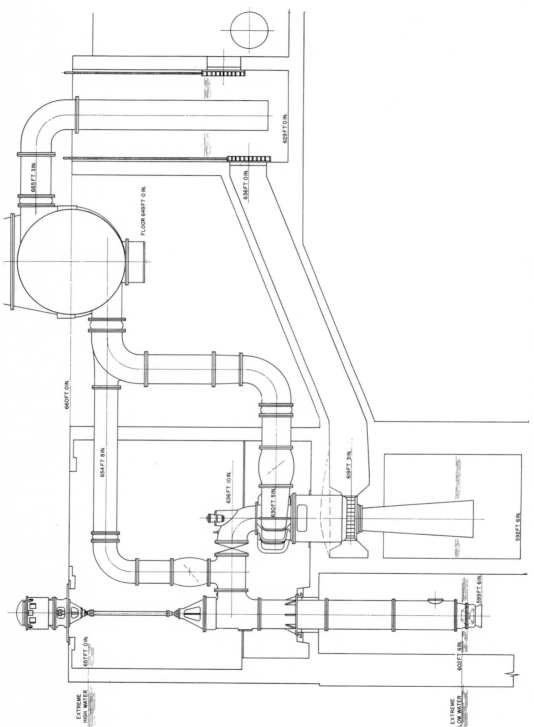

Fig. 22.18 Condenser circulating water circuit

*Circuit utilizes hydraulic turbine-driven booster pump to recover static head be-
yond possible siphon effect.*

have an efficiency of 84 per cent as compared to 220 rpm and 88 per cent for the 54-in. dry-pit mixflo pumps.

The comparison, which does not include the installation cost, considerably favors the dry-pit pump. There will be cases where the vertical dry-pit pump will be too expensive because of the extra cost assessed against it for the pump setting and the intake structure. This may be true for certain midwestern rivers, where water levels vary from 40 to 80 ft between low water and flood stages, and where additional concrete mass is required in the foundation to resist high water level pressures. But in many cases, if a very thorough analysis is made, it can be shown that the higher first cost of the dry-pit foundations and structure would be regained in a very few years of operating by the savings accruing from the lower maintenance and pumping power costs of the dry-pit pump.

Condenser circulating arrangement for high-static heads

Along midwestern rivers such as the Ohio, the Mississippi and the Missouri, where flood stages are as much as 80 ft above low water, it is customary to place the condensers in deep pits to maintain a normal pumping head and at the same time keep the turbine-generator above flood stage. This is very expensive construction to which is added the additional heavy expense of a long and large exhaust connection piece between turbine and condenser.

An arrangement of circulating water pumps has been devised for such an installation that eliminates the costly condenser pit construction and permits locating the condenser directly beneath the turbine as in the conventional type of plant with only a slight increase in the power required for pumping.

This method is shown in Fig. 22.18, which is a diagrammatic sketch of the circulating water circuit of a steam power plant recently constructed on the Ohio

River. The elevations, as shown, correspond to those of the plant; the amount of water involved is 55,000 gpm.

The arrangement, as shown, consists of two circulating pumps operating in series, each at full capacity. The primary pump is a propeller-type, motor-driven pump; the secondary or booster mixflo pump is mounted directly upon and driven by a hydraulic turbine.

Advantage is taken of maximum siphon effect by the discharge drop leg from the condenser. From the sealing or control well all of the circulating water is passed through the hydraulic turbine. Thus, 75 to 80 per cent of the static head from control well level to river level is available for useful work in the hydraulic turbine-driven booster pump.

With its high starting head characteristic, the propeller pump is able to deliver enough water to the condenser for starting purpose and bypasses the mixflo pump. When there is enough water in the control well, the hydraulic turbine is started and the propeller pump discharges directly to the mixflo automatically, and the two pumps operate in series at full capacity.

The booster pump functions until the river rises high enough to make the hydraulic turbine ineffective, at which point the motor-driven propeller pump operates alone, giving full capacity under the reduced head. The system described above is the most feasible for installations with considerable fluctuations in water level.

Another application of this same idea would apply to a contemplated power plant where it would be desirable, for reasons of lower construction cost, investment, accessibility, or geographical location, to build at an elevation considerably above a source of circulating water supply, even though flood conditions do not exist. In such an installation, the two pumps could be designed to operate in parallel, each pump discharging separately against the total head. The total capacity requirements would be proportioned between the

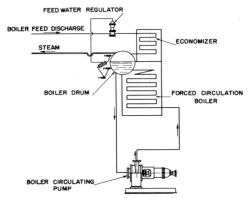

FEEDWATER REGULATOR

BOILER FEED DISCHARGE

STEAM

ECONOMIZER

BOILER DRUM

FORCED CIRCULATION
BOILER

BOILER CIRCULATING
PUMP

Fig. 22.19 Forced or controlled circulation cycle

two pumps, with all water being returned through the hydraulic turbine.

If flood hazards do not exist, horizontal pumps and drivers can be applied, providing other conditions are suitable.

Boiler circulating pumps

The forced circulation or "controlled circulation" boiler, as it is frequently called, has certain advantages over the conventional natural circulation boiler in the opinion of some power plant designers. The major advantages they claim for controlled circulation are:

1. Use of smaller diameter tubes
2. Freedom of tube circuit arrangement and distribution
3. Reduction in number and size of downtakes and risers
4. Lower over-all boiler weight and, consequently, lower structural support weight
5. Greater flexibility of operation.

A negative factor is that the forced circulation must be accomplished by pumps that are subject to most severe conditions of service. The circulating pumps take their suction from a header connected to several downcomers, which originate from the bottom of the boiler drum. They develop a pressure equivalent to the friction losses through the various tube circuits operating in parallel (Fig. 22.19). Thus, in the case

of a boiler operating at 1,900 psig, the boiler circulating pump must handle feedwater at approximately 630°F under a suction pressure of 1,900 psig. Such a combination of high-suction pressure and high-water temperature at saturation impose very severe conditions on the circulating pump stuffing boxes, making it necessary to develop special designs for this part of the pump.

The net pressure to be developed by these pumps is relatively low, ranging from 30 to 100 psi, at most, and therefore, these pumps are all built in a single stage. Because of the difficult conditions imposed on the stuffing boxes, it has always been preferable to use single-suction, end-suction designs, with a single stuffing box. This construction has disadvantages, however; as described in Chap. 5, an extremely severe axial thrust is imposed on the pump, placing a load of several tons on the thrust bearing. In many cases, a thrust-relieving device must be incorporated for start-up.

A second difficulty arises from the fact that the forced circulation pumps have a relatively limited NPSH, whereas their capacity may be from four to eight times the maximum boiler feeding capacity. Although the low design net pressures permit the use of operating speeds much lower than for boiler feed pumps, the suction problem can become relatively difficult. The pumps can take advantage of the fact that required NPSH for satisfactory operation is reduced at high temperatures (Chap. 17) even though they are started before the water in the boiler drum has reached this high temperature and, therefore, before this beneficial effect on required NPSH has taken place. Remedies for this situation will be discussed later.

Two general types of construction are used for boiler circulating pumps: (1) The conventional centrifugal pump with various stuffing box modifications, and (2) the submersible motor pump of either the wet- or dry-stator type (Chap. 20).

In the lower boiler pressure range—up to

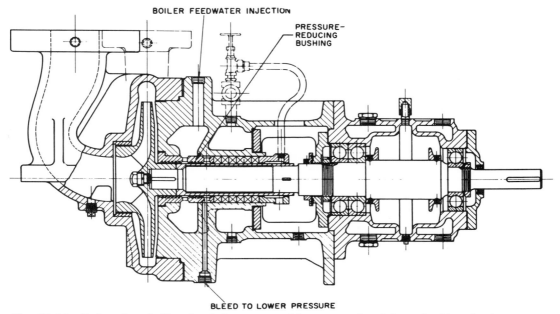

Fig. 22.20   End-suction, boiler circulating pump with pressure breakdown bushing for low-pressure range

500 or 600 psig—a construction as shown in Fig. 22.20 may be used. The pump is of the same general type as is used on high-pressure heater drain service. The packed stuffing box is preceded by a pressure-reducing bushing. Feedwater from the boiler feed pump discharge, at a temperature lower than in the boiler drum and at a pressure somewhat higher than pump internal pressure, is injected in the middle of this bushing. Part of this injected feedwater proceeds towards the pump interior, making a barrier against the outflow of high-temperature water. The rest proceeds outwardly to a bleed portion of the bushing from where it is bled to a lower pressure. The packing need only withstand the lower boiler feed pump temperature and a much lower pressure than boiler pressure.

More sophisticated designs are required for the higher pressures from 1,800 to 2,800 psig (Fig. 22.21). The sealing of the shaft is accomplished by two floating-ring pressure breakdowns and a water-jacketed stuffing box. Boiler feedwater is injected at a point between the lower and upper stack of floating-ring seals at a pressure approximately 50 psi above the pump internal pressure. Here again, part of this injection leaks into the pump interior. The rest leaks past the upper stack of seals to a region of low pressure in the feed cycle. Leakage to atmosphere is controlled by the conventional stuffing box located above the upper stack.

Manufacturers who build both the injection-sealed type and the submersible circulating pump say that there are advantages to both types; the former is somewhat lower in first cost, but the latter eliminates the injection and bleed-off system.

This injection requirement may pose a problem in certain installations. If the boiler feed pumps operate at constant speed, the discharge pressure always exceeds the boiler drum pressure by a very substantial margin, a margin that is throttled in the feedwater regulator. If the boiler feed pump operates at variable speed, however, and the feedwater regulator is eliminated, there is no appreciable excess pressure between the feed pump discharge and the boiler drum at low flows. Consequently, it becomes necessary either to provide a small booster pump for the injection flow, or to install a valve in the feed pump discharge. The valve creates sufficient artificial friction to provide the excess pressure necessary to permit the injection flow to take place, even at low flows to the boiler.

As described previously, the available NPSH may not be sufficient at start-up, when the water in the boiler is cold. Therefore, certain installations include two-speed motors, so that a lower NPSH is required at start-up while operating at the lower of the two speeds. There is an added advantage to this arrangement: Under normal op-

erating conditions, the feedwater will be at boiler saturation temperature and, therefore, will have a specific gravity of as low as 0.60; at start-up, however, the specific gravity will be 1.0. The power consumption on cold water, therefore, would be 65 per cent higher than in normal operation if the pump operates at the same speed, and a much larger motor would be required. If a two-speed motor is used, however, and the pump is operated at lower speed when the water is cold, the motor need be no larger than the maximum horsepower under normal operating conditions.

The pump can be protected against low available NPSH by another means which is not only especially effective against the possibility of fluctuation in boiler pressure and temperature, but also avoids possible long-term damage from operating too close to cavitation range. This method consists of subcooling the feedwater entering the boiler circulating pump and increasing the available NPSH by lowering the vapor pressure of the water. This subcooling is accomplished by allowing all or a portion of the boiler feed pump discharge to mix with the circulated flow ahead of the circulating pump suction (Fig. 22.22). Figure 22.23 gives the temperature depression achieved by this mixing in terms of both the temperature difference between the downcomer and the injected feedwater, and the ratio between the injected and total flows. The effective increase in NPSH for a given temperature depression and at a given boiler drum saturation temperature is plotted in Fig. 22.24. For example, if it is assumed that 100 ft surplus NPSH are desired for an installation involving a 1,900-psig boiler (saturation temperature = 630°F), the required temperature depression is only 2°F. If injection feedwater is available at 330°F, the difference between saturation and injection temperatures is 300°F. The ratio of injected flow to the total circulation flow need be only 0.0067, or less than 1 per cent.

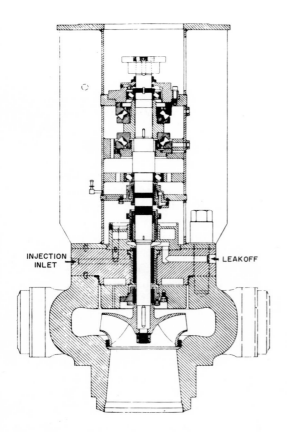

INJECTION INLET

LEAKOFF

Fig. 22.21 Vertical, injection-type, boiler circulating pump for high pressures

(Courtesy Ingersoll-Rand.)

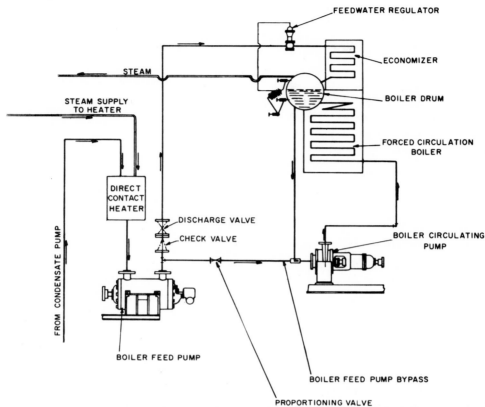

Fig. 22.22 Arrangement of boiler circulating cycle with subcooling

Feedwater is injected directly into circulating pump suction.

Special nuclear power applications

In oversimplified form, the nuclear energy steam power plant only differs from the conventional power plant in the fact that it uses a different fuel. Thus, the so-called "secondary cycle" (consisting of turbogenerator, condenser and auxiliaries, and boiler feed pumps) is not very different from its counterpart in the conventional steam power plant. The main differences are a desire for even greater equipment reliability and a preference for absence or minimum of leakage, to avoid any possibility of contamination with radioactive material.

One other major difference in certain nuclear power installations concerns the boiler feed pumps The rate of heat flow into the heat exchanger in the "primary cycle" (the boiler) is largely independent of station load, because neither the flow on the primary side nor the heat output of the reactor vary. Since this heat must be carried away at a constant rate, a temperature rise is permitted in the boiler. Consequently, the steam pressure generated in such a heat exchanger will rise with a decrease in load. This has the same effect as if conventional power plant were arranged to have a boiler pressure that rises with a reduction in load.

The exact value of the pressure rise from full load to no-load in a nuclear power plant will vary because different reactors will have different pressure rise characteristics. The system-head curve will have the

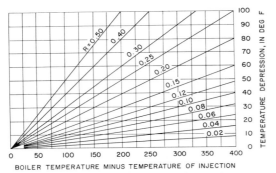

Fig. 22.23 Temperature depression

$$dT = r \ (T_1 - T_2)$$

in which:
  dT = *temperature depression*
  r = q/Q
  q = *injected capacity, in pounds per hour*
  Q = *total flow, in pounds per hour*
  T_1 = *boiler temperature, in degrees Fahrenheit*
  T_2 = *temperature of injection, in degrees Fahrenheit.*

shape shown in Fig. 22.25. Therefore, the boiler feed pump may have to be designed for a higher total head at maximum capacity than otherwise necessary, to insure that it will develop sufficient head at shutoff (Fig. 22.26). A paradox occurs: The discharge of the boiler feed pump must be throttled at full load more than at reduced loads; as an alternate, a variable speed device will slow the pump down at full load.

The primary coolant pump that circulates the heat exchange medium through the reactor, and then through the heat exchanger serving as a boiler, has to meet conditions of service similar to those of a circulating pump with a controlled circulation boiler. Therefore, the problems are similar in both cases, except for the extreme danger of leakage. Until now, canned rotor submersible pumps, which eliminate all external leakage, have been preferred. Too many designs are being currently developed to make it advisable to illustrate them here, as they would be only of passing interest.

The same is true of liquid-metal, primary-coolant pumps used in sodium- or sodium-potassium-cooled reactors. The in-terested reader should contact pump manufacturers, who can give him the latest information on this equipment.

WATER WORKS

The supply of water to industry and residential users is one of the major fields of application for centrifugal pumps. Both ground water from shallow or deep wells and surface water from rivers, lakes, or artificial reservoirs are used as a source of supply. Except for some well water, which may be sufficiently clear to require little treatment, most raw water must be processed to remove suspended matter and bacteria. Thus, raw water is transported by centrifugal pumps from its source to a purification plant. The amount of treatment required for the raw water will depend on the use to which it will be put and the character of the impurities it contains. Raw water may be subject to any or all of the following treatments: Coagulation, sedimentation, filtration, activation, chlorination, fluoridation, and softening. Because pumps that transfer raw water from a surface supply to a purification plant usually operate against a low discharge pressure, they are known as "low-service" pumps.

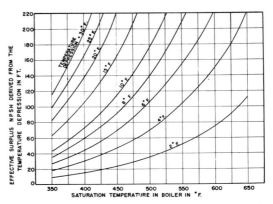

Fig. 22.24 Surplus NPSH obtained by sub-cooling

In terms of temperature depression and boiler saturation temperature.

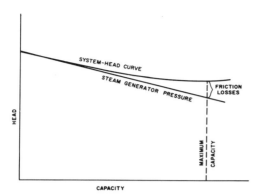

Fig. 22.25 System-head curve of boiler feed pumps in some nuclear power plants

Pumps that discharge into distribution systems after purification must generate higher heads and are known as "high-service" pumps. The distribution system consists of large-size transmission pipes, which feed into a network of medium- and small-size piping.

The following are the types of centrifugal pumps most frequently used in the water works field:

1. Horizontal-shaft, end-suction, single-stage, volute, either frame- (Fig. 2.12) or motor-mounted (Fig. 1.14) pump.

2. Horizontal-shaft, double-suction, volute, side-discharge, side- (Fig. 2.10) or bottom-suction (Fig. 2.19) pump, with casing axially split. Occasionally, a bottom-suction, bottom-discharge pump is used.

3. Horizontal-shaft, single- or double-suction, multistage (usually two, or possibly three, stages) volute pump.

4. Series units, consisting of two, or occasionally three, horizontal-shaft, double-suction, volute pumps, with casings axially split, connected in series with separate or common driver (Fig. 2.20 and 2.21).

5. Vertical-shaft, single-suction, single-stage, dry-pit, volute pump (Fig. 14.11 and 2.7).

6. Vertical-shaft, wet-pit, single-stage or multistage turbine (Fig. 14.13 and 14.14) or propeller pump (Fig. 14.18 and 14.19).

The horizontal, end-suction, single-stage, volute, frame- or motor-mounted pump is an ideal design for small to medium capacities. This pump has not found universal acceptance in the water works field, possibly because many pumps of this type have been light-duty lines made for intermittent service and, therefore, not suitable for water works service. Currently, most manufacturers offer substantial designs as well as light-duty lines in this construction. These frame- or motor-mounted substantial designs are volume built, usually with impeller designs that load up standard motor sizes. They are standardized production line units and, in most cases, are built in lots for stock. As a result, the manufacturers cannot tailor the impellers for specific service conditions, nor can they afford to make special tests to demonstrate the pump performance. Nevertheless, this type of pump has considerable application for small-capacity water works service and can be obtained at reasonable cost, provided it is purchased without voluminous specifications or special guarantee requirements.

In general, the horizontal-shaft, double-suction, volute pump is the most commonly used type for medium capacities up to 20 mgd and for heads up to 300 ft or more. At this capacity the vertical-shaft, single-suction, single-stage pump sometimes makes a

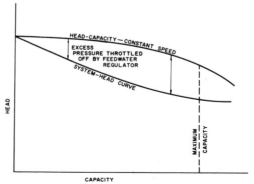

Fig. 22.26 Operation of boiler feed pumps at constant speed in system shown in Fig. 22.25

more economical installation, when all factors are taken into account. In the 20- to 40-mgd capacity range, local conditions may favor either type, but over 40 mgd the probability is that the vertical pump would be the most economical.

The vertical turbine pump is universally used for pumping from deep wells. There has been some tendency to apply this type to shallow wet pits in place of the conventional horizontal-shaft or vertical dry-pit types. Although the vertical turbine pump installation has a lower first cost in many cases, the horizontal-shaft or vertical-shaft, dry-pit pump has a much lower maintenance cost and, therefore, is favored.

SEWAGE

For centuries, inland rivers and coastal waters afforded a convenient disposal of various untreated wastes, such as domestic sewage and storm water drainage. As technological civilization progressed, industrial-process wastes of many kinds, drainage of contaminated mine waters, and liquids such as brine from oil fields were added. The resulting pollution made these river waters unfit for consumption and some coastal waters extremely unpleasant to the nearby population. For a number of years, therefore, local and national government bodies have exerted greater and greater efforts to pass legislation that will force communities and industries to treat wastes before ultimate disposal.

Today, municipal sewage treatment prevents excessive pollution of water caused by the discharge of raw domestic sewage and liquid wastes from cities and towns. In the same manner, individual industries are treating their wastes, either voluntarily or under the pressure of legislation.

The degree of sewage treatment in a particular installation depends primarily upon the amount of pollution to be disposed of and the size of the stream flow into which wastes are discharged. Partial treatment

may involve merely screening of the wastes and chlorination, followed by some sedimentation. Intermediate treatment may include the addition of chemical precipitation, as well as some filtration. Finally, complete sewage treatment may also include multistage high-rate filtration, trickling filtration, and an activated sludge process. For a complete discussion of various types of sewage treatment the reader is referred to the numerous textbooks available on the subject.

All of these processes require the use of pumping equipment to move liquids—either raw or treated sewage—from one part of the process to another, and finally to the ultimate disposal of the effluent.

Pumps suitable for pumping sewage must be capable of passing the solids contained in the sewage. Raggy and stringy solid material gives the most trouble and, except for very large pumps with stationary covers over the shaft, makes it undesirable to use pumps in which the shaft projects through the suction waterways. For the same reason, diffuser vanes with thin edges that will catch rags and stringy material are also undesirable. Thus, almost all sewage pumps are of the end-suction, volute type (Fig. 14.2) and, in general, the impellers are of a special non-clogging design (Fig. 4.8). Most applications of sewage pumps are for low heads up to 70 to 80 ft maximum in the smaller capacity range and up to 40 to 50 ft maximum for large capacities, so that single-stage pumps can always be used.

The standard water closet bore is $2\frac{1}{2}$ in. in diameter and non-clogging impellers are specified to pass objects up to at least $2\frac{1}{2}$ in. in diameter. If screens with finer mesh than $2\frac{1}{2}$ in. or comminutors that macerate and cut up foreign matter in the raw sewage are used ahead of the pumps, some specifications reduce the requirement to 2 in. or even $1\frac{1}{2}$ in., to take advantage of the resulting increase in pump efficiency.

Vertical sewage pumps are preferred because they can be located so that priming is unnecessary (Fig. 22.27). The centerline of the impeller is located at an elevation below the suction supply. Dry-pit construction is generally preferred, although wet-pit pumps are sometimes used for intermittent duty or for sewage sump draining.

The same general type of end-suction pump with special non-clogging impellers is used for secondary and activated sludge pumping in the sewage treatment pumps as well as for sludge recirculation. For recirculation or transfer of settled sewage that contains neither large solids nor stringy material, both wet- and dry-pit vertical pumps are used, with a trend to axial-flow, wet-pit propeller pumps when the total heads are very low.

Both wet-pit, axial- or mixed-flow propeller pumps and dry-pit, vertical or horizontal volute pumps are used for storm water pumpage, whichever type best suits the individual problem.

A trend may exist toward the elimination of multiple sewage pumps or stations in favor of single pumps of larger capacity, because maintenance is costly with many small installations spread over a large area. In contrast to this is a growing interest in small package-type sewage disposal plants with capacities as low as 25 to 100 gpm for small public installations such as schools.

The majority of sewage pumps are electric motor driven, although some gasoline engine or diesel engine installations may be encountered. Where wide variations in pumping capacities and total heads are encountered, variable speed operation is becoming more and more popular. The variable speed is obtained through the use of magnetic drives, and the operation is generally made fully automatic, responsive to level control.

DRAINAGE AND IRRIGATION

Drainage pumping is employed either to recover low-lying lands otherwise suitable for farming, or to maintain areas subject to infiltration flooding in relatively dry con-

Fig. 22.27 Vertical, bottom-suction sewage pump with motor supported directly on pump

dition. In many cases this latter function is essentially similar to flood control of areas lying below water level that are protected by levees. Capacities of drainage and flood control pumps are generally high, although the total head is relatively low—frequently as low as 5 to 10 ft. Vertical and horizontal, mixed-flow and axial-flow pumps are used on this service. When higher heads, up to 50 ft, are encountered, mixed-flow pumps only are used.

Drainage pumps are subject to the same problem as sewage pumps; the water may contain a considerable amount of foreign matter, including small branches, vegetable matter, trash, abrasive sand, and mud. Although a bar screen ahead of the pump suction or coarse mesh strainers on the bell-mouth of the suction piping can eliminate some of these solids, it is still necessary to use the same type of non-clogging impellers as for sewage pumps. When large-capacity pumps are involved, however, the waterways between the vanes of a normal impeller will pass sufficiently large solids. Similarly, vertical-turbine pumps or propeller pumps with diffusers are undesirable because the debris carried in the drainage water could catch on the diffuser vane edges.

Ditch-type or flooded-type irrigation presents the same problems as drainage, namely, heads are low and the water handled may be very dirty and even contain abrasive sand or mud. Capacities of pumps on irrigation service, however, are limited to a lower range than on drainage service. Sprinkler system irrigation, on the other hand, demands a practically clean water supply. Therefore ordinary general-service centrifugal or vertical-turbine pumps are used, depending on the source of supply and the hydraulic requirements.

Pumps on this service may require total heads from 100 to 200 ft and higher.

In some cases, irrigation pumps are installed in permanent locations whereas in others pumps are portable or semiportable. Permanent pumps can be driven by electric motors (Fig. 22.28) or diesel or gasoline engines. Portable pumps are mostly gasoline or diesel engine driven and are mounted on skids, trailers, or trucks.

Except for wet-pit, submerged-type pumps, irrigation pumps must be provided with some form of priming device (Chap. 21). Figure 22.29 shows a diesel-engine-driven irrigation pump mounted on a portable skid; a jet primer, which uses the engine exhaust, is employed to prime the pump. A clutch between the pump and the engine permits the engine to idle with the pump disconnected while it is being primed.

Unless there is a considerable amount of sand or grit in the water, the manufacturer's standard materials for ordinary cold

Fig. 22.28   Propeller type, 42-in., horizontal pump on drainage service
Electric-motor driven through a Multi-V drive

Fig. 22.29   Diesel-engine-driven, mixflo drainage pump with exhaust jet-primer mounted on portable skid-type base

water service are used. Special materials for parts subject to the erosive action of sand may be justified for the larger pumps; this may include nickel cast-iron casings, chrome-steel wearing rings, chrome-steel shaft sleeves, and sometimes chrome-steel impellers.

FIRE PUMPS

Even though the centrifugal fire pump is purchased with the hope that it will never be operated, it is a very sound and profitable investment. Not only is it important to provide a fire protection system, but the reduction in fire insurance premiums will generally amortize the initial cost of the installation in a few years.

Centrifugal fire pumps are classified in two groups from the point of view of function: If the installation to be protected is located where municipal fire protection is not available, fire pumps are installed to provide this protection by themselves; in other cases, inadequate water pressure supply is augmented by booster fire pumps. The latter is probably the more frequent case, and most large buildings in urban areas include such booster fire pumps.

Because of the interest insurance companies have in the satisfactory protection obtained from fire pumps, the size and type of this equipment, as well as many details

Fig. 22.30   Motor-driven centrifugal fire pump with underwriters' approved fire fittings

of its installation, are subject to the approval of several governing bodies connected with insurance companies. These pumps, therefore, must be built in accordance with the requirements of the Underwriters Laboratories, Inc., the National Board of Fire Underwriters, or the Associated Factory Mutual Fire Insurance Companies, Inc. In addition, an installation may have to be locally inspected and approved. These underwriters' organizations issue standards on approved equipment; they should be consulted for specific details not given here. In general, these standards specify both the hydraulic characteristics that must be met before acceptance and the auxiliary equipment that must be furnished with the installation.

Fire pumps are classified in the standards as low-pressure service (40 to 100 psi) and high-pressure service (100 psi and above). Both horizontal and vertical pumps are acceptable for fire service. Horizontal pumps are limited to a total suction lift of 15 ft when delivering 150 per cent of rated capacity. The standards specify a permissible range of heads at shut-off and at 150 per cent of rated capacity and provide standard capacity ratings. Approved ratings range from 500 to 2,500 gpm. Special ratings of 200 to 500 gpm are also listed for limited-service fire protection. These are subject to special consideration in each individual case. Vertical-shaft, turbine pumps are listed in the underwriters' standards for capacities between 500 and 2,500 gpm.

Where the power supply is adequate and reliable, electric-motor drive is preferable for fire-pump service (Fig. 22.30). The electrical equipment must comply with the provisions of the National Electrical Code, except as modified by the standards described above. Steam turbines and internal combustion engines may be used to drive fire pumps where other sources of power are unreliable or not available (Fig. 22.31).

Any number of the following accessories may have to be supplied with a fire pump installation, depending on the particular local requirements:

CHEMICAL INDUSTRY

1. Automatic air release valve

2. Circulation relief valve (for horizontal pumps)

3. Eccentric tapered reducer at suction nozzle (for horizontal pumps)

4. Hose manifold with hose valves

5. Suction and discharge pressure gages

6. Priming connections (for horizontal pumps operating with suction lift)

7. Relief valve and discharge cone

8. Splash shield between pump and motor (for horizontal pumps driven by open motors)

9. Test valves with piping connections

10. Umbrella cocks (for horizontal pumps)

11. Suction strainer (for vertical submerged pumps).

The chemical industry presents the widest variety of pumping problems and different liquids to pump of any industry today. Raw materials in liquid form are generally delivered to a chemical plant in tank cars and must be transferred into reservoirs. They are later transferred to the plant and pumped from one part of the process to the next. The semifinished and finished products undergo many changes in chemical composition, consistency, temperature, viscosity, and corrosiveness.

Centrifugal pumps are used in approximately 90 per cent of all the applications that involve the handling of corrosive liquids in chemical plants. The most compelling reason for this preference lies in

Fig. 22.31 Gasoline-engine-driven centrifugal fire pump with underwriters' approved fire fittings

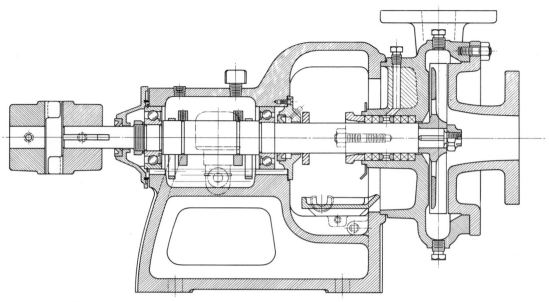

Fig. 22.32   Single end-suction open impeller pump used in chemical service

the much lower cost of centrifugal pumps made of special alloys.

The most important factor in the selection of pumps for chemical service is the selection of materials to withstand the corrosive or corrosive-abrasive liquids that have to be pumped. As described in Chap. 16, centrifugal pumps have been made of every imaginable material, including graphite, rubber, plastics, glass, and ceramics. The Standards of the Hydraulic Institute recommend a very extensive listing of materials and their suitability for pumps

Fig. 22.33   Chemical pump shown in Fig. 22.32

Antimonial lead drip pan is used to carry away stuffing box leakage.

handling a variety of chemically active liquids. Although these recommendations are followed exactly, as little as 1 per cent variation in the constituents of an alloy may alter the rate of corrosion with a given liquid by as much as several hundred per cent. Variations in foundry techniques may also have the same effect without any change in the alloy composition. Finally, the unexpected presence of even minute amounts of alien components in the liquid pumped, a slight increase in temperature, or the presence of dissolved gases may seriously alter the ultimate life of a given material. In many cases, therefore, the operators who are using the equipment are the only ones who have a firm basis for the proper selection of the material to be used.

Most manufacturers specializing in centrifugal pumps for chemical service maintain extensive corrosion data files which are the result of many years of field experience and corrosion testing. As a result, it is generally sound policy to give the manufacturer as much information on the liquid as possible even when a particular material is being specified.

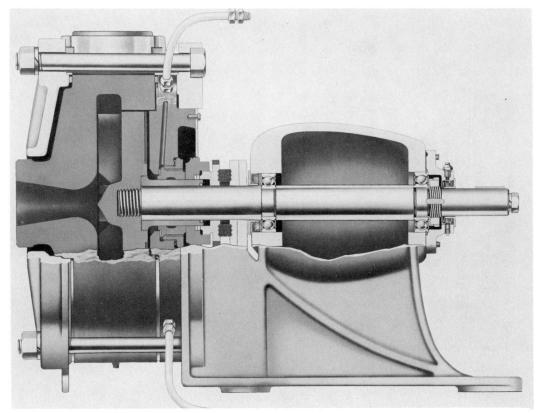

Fig. 22.34 Chemical pump with graphite liquid-end parts

"Karbate" impervious graphite pump. (Courtesy National Carbon Co.)

The extreme severity of chemical-service applications has resulted in the acceptance of very different criteria of satisfactory life as compared to most centrifugal pump uses. A life of six months to two years for an impeller, for example, may be quite acceptable in many cases, although this would not be tolerated on boiler feed service.

Because so many applications require pumps made of special alloys or special materials, designers tend to use the simplest possible constructions and configurations. The stuffing box problem is a particularly difficult one in many chemical process applications, with the result that the most favored style of chemical pump is the single-stage, single-end-suction, overhung-impeller type, generally with an open or semiopen impeller (Fig. 22.32 and 22.33). The single-stage, double-suction pump with an axially split casing is widely used in the chemical industry, but it is restricted to general service, cooling water service, or the handling of liquids that have the same corrosion and abrasion qualities as water. This pump is not suitable to corrosive service because: (1) It is too expensive to build with the special materials required, and (2) it has two stuffing boxes, whereas single-end-suction, single-stage pumps have only one.

This discussion will be limited to general remarks on certain specific features of centrifugal pumps intended for chemical service that can assist the user in evaluating any particular design. Many special design refinements can be furnished that

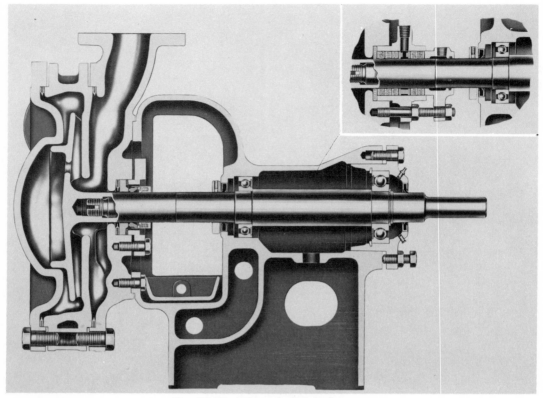

Fig. 22.35   Glassed chemical pump

(*Courtesy Goulds Pumps, Inc.*)

would prove to be an asset to a plant with well-trained mechanics, but these same refinements may prove to be a disadvantage if certain adjustable or complicated features are improperly used. As a general rule, therefore, the simplest construction will prove to be best.

The serious problems that stuffing boxes frequently present in chemical service can often be remedied by thorough understanding and intelligent handling.

Packing used on chemical service can be impregnated with various lubricants. Unfortunately, these lubricants do not last very long under corrosive conditions. Furthermore, the abrasive nature of many liquids shortens the life of the packing and of the shaft sleeves. In most cases, therefore, it is necessary to use seal cages with

an independent supply of sealing liquid that is non-corrosive and free of abrasive particles. The easiest and cheapest liquid to apply is water, and it can be used in most cases. If the supply pressure is properly regulated, the amount of sealing liquid that will enter the pump and dilute the pumped liquid is insignificant. For example, if 10 gal per day of sealing water enters a pump handling 100 gpm, the dilution is only 70 ppm. If even this amount is objectionable, a suitable insoluble grease can be fed to the stuffing box seal cage under continuous pressure. There are standard, low-cost grease sealers available on the market for this.

Mechanical seals have been used very successfully on chemical pumps, provided they are properly installed and properly op-

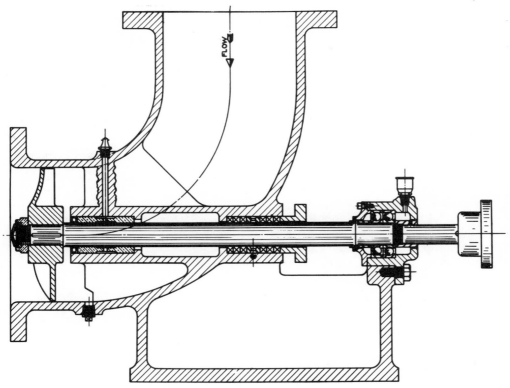

Fig. 22.36 Elbow-type propeller pump

Fig. 22.37 Elbow-type propeller pump shown in Fig. 22.36

Fig. 22.38   Rubber-lined pump

(Courtesy Allen-Sherman-Hoff Pump Co.)

erated. For example, if the liquid pumped contains abrasives, they must be prevented from penetrating the seal. Means for cooling the seal gland are generally necessary. If the pump handles liquids at high temperatures, cooling must be supplied to the chamber around the seal as well.

Not all pumps can be converted from a packed box construction to the use of mechanical seals. In many cases, the stuffing box packing assists the bearings in supporting the shaft. Elimination of the packing and of this assistance may lead to an increase in the deflection, which would make a mechanical seal impossible to maintain.

A few special pump constructions are shown in Fig. 22.34–22.39. These figures illustrate some of the various designs made available by manufacturers to answer special needs of the chemical industry.

Because graphite is impervious to a wide variety of corrosive liquids, pumps are often made of this material. However, because of the possibility of breakage, the graphite

liquid end parts are encased in cast steel armoring (Fig. 22.34).

Glass is another material with many applications in chemical service, but which breaks easily. One design utilizes "glassed" metal parts (Fig. 22.35), so constructed that every part of the pump coming into contact with the liquid is glassed. The glass is permanently fused to the metal in successive applications and firings, and the casing is made in three pieces to facilitate the glassing operation. The joints between the suction piece, the casing, and the casing cover are sealed by retained Teflon gaskets. A special feature of this design is that the impeller is so arranged in the casing that the stuffing box is under suction pressure. The design illustrated in Fig. 22.35 is fitted with a mechanical seal.

Figures 22.36 and 22.37 show the sectional and external views of an elbow-type propeller pump frequently used in evaporator transfer service or as an external circulator pump. Agitators use a modification

Fig. 22.39   Method of inserting rubber inner casing into the rubber-lined pump shown in Fig. 22.38

(Courtesy Allen-Sherman-Hoff Pump Co.)

of this design, eliminating the casing. The pump manufacturer furnishes the shaft, the propeller, and a stuffing box; the user mounts this assembly in a tank where mixing and agitation is required, locating the propeller inside a draft tube.

Rubber-lined pumps are widely used for chemical service because they offer excellent resistance to abrasive corrosion and erosion. The pump shown in Fig. 22.38 has a radially split casing, entirely protected by Maximix rubber, Neoprene, or other synthetic material, according to the application. The manner in which the synthetic inner casing is inserted in the pump is shown in Fig. 22.39. The rubber parts of the split casing lock into their proper position and are replaced in the same manner as changing the tube in a tire. The impeller is made of the same material. These pumps are also used in a variety of other services requiring protection against abrasives, such as pulp concentrations in paper mills, and water and sand mixtures in dredge service.

PETROLEUM INDUSTRY

The pumping requirements of the petroleum industry may be divided into three general categories, as follows:

1. Production includes well pumping, gathering, and waterflooding, as well as various auxiliary services such as fire pumps, and contractors' pumps.

2. Transportation includes transportation of the crudes, the refined oils, and gasoline. The transportation may be by pipeline, loading to tank cars or tankers, and cargo unloading.

3. Refining includes pumping for all processes and auxiliary plant services such as cooling.

Production

The only type of centrifugal pump used in oil-well pumping is the submersible pump described in Chap. 20.

Waterflooding, or secondary recovery, consists of injecting water in wells that have stopped flowing as primary production wells. The repressuring of exhausted areas permits recovery of large quantities of oil, otherwise out of reach. Whereas the majority of repressuring installations employ reciprocating pumps, some centrifugal pump installations have been made for medium- and high-capacity for pressures up to 1,500 psig. These pumps are of the vertical-can type and are sometimes operated with two pumps in series. Where natural gas is available at low price, gas engines may be the most desirable driver, if the gas is not sour. However, despite the higher price of electricity, there are many advantages in using electric motors: The initial

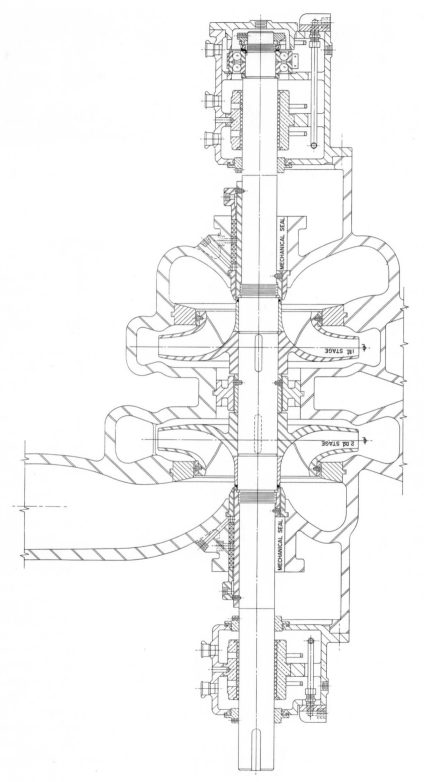

Fig. 22.40 Two-stage volute pipeline pump with axially split casing showing external inter-stage passages

(Courtesy United Centrifugal Pumps.)

cost is lower, automatic operation and control may be more practical, and the cost of maintaining the driver is low.

Salt water is the most widely used medium for waterflooding, as fresh water is conserved for irrigation and other purposes. This water comes either from independent water wells or comes up with the oil in primary production. Most oil-field salt water is brackish and strongly corrosive to bronzes and copper alloys. All-iron pumps have to be used, and stainless alloys are required in many cases.

Transportation

Since the middle 1920's, centrifugal pumps have almost completely replaced reciprocating pumps for pipeline transportation service. Pipelines are used to transport crudes, as well as refined or finished products, including gasoline. Pipelines for motorized transport and aviation gasoline serve an important strategic purpose in Europe, the Middle East, and Africa. In certain cases, the same pipeline may serve to

deliver several different products at different times.

Pipeline stations are installed at reasonable intervals, but still the total heads involved are high enough to require multi-stage pumps. In the past, pipeline pumps were frequently built in six, eight, and even ten stages. Stations were installed where two such pumps may have operated in series, as well as in parallel. Series operation made it necessary to provide casings suitable for the high pressures developed and stuffing boxes capable of withstanding the high-suction pressure of the second pump in the series. Recent installations have been made with pumps of fewer stages, partly because larger size installations are involved and higher heads per stage can be developed in the higher capacity pumps. Figure 22.40 shows the sectional view of a two-stage pump design frequently applied in pipeline service, with external interstage passages. This pump can be either furnished with a conventional packed stuffing box or arranged with a mechanical seal. Figure 22.41

Fig. 22.41 Installation of two-stage motor-driven pipeline pump with axially split casing
(*Courtesy United Centrifugal Pumps.*)

is an installation view of a two-stage pipe-line pump. Where water-cooled bearings are required and water is not available, as is common in the Middle East, the product pumped is used for bearing cooling.

The trend today is to larger pipelines (sometimes obtained by "looping" or paral-leling an existing line), outdoor stations, and automatic remote control. Whereas both gas engines and electric motors have been used for pipeline pump drives in the past, a number of gas turbine installations have been made recently.

A vertical, "in line," single-stage, volute pump used for pipeline booster service is shown in Fig. 22.42 and 22.43. The pump casing is designed with same-size suction and discharge nozzles on the same vertical and horizontal centerlines, 180 deg apart. In the smaller sizes, the pump has no sup-porting feet or foundation, as the pump and its driving motor are supported by the pipe itself. Larger sizes are provided with supporting feet. If flood conditions exist, an extension bracket is furnished between the pump casing and the motor. The same construction is used if the pipe is buried underground, in which case the pump it-self is either buried to match the piping or installed at pipe level in a dry pit.

Pipelines terminate either at refineries or bulk-loading stations for tankers, or at storage tank stations from where the crude or refined product is later loaded to tank trucks or tank cars. Pumps used at these terminal points may be either single-stage or multistage and either horizontal or ver-tical. Vertical pumps can be used for "strip-ping," that is transferring oil or gasoline from the very bottom of the tanks.

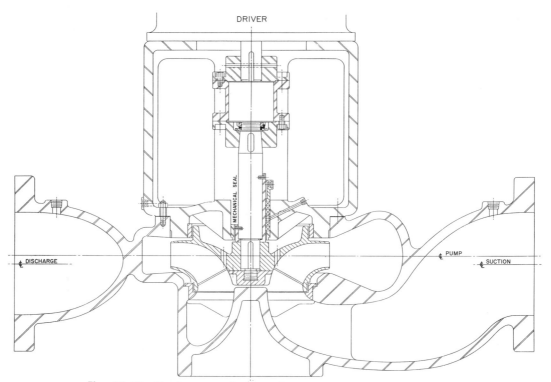

Fig. 22.42   Single-stage, single-suction, vertical in-line booster pump
(*Courtesy United Centrifugal Pumps.*)

Fig. 22.43 Vertical, in-line, booster pipeline pump
(*Courtesy United Centrifugal Pumps.*)

Refining

This discussion will not present a detailed catalogue of all the pumping applications encountered in refining petroleum. Engineers in this industry are quite familiar with these applications, and a complete textbook would be necessary to acquaint others with them. Briefly, the range of products handled in a refinery is quite extensive; specific gravities as low as 0.60 to as high or higher than water occur; viscosities range from lower than water to such high values that centrifugal pumps can no longer handle them and rotary pumps must be used; temperatures vary from cold to as high as 850°F; small to medium to high capacities are encountered; pressures range from as low as a few psi to as high as 1,200 psi or higher; the product may be as inert as water or extremely corrosive and require high stainless alloys.

The petroleum industry has always tended toward a high degree of standardization in its pumping equipment. This may be due partly to the fact that processes change frequently, and users wish to be able to switch pumping equipment from one service to another without requiring too many modifications. Each oil company has generally prepared its own standard specifications for various services in the refinery, and manufacturers have developed lines of pumps that can readily fit these specifications. Recently, a combined specification has been prepared by the American Petroleum Institute for "Centrifugal Pumps for General Refinery Services" (API Standard 610, Third Edition, January 1960). Most oil companies refer to this standard in their specifications, qualifying it sometimes by saying: "As per API Standard 610, except as modified or supplemented by . . ." These standards detail the design, materials, shop inspection, tests, and even such items as preparation for shipment, drawings, and preparation of proposals of pumps intended for refinery service.

For most applications where single-stage pumps can be used, preference is for end-suction (Fig. 22.44 and 12.11) or top-suction (Fig. 22.45), frame-mounted pumps with overhung, single-suction impellers. For light-duty service (low temperatures, low suction pressures, low discharge pressures) water-cooled stuffing boxes are not required. Heavy-duty services require water-cooled stuffing boxes and bearings. Both packed boxes and mechanical seals are encountered in refinery process work. Some designs are adaptable for either construction. Others use an integral seal construction, with no provision for conventional packing, to take advantage of the resulting shorter shaft overhang (Fig. 22.46). Most manufacturers standardize on a small number of bearing frames, each size of which is suitable for a large number of different liquid ends.

Two-stage pumps of the overhung-impeller design are used when heads are higher than can be met in the single-stage design (Fig. 22.47).

A wide range of materials is encountered in the construction of refinery process pumps, from cast-iron casing and internal parts, through cast steel, Ni-Resist, and stainless alloys. Recently, nodular cast iron has been used in certain services previously requiring cast steel casings.

Suction conditions are one of the most important aspects of the application and operation of refinery pumps for two reasons: (1) In many applications the pump must handle hydrocarbons at temperatures corresponding to the boiling point, and (2) it is expensive to locate the vessels from which these pumps take their suction at such an elevation that ample NPSH is available. The hydraulic design of refinery pumps, therefore, is directed at requiring very low NPSH values.

As described in Chap. 17, the NPSH required for a given capacity and a given pump is frequently lower for hydrocarbons than for cold water. At present, there are no authoritative universally applicable data that can allow the correction of cold

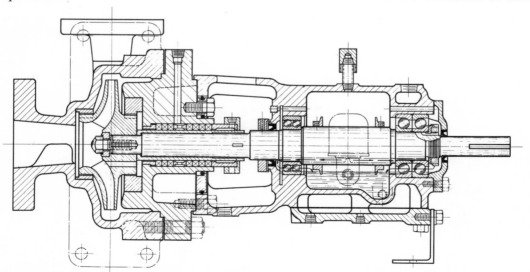

Fig. 22.44   End-suction, single-stage, frame-mounted refinery pump with radially split casing

Fig. 22.45   Top-suction, single-stage, frame-mounted refinery pump with radially split casing

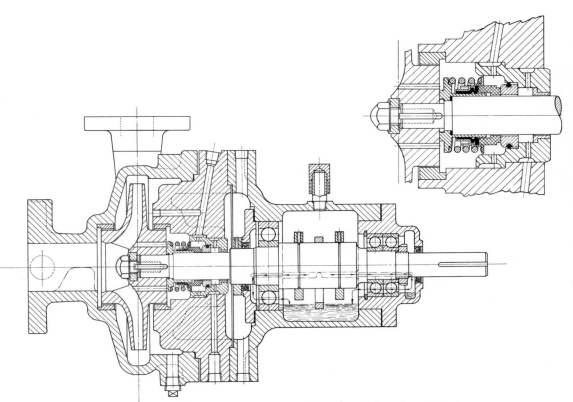

Fig. 22.46   Refinery pump with internal mechanical seal construction

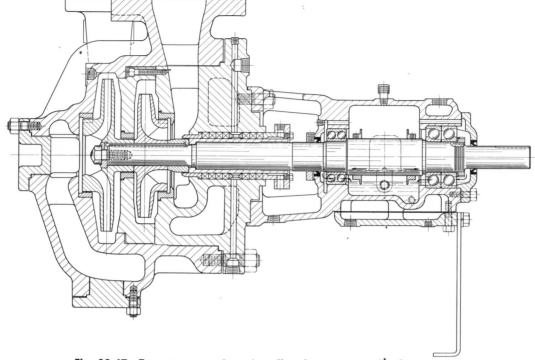

Fig. 22.47 Two-stage, overhung-impeller, frame-mounted refinery pump

water tests for application on hydrocarbons. A number of designers or application engineers have been applying their
own correction curves, established empir-
ically. In most cases these curves are reasonably conservative since they are based on
field experience with trouble-free units.
The most logical solution would be for the
manufacturer to present his NPSH requirements on the basis of cold water tests
and for the user to apply his own correction
factors based on his experience. It is quite
possible that at some future time the accumulation of a large number of tests on
many different designs will permit the establishment of a more general method of
correction.

Because of the severe suction conditions
encountered in these services, some refineries have tended toward the use of vertical-
can pumps (Fig. 14.17).

Hot oil charge pumps may develop pressures of more than 1,000 psi and are gen-

erally constructed with a radially split
double casing. Pumps similar to those employed in boiler feed service (Fig. 3.14 and
3.15) are used in this service, modified
where needed, but especially at the stuffing
boxes. Outward leakage of oil at these extreme temperatures (700 to 900°F) cannot
be tolerated because of fire hazard, and cold
oil must be injected ahead of the boxes.

De-coking is another special service in
refineries where high-pressure, multistage
pumps are used. Water discharged through
special nozzles under high pressure is used
to cut the coke inside coking stills. Depending on the range of pressures involved,
these pumps may have axially split casings
(Fig. 5.8, 5.11, 3.6, or 3.7) or radially split
double casings (Fig. 3.14 and 3.15).

Both electric motors and steam turbines
are used to drive refinery service pumps.
The latter type of driver is preferred in
certain applications because a refinery uses
a large amount of steam in heat exchange

processes and, as a result, steam turbines operating at back pressures to provide steam for the process are very economical.

PULP AND PAPER MILLS

The manufacture of pulp and paper is a complex series of processes, each of which employs a great number of centrifugal pumps. The paper industry is a separate industry from pulp making, as many mills making paper do not make pulp. Pulp manufacturing consists of separating cellulose fibers from any foreign matter present in the raw material and preparing these fibers into a pulp suitable for the manufacture of paper, paper board, rayon, cellophane, explosives, and a variety of other products.

Pulp can be manufactured either by a strictly mechanical process or by breaking up wood or other vegetable matter and cooking the broken-up material with chemicals in a digester. The products of chemical pulp making may be an alkaline pulp (soda or sulphate) or an acid pulp (sulphite), depending on the chemicals used and on the desired end product. Thus, a pulp mill uses pumps to handle liquids such as clear water, acids, caustics, white liquor (the cooking liquor in the digester), black liquor (the spent liquor separated from the stock after cooking), and green liquor (chemicals recovered from the recovery furnace and dissolved in water). Therefore, in addition to general-service pumps, which handle clear water, pulp mills use a variety of specialty pumps, generally similar to chemical pumps.

In paper mills pumps must handle stock with a consistency ranging from 1 per cent to as much as 6 per cent by weight, as well as clear water. Pumps handling high-consistency stock are generally built with special non-clogging impellers, with the fewest number of vanes possible, and sometimes with a special booster or screw-feeder that helps the high-consistency stock to en-

Fig. 22.48 Paper stock pump shown in Fig. 2.11

(Courtesy Allis-Chalmers Corp.)

ter the impeller without separating or dehydrating. Horizontal paper stock pumps are frequently built with diagonally split casings (Fig. 2.11). Figure 22.48 shows the cross-section of such a pump, with conventional stuffing box packing and with a seal connection to the stuffing box ahead of all the packing. An external source of clear sealing water prevents any of the fibers in the stock from entering the stuffing box and interfering with the life of the packing.

The friction losses of stock vary widely with the stock consistency. Friction losses of groundwood stock and of sulphite stock are detailed in the standards of the Hydraulic Institute, and this information should be used in estimating head requirements of paper stock pumps.

TEXTILES

Textiles is another industry that requires the handling of a wide variety of liquids in addition to clear water. General-service pumps are used for water or cold non-corrosive liquids. Chemical pumps are widely used to handle liquids such as acids, alkalis, acetates, and dyes. One very important requirement in textile manufacturing is the complete assurance of the absence of foreign matter that could affect the coloring of the end product. As a result, stainless alloys are generally used—more for eliminating any corrosion products whatsoever than for increasing the life of the pumping equipment.

NATURAL AND SYNTHETIC RUBBER

Centrifugal pumps are used to handle a variety of liquids such as acids, sodas, hydrocarbons, latex, and solvents. The problems of pump selection for these applications are similar to those encountered in the chemical industry.

FOOD AND BEVERAGE PROCESSING

Pumps handling foods or beverages must be of a special sanitary design. The main requirements are a complete absence of corrosion and ease of dismantling for cleaning. Most food-processing pumps are made of stainless steels, glass, or porcelain, although some all-bronze pumps are used, especially in the beverage industry. The designs must be free of any pockets where particles or liquids could accumulate and .where bacteria might breed. Neither oil nor grease lubricant can be permitted to enter the pump, and, as a result, special packing must be used in the stuffing boxes or mechanical seals applied.

MINING

Two general categories of application of centrifugal pumps are encountered in mining operations. The first involves the dewatering of different types of mines, such as coal and metal ore. The second category includes all applications where pumps are used in the processing of the mined product, such as in leaching operations, coal washing, transfer of precipitates, and waste disposal.

In almost all cases mine dewatering pumps handle corrosive or abrasive waters against relatively high heads. Multistage pumps, built of corrosion-resistant materials, ranging from all-bronze to chrome or nickel stainless alloys are, therefore, used most frequently. Vertical turbine pumps have become a favorite style in recent years as they require very little space. In general, the difficulty of introducing pumps into their ultimate location within the mine calls for special construction features. In some cases, it is necessary to dismantle the pump, transport the parts into the mine, and reassemble the pump when it is in place.

As mines are always in danger of becoming flooded, special sinking pumps, mounted on small trucks for portability, are employed. These units can use the mine tracks to move from place to place. Frequently the pumps must be capable of operating in an inclined position. These pumps may also be mounted in cages that are lowered into the shaft as dewatering progresses. Lengths of discharge piping are added each time the pump is lowered further.

As described in Chap. 18, the problem of dewatering a flooded mine is unlike most applications of centrifugal pumps in that the pump must work against a total head that varies from practically zero to the maximum that occurs when the mine is almost clear of water. A dewatering pump designed to work as two two-stage pumps in parallel, initially, and one four-stage pump when the total head requirement exceeds that developed by the two-stage combination is shown in Fig. 22.49. Its hydraulic performance was shown in Fig. 18.36. The pump with its 600-hp, 1,450-rpm motor is mounted in a welded channel frame. A cable sheave is mounted in the frame to permit the unit to be suspended vertically.

The frame and sheave are capable of supporting the discharge pipe full of water weighing up to 75,000 lb.

Pumps for coal washing, slurry handling, and leaching generally handle water containing a considerable amount of abrasive solids. Corrosion-resistant materials with a great degree of resistance against abrasion are required. Manganese steels, Ni-Resist and similar alloys are frequently applied. The rubber-lined pump shown in Fig. 22.38 is another type used frequently because of its resistance to abrasion. In general, operating speeds are held to much lower limits than in handling clear water. Because the abrasives contained would quickly destroy both packing and shaft sleeves, the stuffing boxes are sealed with clear filtered water, wherever possible.

CONSTRUCTION

Centrifugal pumps used in construction work include units for drainage or jetting. Drainage units are generally small, engine-driven, portable units, self-priming, if horizontal. Jetting pumps may develop pressures from 75 to 150 psi and are usually designed as portable, engine-driven units.

STEEL MILLS

The steel industry uses a large number of centrifugal pumps for a wide variety of services, such as cooling of rolls and furnaces, general water supply service, hydraulic-descaling, scale-pit recirculation, and many others.

Service water pumps handle clean water at pressures up to 150 psi and for capacities up to 10,000 gpm or more. Standard horizontal or vertical, general-service pumps, both single-stage and multistage are used on this service. In a few cases, where river water is contaminated by industrial wastes and is quite corrosive, special materials may be required, but, generally, standard fitted pumps are used.

Fig. 22.49 Mine dewatering pump arranged to operate as either two- or four-stage unit

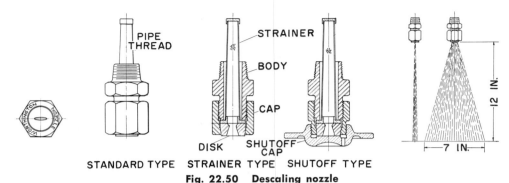

Fig. 22.50   Descaling nozzle

Various cooling services may use pressures up to 250 psi. Here again, unless special water conditions exist, standard fitted pumps are satisfactory.

Hydraulic descaling is the process in which scale is removed from metal by using the impact of a jet of water directed at the metal surface through special nozzles (Fig. 22.50). In the 1930's, when descaling was first developed, pressures of 1,000 psi were considered quite sufficient, but the recent trend has been toward raising the pump discharge pressure to 1,500 or even 1,800 psi. A typical hydraulic-descaling installation is illustrated in Fig. 22.51. Preference generally has been given to multistage pumps with axially split casings, as shown in Fig. 3.6 and 3.7, although some installations of pumps with radially split double casings (Fig. 3.14 or 3.15) have been made. Because of their long-time familiarity with gears, steel mills have often used step-up gears with 1,800-rpm motors to drive centrifugal pumps at 3,600 rpm or even higher, years ahead of steam power plant practices.

Descaling pumps are designed either to pump directly to the system or to be assisted by accumulator air bottles. An accumulator system is advantageous because smaller pumping equipment may be used; the accumulator is replenished during the time when water is not required. The cost, however, is considerably higher than that of a direct pumping system, and a thorough analysis must be made of the average and peak demands before choosing between the two systems.

Scale pit recirculating pumps are intended to recover the water used in hydraulic descaling and return it for further re-use after it is settled out. As a result, these pumps handle water containing considerable amounts of scale; special, hard, abrasion-resistant materials are generally necessary in this service.

HYDRAULIC PRESSES

Applications for hydraulic presses occur in the metal-forming industry and in many other industrial processes. Installations include such applications as forging, die-forming, extrusion, injection-molding, and die-casting. Unless a large number of individual operations are served by a single pump, the tendency is to reduce the size of the pumping equipment by using accumulator systems, as in the descaling process. The same type of pumps used in descaling service are employed for the higher pressures, whereas two- and four-stage pumps with axially split casings are used for the lower range of pressures.

The water handled by these pumps is generally pure and non-corrosive, with occasionally a slight admixture of water-soluble oil. Corrosion problems are, therefore, seldom encountered, and materials are selected for their suitability for the particular pressure requirements.

AIR CONDITIONING

Centrifugal pumps used in air conditioning include general water supply units, chilled water pumps to circulate water from the chillers through the air-cooling coils, air washing pumps in some installations, and circulating pumps for the refrigerating condenser. The water handled is clean, and pressures in most cases such that single-stage, standard, fitted pumps can be used. Depending on size of the unit, end-suction, frame-mounted (Fig. 2.12), close-coupled (Fig. 1.14), or axially split casing pumps (Fig. 2.10) are used. Some vertical-turbine pumps are used for the water supply application. The availability of me-

chanical seals in recent years has made this construction of value in some cases because of the reduced amount of overseeing required.

REFRIGERATION

Brine pumping is the one special application required in refrigeration systems, cold storage, or ice making. Both sodium chloride and calcium chloride brines are used, the latter much more often than the former because it has a lower freezing point and is less corrosive. The specific gravity of the brine solution that the pump has to

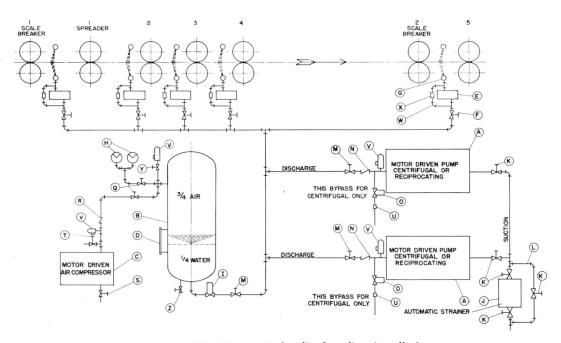

Fig. 22.51   Hydraulic descaling installation

KEY:
A = pump and motor
B = accumulator
C = air compressor
D = gage glass
E = two-way automatic valve
F = stop valve
G = nozzles and header
H = pressure regulators
I = automatic stop valve
J = automatic strainer
K = stop valves
L = bypass
M = stop valve
N = check valve
O = automatic bypass
Q = stop valve
R = check valve
S = stop valve
T = bypass for starting
U = orifice nipple
V = relief valves
W = bypass
X = orifice nipple
Y = stop valve (lock type)
Z = drain valve.

Fig. 22.52   Condensate return unit

handle varies between 1.05 and 1.28, depending on the minimum temperature maintained in the expansion coils.

The viscosity of brine solutions increases with specific gravity and with decreasing temperatures to a degree that it definitely affects pump efficiency. If this effect is not considered, a strong probability exists that a motor driver selected on the basis of water efficiency will be overloaded. It is necessary, therefore, to establish the viscosity of the brine at the minimum pumping temperature and determine the efficiency correction factors as described in Chap. 17 (Fig. 17.54).

All-iron pumps are used for calcium chloride brine, and all-bronze pumps are preferred for sodium chloride brine. Special attention is required for the bearing construction; since the temperature of the

brine is below that of freezing water, adequate seals must be used to prevent the moisture present in the air from condensing inside the bearings. In some installations, it is necessary to circulate warm water through the cooling jackets of the bearings to avoid lowering the temperature of the lubricants excessively.

HEATING

Building heating systems require hot water circulating pumps and, in some cases, condensate return units. The first are generally small end-suction, close-coupled pumps built for that special purpose by several manufacturers. Regenerative pumps (Chap. 15) are also used for this purpose, as they are suitable for the range of low capacities and high heads encountered. There is a tendency to supply these pumps with mechanical seals to reduce maintenance.

Various condensate return units are available on the market. They consist of a condensate returns receiver and one or two small, close-coupled centrifugal pumps (Fig. 22.52). An automatic float switch starts and stops the pump, which discharges the returns back into the boiler.

A definite trend exists toward high-temperature heating applications, with temperatures ranging from 240 to 400°F. A high-temperature, hot-water system consists

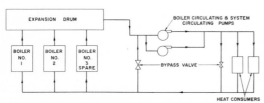

Fig. 22.53   High-temperature, hot-water heating system

Single-pump system.

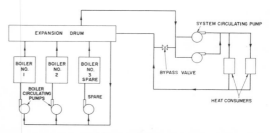

Fig. 22.54   High-temperature, hot-water heating system

Separate-pump system.

of a hot-water boiler, an expansion drum, a pumping system and the heat-using system. Two different arrangements are used: (1) The single-pump system (Fig. 22.53), where the same pump or pumps are used to provide circulation through the boilers and to distribute the hot water through the system, and (2) the separate-pump system (Fig. 22.54), where two separate sets of pumps are used, one for boiler circulation and one for hot-water circulation to the heat users. In either case, the pumps take their suction from the expansion drum, which is used to pressurize the system and to store hot water.

Because of the relatively high operating temperatures and suction pressures, end-suction pumps with only one stuffing box of the same type as high-pressure heater drain pumps or low-pressure steam power plant boiler circulating pumps (see Fig. 22.20) are used. There is a trend to the use of mechanical seals with cold water injection at the seal faces.

Fig. 22.55 Self-contained condensate return centrifugal pump

(Courtesy Cochrane Corp.)

PROCESS CONDENSATE RETURNS

In a number of processes that reclaim condensate returns, a certain amount of air returns with the condensate and has to be adequately vented to avoid air binding the return pump. One special design is illustrated in Fig. 22.55. No return tank is used, as the unit is self-contained. The centrifugal pump draws water from the priming loop and discharges it at high velocity through a nozzle. The mixture of hot condensate, steam, and air from the returns flows into the low-pressure area surrounding the nozzle and is drawn into the mixing tube where it is combined with the circulating water. This additional amount of condensate discharges an equal amount through the air separator, either directly to the boiler or to the reservoir from which a boiler feed pump delivers it to the boiler. The fins that surround the priming loop act to subcool the liquid-vapor mixture in

it and reduce the submergence required at the pump suction by reducing the vapor pressure of the condensate entering the pump.

MARINE PUMPING

The wide range of requirements for pumping services aboard ship calls for a correspondingly extensive selection of types and sizes of centrifugal pumps. The most important applications for marine service are listed in Table 22.4. Although it is impractical to discuss the requirements of each of these applications in detail, some of them will be described in this chapter.

There are, first, certain general considerations that apply to all marine services. Centrifugal pumps for ship use must operate at or near the safe maximum speed at which a particular condition can be met, as the use of the highest rotative speed permitted by the suction conditions will result in the smallest pump and driver and, therefore, the lowest weight and smallest total size unit.

Marine service requires the greatest possible degree of reliability. Standards have been developed to guarantee acceptable designs, and numerous agencies have prepared specifications for centrifugal pumps in this service. Some of these agencies are:

1. American Bureau of Shipping
2. US Maritime Commission
3. US Coast Guard Marine Inspection Service
4. US Navy Department Bureau of Ships
5. US Navy Department Bureau of Yards and Docks
6. Army Transport Service
7. US Corps of Engineers
8. Lloyds.

In most cases, special inspection procedures made by a customer's representative or by a navy inspector are required at the point of manufacture. Witnessed performance tests are generally required. Type approval of a particular pump is generally required in the case of pumps for US Navy

vessels. This involves a thorough analysis of the design as well as extensive tests at navy laboratories.

Boiler feed pumps

With the exception of some of the latest ships, which operate at higher steam pressures and require double-casing pumps, most marine feed pumps are multistage pumps with axially split casings. They are primarily designed for steam-turbine drive, which permits the desirable higher speed operation and variable speed control.

Although some standard commercial designs adapted for marine service are occasionally used, special boiler feed pumps are generally developed by manufacturers for this service, especially for navy application. Figure 22.56 shows a two-stage boiler feed pump connected to a steam turbine through a flexible coupling. Forced feed lubrication for pump and driver bearings is provided by means of a submerged oil pump. Figure 22.57 shows a sectional view of a three-stage navy boiler feed pump.

Figure 22.58 shows a three-stage boiler feed pump with its steam turbine. A single shaft, supported in three bearings, is used

---

TABLE 22.4   INDEX TO CENTRIFUGAL PUMP APPLICATIONS IN MARINE SERVICE

Atmospheric drain	Elevator
Automatic priming	Engine cooling
Ballast	Evaporator system
Bilge	Fire
Boiler feed	Flushing
Booster	Fresh water
Brine circulating	Gasoline
Butterworth system	General service
Caisson gate dewatering	Gun cooling
Cargo	Heater drains
Condensate	Hot-water circulating
Condensate return	Ice-water circulating
Condenser circulating	Jetting
Damage control	Make-up feed
Diesel fuel oil	Potable water
Distiller plant	Refrigeration brine circulating
Drainage	Refrigeration condenser circulating
Drinking water	Sanitary
Drydock	Wash water

Fig. 22.56 Two-stage navy boiler feed pump connected to steam turbine by flexible coupling

for pump and turbine, thus eliminating the connecting coupling.

A four-stage, double-casing, barrel-type boiler feed pump used on navy vessels operating with high-steam pressures is shown in Fig. 22.59.

A different type of boiler feed pump is shown in Fig. 22.60. It is a turbine-driven unit of Monobloc construction, designed so that there is no stuffing box between the pump and the turbine. Leakage from the pump proceeds into a mixing or quenching chamber, from which it is fed back into the feedwater system.

Some vertical, multistage boiler feed pumps have been installed aboard ships to save space. One such unit is shown in Fig. 22.61.

Except for low-pressure service and some commercial applications, boiler feed pump casings are made of chrome or chrome-nickel steel. Impellers, wearing rings, diaphragms, shaft sleeves, and shaft nuts; that is, all parts in contact with the feedwater, are usually made of chrome steel or monel metal.

Condensate pumps

Condensate pumps are almost always arranged for vertical operation. Figure 22.62 shows a vertical, single-stage, single-suction condensate pump with an internal water-lubricated bearing, an external ball thrust bearing, and a mechanical seal. Figure 22.63 shows a vertical, two-stage, single-suction condensate pump. The inlet of the first-stage impeller points upwards to facilitate venting. Figure 22.64 shows a photograph of this pump, with the removable casing half lifted and the rotor sectionalized for display.

Condensate pumps must always be located close to the condenser hotwell, to minimize the effect of the ship roll. Casings are generally made of bronze while impellers are bronze or monel.

Condenser circulating pumps

Commercial designs of horizontal, double-suction, single-stage pumps with axially split casings are frequently adapted for vertical service on marine application.

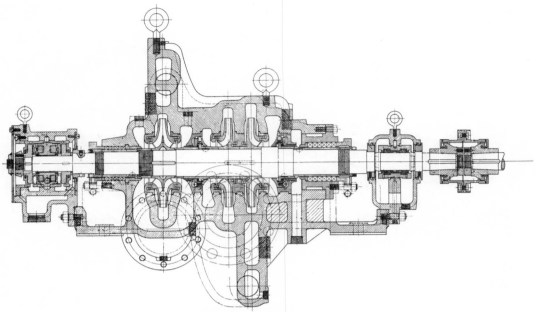

Fig. 22.57   Three-stage navy boiler feed pump

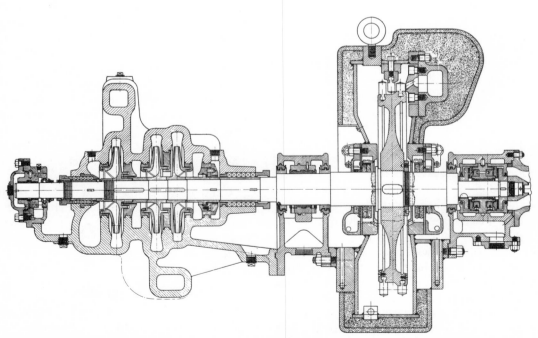

Fig. 22.58   Turbine-driven, three-stage navy boiler feed pump with common turbine and pump shaft

Fig. 22.59 Four-stage, double-casing, barrel-type navy boiler feed pump

Figure 22.65 shows such a pump with external ball bearings and a flexible coupling. This is essentially a low-speed pump and requires more head room than the mixflo or propeller pump would under identical operating conditions. This type of pump has also been built with an internal grease- or water-lubricated bearing and a rigid coupling; the motor or turbine thrust bearing is designed to take the weight of the pump rotor and any unbalanced axial hydraulic thrust.

Vertical mixflo pumps are frequently used for marine condenser circulating service. They operate at higher speeds than double-suction pumps. Still higher rotative speeds can be used with propeller pumps. Figure 22.66 shows a vertical propeller pump of a special short design with a water-lubricated bearing and a rigid coupling,

Fig. 22.60 Turbine-driven, Monobloc boiler feed pump without stuffing boxes

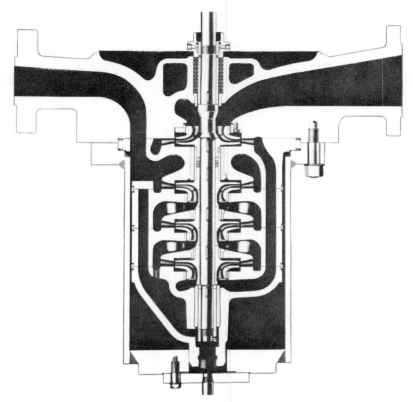

Fig. 22.61   Vertical, multistage boiler feed pump

(Courtesy Ingersoll-Rand.)

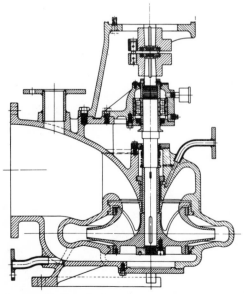

Fig. 22.62   Vertical, single-stage, single-suction marine condensate pump

and Fig. 22.67 shows a photograph of the same pump driven by a geared-down steam turbine.

Materials generally used for condenser-circulating service are: bronze for the casing, impeller, and wearing rings; monel metal for the shaft; bronze or monel for the shaft sleeves; and leaded bronze for the water-lubricated bearings.

Cargo pumps

Centrifugal cargo pumps are designed to pump large volumes of crude oil, fuel oil, and other liquid hydrocarbons, as well as salt water ballast, to give fast cargo unloading, so essential to profitable tanker operation.

Figure 22.68 shows a typical single-stage, double-suction pump used in this service.

Two large flanged vent connections on top of the casing suction spaces vent off vapors and air that would otherwise cause the pump to lose suction when the liquid level in the cargo tanks lowers. This feature permits pumping down each cargo tank to the lowest possible level with the high-capacity centrifugal cargo pumps, leaving a minimum of cargo to be stripped.

Vertical turbine pumps (Chap. 14) are also widely used in cargo service.

Drydock pumps

Drydock service requirements are very special, in that the head varies from a low value, made up principally of friction head at the beginning of pumping, to a higher head composed largely of static head at the end of the pumping period. The head-capacity curve of drydock pumps should, therefore, be quite steep, to provide the minimum of capacity reduction with this

Fig. 22.64 Condensate pump shown in Fig. 22.63, with casing half lifted and rotor sectionalized

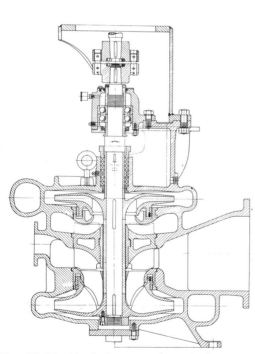

Fig. 22.63 Vertical, two-stage, single-suction marine condensate pump

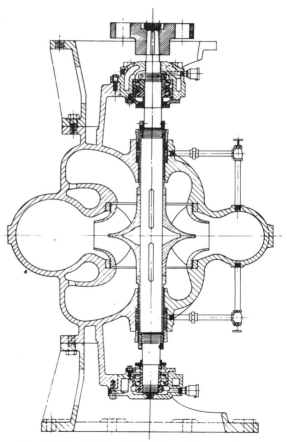

Fig. 22.65 Vertical, double-suction marine condenser circulating pump with axially split casing

increase in head and a minimum of reduction in efficiency. As the characteristics of mixflo pumps are well-suited for this requirement, most of the drydock pumps are this type (Fig. 14.11). Other pumps used as main drydock pumps are the vertical turbine pump, or the vertical hiflo type (Fig. 14.19).

All of these general types of pumps may be used for graving docks or in floating drydocks, with special construction in each case to suit the installation. Graving docks generally have a dry-pit pump room on one side of the dock, with the main pumps located on the floor that is approximately at the same level as the dock floor. A flared suction bell below the pump opens into the suction well connected to the dock by means of a large culvert. The pump is connected to the motor near the ground surface level by a long shafting with guide bearings.

Floating drydocks invariably require submersible pumps located in the flooded sections of the dock. The pump shafting is en-

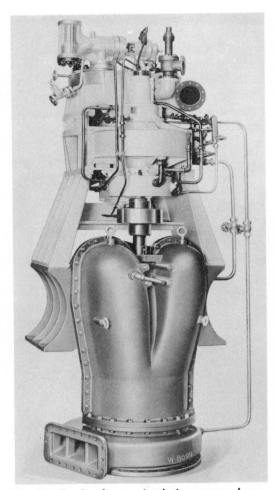

Fig. 22.67   Condenser circulating pump shown in Fig. 22.66

closed in a cover pipe to protect the shaft, bearings, and couplings from water. It is usually desirable to have a split section of cover pipe adjacent to the pump, enclosing a removable short piece of shaft. This permits dismantling the pump without disturbing the intermediate shafting.

PUMPED STORAGE

A pumped storage installation is, in a manner of speaking, a hydraulic storage battery. In a pumped storage system, electric power is used to lift water into a reservoir during off-peak periods. Later, the

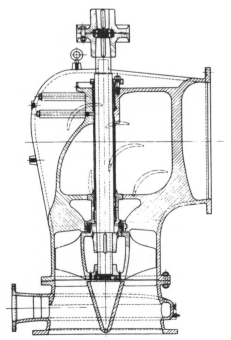

Fig. 22.66   Vertical, propeller-type, condenser circulating pump

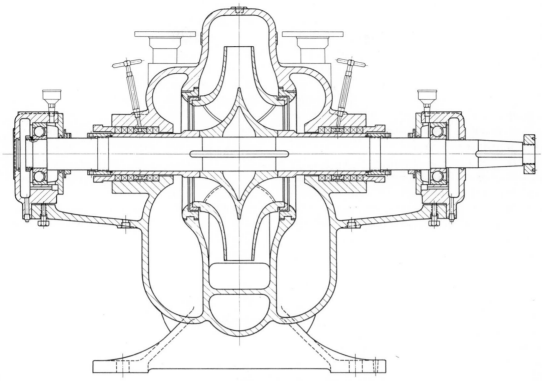

Fig. 22.68   Single-stage, double-suction cargo pump

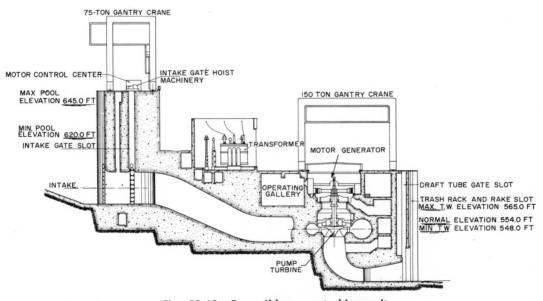

Fig. 22.69   Reversible pump-turbine unit

Tuscarora Pump Generating Plant, Niagara Falls, N. Y. (Courtesy Allis-Chalmers Corp.)

water is drawn from the reservoir to drive a hydraulic turbine that provides incremental power during peak load periods. The electricity used to drive the pump may be obtained from a hydroelectric installation or it may be available from a neighboring steam electric station. This latter arrangement is presently creating considerable interest among utilities who see in it a means for averaging out their steam electric load. In a number of cases, the building of special reservoirs may be justified for storage of night-time or weekend off-peak power to be utilized in the daytime for generating additional electricity.

Several arrangements are possible. A separate motor-driven pump and a water-turbine generator may be installed, or a vertical tandem unit, which consists of a motor-generator, a conventional water turbine, and a centrifugal pump, can be used. When delivering power, the electric component acts as a generator, driven by the turbine and producing electricity; the pump is declutched and remains idle. At off-peak load periods, the electric generator draws electricity and operates as a motor driving the pump, which discharges water into the reservoir. During this period, the water turbine is filled with air under pressure and is idling. One special arrangement involves a reversible pump, that is, turbine-driven or driving a motor-generator. The pump-impeller runner operates within a casing with a stay ring and movable wicket gates. It operates as a pump when rotating in one direction, driven by the motor, and discharges water into the storage reservoir. Rotating in the opposite direction, it operates as a water turbine and the motor becomes a generator. Figure 22.69 shows one of twelve reversible pump-turbine units at the Tuscarora Pump Generating Plant, Niagara Falls, N. Y. It is rated 28,000 hp at 75 ft head as a turbine, 25,000 kva as a generator, 37,500 hp as a motor, and 3,400 cfs at 85 ft head as a pump. Its operating speed is 112.5 rpm.

23 *Preparing Centrifugal Pump Inquiries*

Before purchasing centrifugal pumping equipment, a customer will make a careful analysis of all factors relating to the installation. He will assemble these data into a formal inquiry and send it to several pump manufacturers, asking them to prepare detailed descriptions of the equipment they would recommend to meet his particular needs.

It is impossible to present an exhaustive survey of each one of the factors to be studied when preparing such an inquiry; experience and intimate contact with the particular installation are the only guides to complete knowledge. Nevertheless, it is possible to outline the essential data required to allow the manufacturer to select a centrifugal pump for any given installation intelligently (Fig. 23.1).

ESSENTIAL DATA REQUIRED

The following represents the required essential data:

1. Number of units required
2. Nature of liquid to be pumped
   a. Fresh or salt water, acid or alkali, oil, gasoline, slurry, or paper stock
   b. Vapor pressure of liquid at pumping temperature
   c. Specific gravity
   d. Viscosity conditions
   e. Amount of any suspended foreign matter present, and size, nature, and abrasive quality of solids—if liquid is pulpy, consistency either in per cent or in pounds per cubic foot of liquid
   f. Chemical analysis, including pH value, predicted variations from analysis, impurities, oxygen content, past history, and scaling tendency, if any
3. Required capacity, as well as minimum and maximum amount of liquid pump must deliver
4. Suction conditions
   a. Suction lift or suction head
   b. Constant or variable suction conditions
   c. Length and diameter of suction pipe, fittings and valves involved
5. Discharge conditions
   a. Static head description—constant or variable
   b. Friction head description and how estimated
   c. Maximum and minimum discharge pressures against which pump must deliver liquid
6. Type of service—continuous or intermittent

403

7. Pump installation—horizontal or vertical position (if vertical, type of pit—wet or dry)
8. Type and characteristics of power available to drive pump
9. Space, weight, or transportation limitations
10. Location of installation
11. Special requirements or marked preferences with respect to pump design, construction, or performance.

Number of units

The number of units is important primarily to increase pump reliability; standby units are often necessary, especially in cases where the life of the pump may be jeopardized by the severity of the service. It is important to determine whether one or more units can be operated in parallel, as the hydraulic performance of the individual units may need adaptation for this purpose.

The choice between the use of a single pump and the installation of several pumps in parallel for the full demand is influenced by the expected load factor. When the demand is more or less constant, the tendency is to select a pump for the full demand, adding a safety margin as provision against pump wear. If, on the other hand, the load is of a fluctuating nature, two or more pumps may be operated in parallel. At periods of light load, one or more pumps can be taken off the line for more efficient operation. The capacity of the individual pump is selected with this type of operation in view. For instance, if the demand remains at 65 per cent of the maximum capacity most of the time, two pumps, each designed for about 70 per cent of maximum flow, could be installed. One pump alone would be able to carry the load most of the time. When the total required flow exceeds 65 per cent of maximum, the second pump is put on the line to share the load.

Exceptions to this general practice are as follows:

When the total demand is too low to be divided efficiently between two pumps, a single pump may have to be used, regardless of the nature of the load factor.

When the efficiency of operation is inconsequential—intermittent service or turbine-driven pumps used to provide exhaust steam for process work or for heating—low first cost will generally determine the number of pumps to be used in the installation.

When the maximum demand is too great to permit the use of the most economical pump or of the most economical operating speed, the demand may be split up between two or more units, regardless of the nature of the load factor.

If more than one pump is used to provide the maximum demand and if a partial reduction in available capacity is permissible, the installation may not require a spare pump. Enough spare parts must then be stocked to restore full available capacity rapidly. If, however, no reduction in available capacity is permissible, a spare pump must be provided, which can be put into service immediately upon failure or stoppage of one of the main pumps. Because this substitution must provide against the interruption of the prime mover as well as for pump failure, a different source of power is frequently used to drive the spare pump. Thus, if the normal operation is based on motor drive, the spare pump may be driven by a steam turbine or an internal combustion engine. The switch over to the spare pump can be accomplished automatically when the pressure in the discharge header decreases.

Nature of the liquid

Type. To a certain extent the nature of the liquid pumped determines the types of pumps most frequently used for the service involved. The widest range of selections is available for water service, but this must be divided into fresh and salt water. The selection of materials for the salt-water service varies greatly from the point of view of first cost, length of life, and customer preference and experience. Apparently insignificant impurities in the liquid may be a highly important factor in the selection of the proper materials. Similarly, the nature of the liquid to be pumped will greatly affect not only the pump material, but possibly even the mechanical construction best suited to the purpose, depending on whether the liquid is an acid, alkali, or oil. For example, if the pump is to handle highly corrosive liquids, not only must the

CENTRIFUGAL PUMP DATA

WORKS NO. _____

DIST. OFFICE NO. _____

PAGE NO. _____ OF _____

ADDITIONAL NO. _____

SOLD
TO

ITEM NO. _____

TYPE OF PUMP_____ QUANTITY_____ TYPE OF DRIVE _____ RATING _____ HP

RATING CURVE _____ PRELIMINARY CURVE # IF SUBMITTED TO CUSTOMER E _____

MATERIALS	CONDITIONS OF SERVICE	
STANDARD FITTED ☐ ALL IRON ☐	** LIQUID	SUCTION LIFT
ALL BRONZE ☐		
* SPECIAL (SPECIFY) ☐		SUCTION HEAD
* CASING		
* IMPELLER	US GPM	SUCTION PRESSURE
* CASING RINGS		
* IMPELLER RINGS	TOTAL HEAD FT.	NPSH AVAILABLE
* SHAFT		
* SHAFT SLEEVES	** SP. GR. @ P. T.	** pH
* STAGE PIECES		
* GLAND	RPM.	** VISCOSITY
* BEARINGS — TYPE		
GREASE LUBE ☐ OIL LUBE ☐	EFF.	** SOLIDS
* COUPLING		
* COUPLING GUARD	BHP	TYPE OF SERVICE:
* BASEPLATE		
* MECHANICAL SEAL		
* MISCELLANEOUS	** PUMPING TEMP. (°F)	** OTHER DATA

DRIVER

FURNISHED BY   WORTHINGTON ☐   CUSTOMER ☐

MOUNTED BY   WORTHINGTON ☐   CUSTOMER ☐

CUSTOMER'S CERTIFIED PRINTS   ATTACHED ☐   LATER ☐

IF MOTOR FURNISHED BY CUSTOMER

MAKE_____ TYPE_____ NEMA FRAME #_____

VERTICAL PUMPS

DISTANCE MOTOR FLOOR

TO SUCTION BELL   (WET PUMP)_____

DISTANCE MOTOR FLOOR

TO PUMP FLOOR   (DRY PUMP)_____

MOTOR STAND

A - SECTION { VERTICAL _____

              SHAFTING _____

B - SECTION { VERTICAL _____

              SHAFTING _____

CONNECTING FLANGES _____

GUIDE BEARING & TYPE _____

CUSTOMER REQUIREMENTS

WITNESS RUNNING TEST   YES ☐   NO ☐

WITNESS HYDRO TEST   YES ☐   NO ☐

INSPECTION (DESCRIBE)

CERTIFIED PRINTS   TIME_____ COPIES_____

TO BE SENT TO

IS DRAWING APPROVAL REQUIRED   YES ☐   NO ☐

PARTS LIST_____ COPIES_____

CONSTRUCTION

HORIZONTAL SHAFT ☐   VERTICAL SHAFT ☐

ROTATION LOOKING FROM_____

   CLOCKWISE ☐   COUNTERCLOCKWISE ☐

NOZZLE POSITION _____

* WORTHINGTON STANDARD UNLESS OTHERWISE APPROVED BY CENTRIFUGAL DIVISION AND COVERED BY AN EXTRA PRICE.

** IF DATA NOT GIVEN WE WILL ASSUME EQUIVALENT TO CLEAR COLD WATER.

Fig. 23.1   Data sheet listing information required to select and build a pump

pump be built of corrosion-resisting material, but the acid must be prevented from leaking out into the atmosphere through the stuffing box.

Temperature. The temperature of the pumped liquid is a very important factor. A standard line of general-service pumps has definite temperature limitations. Higher temperatures may dictate the use of special materials, water-cooled stuffing boxes, or special mechanical features such as center-line support of the casing. For extremely low temperatures, such as in brine service, the use of a nickel cast iron may be warranted in preference to standard gray cast iron, in order to obtain a more refined crystalline structure and thus prevent fractures.

Any wide variations in the operating temperatures should be known, as they will affect the specific gravity and viscosity range of the liquid handled.

If the liquid is water, the vapor pressure can be easily determined from steam tables. If some other liquid is involved, the vapor pressure at the pumping temperature must be carefully noted, because it figures importantly in determining whether or not the prevailing suction conditions are satisfactory.

Specific gravity. The specific gravity must be known in order to determine the power consumption at the design conditions and to select the proper driver size. A pump inquiry frequently expresses the required discharge or net pressure in pounds per square inch, which must be converted into feet of the liquid handled. If a net pressure of 100 psi is specified, the total head is 231 ft of fresh cold water, 193 ft of brine with a specific gravity of 1.2, or 308 ft of gasoline with a specific gravity of 0.75 (Fig. 17.2).

Viscosity. When the viscosity of the liquid handled differs from that of water, the pump capacity, head, and power consumption are all affected appreciably, so that correction factors are necessary.

Foreign matter. The size and nature of the solids suspended in the liquid will de-termine both the type of impeller best suited for the purpose and the materials to be used for the pump construction. If the solids are very abrasive, an open impeller is generally used, and, where necessary, special and more expensive wear-resisting materials can be applied. When solids reach a certain size, or when they are of a stringy nature, special nonclogging impellers are required.

Chemical analysis. Special consideration should be given to the chemical analysis of the liquid if its corrosive or electrolytic properties are not readily apparent from the description of the liquid itself. Thus, pH (measure of acidity or alkalinity) of the water should always be stated if indications show that the water is not neutral.

Required capacity

The capacity required for the installation should be stated in US gallons per minute at the pumping temperature. Any expected variation in the range of capacities should be clearly stated, because centrifugal pumps do not permit as much flexibility in capacity variations without affecting pump efficiency as other types of pumps, such as reciprocating steam pumps. Furthermore, it is generally preferable to have the maximum efficiency of the pump occur at or near the normal capacity conditions.

Any centrifugal pump can run occasionally at much more than its rated capacity, but this may not always be practical or permissible. An increase in capacity means a decrease in the head generated; this may prevent operation of the pump at emergency overloads if excess capacity was not included in the design and if the pump operates on a system-head curve, since the friction losses which make up part of the required head will increase with the capacity. Emergency over-capacity pumping may also be prohibited if the prevailing suction conditions leave no margin over those required for the normal rated capacity. Finally, if the power consumption increases with the capacity, as it does with the majority of centrifugal pumps, operation at capacities greater than originally expected may seriously overload the pump driver.

Minimum operating capacity information is also of great importance. In certain cases, operation at extremely reduced capacities, even for very short periods, presents a definite danger and must be avoided. In other instances, the only disadvantage from operating at reduced capacities is poor economy, and a thorough analysis of the problem may result in the installation of additional smaller units that would be operated during periods of light loads.

Suction conditions

Correct suction conditions for centrifugal pump installations are of paramount importance (Fig. 23.2). Unless the available net positive suction head (NPSH) is equal to or more than that required by the pump selected at the capacity involved, the pump will be unable to meet its capacity design conditions. In addition, the resulting cavitation will damage the pump. If cold liquids are handled, it is necessary to know whether there is head on the suction or whether the pump will operate on a suction lift, and, if the latter, what the maximum lift will be. If the liquid is hot or under a pressure at or near its vapor pressure, the pump must be installed with head

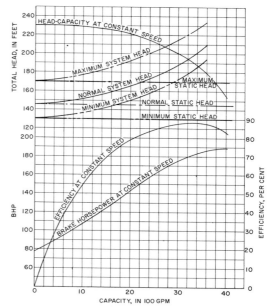

Fig. 23.3 Static head curve

When static head changes considerably, pump may have to operate over wide capacity range.

on suction and the available submergence must be described. In all cases, it is advisable to determine separately the static difference between the liquid level and the pump centerline, and the friction and entrance losses in the suction piping. If these losses have not been determined, it will generally suffice to describe accurately the suction layout, listing all lengths, pipe sizes, and valves.

Discharge conditions

The discharge head for the design conditions should be stated with the understanding that it is generally composed of static elevation (or pressure) and frictional losses in the discharge piping. Any variations in the static head must be known in order to determine the maximum and the minimum head against which the pump is to operate (Fig. 23.3). Specifying an excessive total head has actually the same effect as specifying excessive capacity. Since

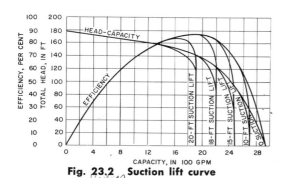

Fig. 23.2 Suction lift curve

Capacity and efficiency fall below normal when suction lift exceeds a given value.

a centrifugal pump will always operate at the intersection of its head-capacity curve and of the system-head curve, a pump which develops an excess of head will, unless artificially throttled, deliver an excess in capacity as its curve will intersect the system-head curve at a greater flow.

By breaking up the discharge head into static and friction head, excessive friction losses can be determined. If the piping to be used is too small, the required pump and its driver will be more expensive than necessary and the cost of operation greater than if the proper size pipe were used.

When the total cost involved is high and justifies an extremely detailed analysis, it is possible to determine the most economic size of pipe by plotting the sum of the amortization of initial pump, driver, and piping cost plus the operating cost against pipe sizes.

In certain special applications, the pump may be required to develop discharge pressures in excess of the design conditions for short periods. A typical example is a condenser water circulating pump that is to be installed in a system utilizing a siphon effect. The head to be developed consists only of the frictional losses through the piping and through the condenser itself and is, therefore, fairly small. However, in order to enable the condenser circulating pump to deliver water through the condenser, the system must be primed, that is, water must be discharged over the siphon. The design of the pump, therefore, must permit it to develop a head at shutoff considerably higher than the normal operating head.

In some condenser circulating pump installations, the head required before the siphon is established might so affect the required characteristic curve that the normal operating head occurs at a point of the curve where the pump efficiency is relatively poor. In a siphon system, however, it is generally possible to eliminate the higher starting head requirement by incorporating a vacuum-producing device.

This device is connected to the top of the siphon, so that the air at the high point can be evacuated and the siphon established with the pump producing a head no greater than that required for normal operation. In many cases, a little study or the installation of relatively inexpensive additional equipment will often simplify a problem of this nature.

Type of service

The expected load factor of the contemplated installation will play an important part in the selection of the best pump to be used. As described previously, the type of service will affect the number of units that will best meet the capacity requirements. If the service is intermittent, it is unnecessary to use the most efficient pump available; the selection is generally made on the basis of lowest first cost. A pump intended for continuous service, however, should be selected for efficiency, reliability, and long life.

Position of installation

Whereas the majority of centrifugal pumps are horizontal units, conditions occur to make a centrifugal pump with a vertical axis or rotation more suitable. Axial-flow propeller pumps and certain mixed-flow type pumps are frequently arranged for vertical operation. This results in a very compact installation and permits submerging the impeller so that the pump is always primed. At the same time, the motor driver can be placed above the pump at a level sufficiently high to avoid accidental damage to the driver if the water level rises excessively. Large and small sewage pumps are frequently installed in this manner.

Vertical pumps can be arranged either for wet- or dry-pit installation (Fig. 14.1 and 14.19). Since the pumps used for these two categories are quite different, it is very important to determine which of the two installations is preferable.

Characteristics of driving power

Whereas modern centrifugal pumps are usually driven by electric motors, steam turbines, or internal combustion engines, many other types of drives and means of power transmission are used. A centrifugal pump, its drive, and its method of operation should form an integral and harmonious unit. The specific application and service for which the pump is intended will dictate not only the choice of the pump itself but also of its driver. Centrifugal pump drivers are described in detail in Chap. 20.

Space, weight, and transportation limitations

Sometimes pumps must be installed in very cramped quarters. Vertical pumps may be preferred in such cases, as they need only a small fraction of the floor space required by a horizontal pump of equal capacity. A pump operating at the highest speed compatible with the conditions of service will also reduce space requirements. Although the use of close-coupled pumps (Fig. 1.14) was introduced primarily from consideration of first-cost economy, the application of such pumps presents definite space-saving advantages. Finally, in a number of cases, the use of horizontal pumps with bottom suction (Fig. 2.19) may simplify considerably the problem of spacing and suction piping accommodation.

The weight of a pump usually does not matter, but aboard ships, for instance, light weight is a deciding factor. For this reason, it is the practice in navy and merchant marine installations to use extremely high-speed units of special design, operating at speeds up to 10,000 rpm (generally turbine-driven).

The facilities available for transportation should be outlined if there is anything unusual about them. Sometimes it is a long journey from the factory where a pump is built to the place where it is needed and the transportation facilities may be poor. For transportation over such routes, lightweight pumps must be supplied, preferably of sectionalized construction.

In other cases a pump must be brought to its ultimate destination through tunnels or shafts of limited dimensions, and the pump selection must be made with such restrictions in mind. All such limitations must be carefully listed, or a fully built pump transported part way may prove to be undeliverable.

Location of installation

The geographical location of the installation has a great bearing on proper pump selection and on its maintenance.

The elevation above sea level affects the pump, as there is a decrease in atmospheric pressure of about 1 in. of mercury per 100 ft of elevation. At an elevation of 4,000 ft, therefore, the atmospheric pressure is 4 in. mercury or about 4.5 ft less than at sea level, which means that a centrifugal pump can handle 4.5 ft less suction lift than at sea level.

The geographical location must also be taken into consideration when recommending spare parts, as a pump that is to operate in an out-of-the-way place should be supplied with sufficient spares to avoid interruption of service if parts become worn and cannot be readily replaced.

The immediate surroundings of a pump will affect its accessibility after installation. A pump located in a cramped, dirty and wet, or poorly lighted position will be neglected by the operators, will not give satisfactory service, and will be difficult to inspect, dismantle, and repair.

Special requirements

The pump manufacturer must become familiar with any special requirements and preferences of the personnel who are going to operate the centrifugal pumping equipment. Some of these preferences may be based on insufficient knowledge of modern practices, or they may originate from experience with pumps operating under the same conditions the new pumps will have to meet. This information may be valuable to help the pump manufacturer in furnishing equipment that will give the longest and most reliable service.

Customer recommendations, however, should be limited to experience with pumps operating under similar conditions, rather than a list of his preferences, which may result in the purchase of very special equipment. The manufacturer's standard construction is preferable to specially built units, both from the point of view of the original cost and of obtaining repair parts later.

It is also necessary to take the life expectancy of the installation into consideration. If the design and materials of construction are selected for a life span much in excess of the life expectancy of the process or installation for which the equipment is intended, the original cost of the equipment will be out of proportion. In other words, it is a short-sighted policy to purchase equipment with a 12-month life for an installation designed to last 15 years; it is just as wasteful to buy a pump to last 20 years if only a temporary 6-month use is expected.

INQUIRIES FOR SPECIFIC SERVICES

A knowledge of the factors outlined above is a prerequisite to an intelligent analysis of the problem involved and to its solution. The factors are not equally important, however, for all the various services under which centrifugal pump applications can be classified. Some representative services most commonly encountered and the most important points to be examined for each of these are described below.

As most pumps are constant-speed motor-driven, these lists have been prepared for this type of drive, with one exception. If a variable-speed drive is used, the operating speed can be adjusted to allow the unit to cover a much wider range of conditions than is feasible with a constant-speed drive. Except for the one item, namely, turbo-centrifugal pumping units, these lists do not describe information needed to make driver selections. If drivers are to be purchased with the pumps, full information as to type to be furnished must be included in the inquiry.

Water works services

Raw water—source to treatment plant

1. Operating head range. Indicated in form of system-head curves for minimum, normal, and maximum static heads and supplemented by information as to the duration or importance of minimum and maximum head pumpage.

2. Capacity rating at design head. When local conditions require obtaining certain variations in capacities with variation in total head, it is advisable, before issuing a formal inquiry, to determine from one or more manufacturers, if the desired characteristics can be obtained with constant speed. If not, it may be necessary to use a variable-speed drive.

Centrifugal pump guarantees are limited to one head-capacity condition with very few exceptions, such as large-capacity, specially built units. Even in such cases the manufacturers expect a reasonable tolerance on head or capacity for any secondary guarantee point.

3. Suction conditions. Detailed information how these are interrelated with total head or capacity.

4. Character of water. Dissolved air or gases and foreign material such as grit or leaves; also chemical contamination, if any.

5. Type of installation. Horizontal or vertical with full data on local conditions.

Treated water—direct service into mains

1. Rated capacity at design head with data on operating range of capacity either alone or in parallel with other units. If the new unit or units are to operate in parallel

with existing units, the characteristics of these existing units should be given.

2. Rated head.

3. Suction conditions.

4. Type of installation. Horizontal or vertical.

Treated water—direct service into mains with stand pipe or reservoir on line

1. Capacity at design head. Information as to desired capacity at other heads if head varies with rate of pumpage or for other reasons.

2. Rated head. Head variation data, if any, described in detail.

3. Suction conditions.

4. Type of installation.

Treated water—long distance pumpage through pipeline to distribution system or reservoir

Such systems necessarily involve considerable friction heads which vary with capacity. Unless pumpage is to be at a constant rate to a distribution reservoir, full data on the system should be given with information as to what is to be accomplished.

Treated water—booster services

Pumps for booster service may either take water from mains under nearly constant pressure and discharge it into a distribution system in which a higher pressure is needed or may take water from mains when the pressure falls below the minimum permitted and add more pressure for the area served by the pump. Without a stand pipe on the system, pumps for the first service should generally have a reasonably small head rise so that excess pressure is kept to a minimum. Depending upon local conditions and the inclusion of a stand-pipe, pumps intended for the second type of installation may require anything between a reasonably flat and a very steep head-capacity characteristic. The purpose, therefore, must be clearly outlined in detail if the manufacturer is to make the proper selection of pumps to be used for booster service.

Treated water—turbo-centrifugal pumping units

Turbo-centrifugal pumping units, that is, steam turbine, reduction gear, centrifugal pump, circulating pump, condenser, and condensate pump with interconnecting piping and auxiliaries, involve various items that must be carefully selected individually to make a proper operating whole. This type of unit is seldom purchased, and the manufacturer must be given very complete information for this type of unit, both for the hydraulic data and the physical location in which the unit will be installed. The data needed is summarized as follows:

1. Rated capacity and head. Range of capacity and heads to be met with data on suction conditions for these ranges.

2. Steam pressure and superheat. Range in variation for both.

3. Water temperature range. Temperature specification at which duty guarantees are to be made.

4. Discharge head for condensate pump.

5. Drawing showing space available for unit, location of basement, operating floor, source of suction supply or location of suction main, and point to which discharge is to be connected.

Sewage

1. Operating head range. Whenever two or more pumps are involved, a curve showing the system head-capacity characteristics. For operating head range specification, reduction in friction head losses with decreased capacity, resulting from an increased static head and the reverse with increased capacity, resulting from a decreased static head must be considered.

2. Capacity at average or design head with limitations, if any, at other heads. Unnecessary restrictions of capacities at other than design heads may require special designs with unnecessary high cost.

3. Suction conditions. Full information on how they vary with total head, capacity, or number of units in service.

4. Type of pump installation. Horizontal or vertical. If vertical in dry pit with shafting between pump and motor or gear, elevation of pump pit floor, centerline of suction (if fixed by existing construction), and motor supporting floor must be described. If the vertical distance is such that steady bearings may be required for the vertical shafting, location of bearing supporting beams or of floors, if fixed by some local conditions, must be described. For vertical wet-pit pumps, distance from supporting floor to pit bottom must be stated. This, with information as to water levels in the pit, will permit the manufacturer to select proper length of pump.

5. Size of solids. The maximum size of solids to be expected in straight domestic sewage is $2\frac{1}{2}$ in. in diameter, as could be flushed down a water closet. Unless the sewage is screened or comminuted, it is desirable that sewage pumps that are on straight domestic sewage service be capable of passing $2\frac{1}{2}$-in. solids. On storm water or combined domestic and storm water systems, larger solids can be expected. It is usual to protect pumps from solids larger than they can pass by a bar screen ahead of the pumps. A pump design capable of passing very large solids for the capacity involved is not desirable as it often forces the manufacturer to offer larger, more expensive, and less efficient pumps than if a more reasonable smaller size solid limitation was involved.

The specific information required for drainage and irrigation installations is generally the same as for sewage installations.

Boiler feed service

1. Required capacity in gallons per minute or in pounds per hour, including maximum, minimum, and normal. If expressed in gallons per minute, the conversion from pounds per hour should include specific gravity corrections. The total pump capacity installed, exclusive of spare equipment, should exceed the maximum steaming capacity of the boilers by 8 to 20 per cent. This margin is intended to provide against boiler swings as well as against the eventual reduction of pump capacity through wear before internal parts are renewed.

2. Temperature of feedwater and possible variations.

3. Suction conditions. Variations in suction pressure and information on the minimum available NPSH. If the pump takes its suction from a direct contact feedwater heater, the NPSH (available energy over and above the vapor pressure) is the static submergence between the water level in the storage space and the pump centerline, less the friction losses in the suction piping. If, however, the pump takes its suction from the discharge of some other pump such as the condensate or a booster pump with closed heaters located between them, the available NPSH is the difference between the suction pressure and the vapor pressure at pumping temperature, converted into feet of water at the pumping temperature.

4. Discharge pressure. If not available, boiler pressure, as this will permit estimation of the required discharge pressure. Experience shows that the required discharge pressure will vary from 115 to 125 per cent of the boiler pressure. For a more accurate determination, it is necessary to know the maximum drum pressure, the static elevation of the boiler drum above the pump floor, the friction losses in the discharge piping and in the closed heaters in the pump discharge, and, the losses in the feedwater regulator.

5. Chemical analysis of feedwater. Description of pH value, if analysis is not complete. Information, if available, on the past experience with boiler feed pump materials handling the same feedwater.

6. Type of regulation.

7. A complete layout of the feedwater system and heat balance diagram should be supplied to the pump manufacturer, if possible.

Condensate service

1. Capacity in gallons per minute or in pounds per hour, maximum, minimum, and normal. If condensate pumps discharge through some auxiliary cooling heat exchangers, such as air ejector aftercondensers or oil coolers, specify the minimum flow required as this may necessitate the inclusion of automatic recirculation controls.

2. Suction conditions. Vacuum in hotwell, minimum available submergence, and expected variations.

3. Discharge pressure. Maximum and normal. If prevailing conditions are complex, supply a system-head curve.

4. Type of system. Open or closed. In other words, does the condensate pump discharge into a direct contact heater or surge tank or does it deliver the condensate into the suction of a boiler feed pump?

5. Type of control contemplated. Will the capacity be automatically regulated by variations in hotwell level through operation in the break, will the discharge be throttled by means of a float-regulated valve, or will the float control regulate the bypassing of a certain quantity of condensate back into the condenser hotwell?

Condenser circulating service

1. Rated capacity. In case of parallel operation of several or more units, desired capacity with fewer units in service.

2. Rated head. In case of parallel operation of several or more units, detailed information so that heads at reduced flows can be determined.

3. Suction conditions. Detailed information if conditions will result in increased capacity at reduced head operation.

4. If the pump will operate on a siphon system, the method of establishing the siphon is either to evacuate the entrapped air or to fill the siphon with the pump. If the pump must fill the siphon, the static head from the suction supply to the top of the siphon must be specified. (Usually it is better to establish the siphon by evacuating the entrapped air.)

5. Character of water. Information as to materials of construction found suitable for existing units.

6. Type of installation. Horizontal or vertical. If the latter, whether wet or dry pit with information as to local conditions.

Industrial water supply

The requirements for water supply to industrial plants vary widely with small to large capacities and low to very high heads handling all types of water—fresh, salt, brackish, often contaminated with chemicals. The data described above under *Water works services* and *Condenser circulating service* will serve, generally, as a proper guide.

Chemical service

1. The nature of the liquid. Acid or alkaline, concentration of the solution, and impurities present in the liquid to be handled, if any. This last item is of paramount importance, as experience has shown that presence of various impurities has a marked effect on the relative resistance to corrosion of various pump materials.

2. Amount of solids in suspension and their abrasive quality.

3. Pumping temperature and vapor pressure or boiling point. Possible or expected variation.

4. Specific gravity and viscosity at pumping temperature.

5. Capacity. Maximum, minimum, and normal.

6. Discharge conditions.

7. Suction conditions. Method used to prime pump if there is a suction lift. Possible change in pump location to arrange for operation under submergence if suction lift is impractical or to reduce positive suction head if sealing stuffing box is impractical.

8. Special stuffing box requirements. Local practice with respect to packing, sealing methods, and effect of dilution by sealing fluids.

9. Past experience with various materials or combination of materials handling this liquid. In many cases, dissimilar materials of the reservoir from which the liquid is drawn and of the pump itself set up a galvanic action which may be harmful to one or the other material, thereby requiring proper isolating precautions.

Petroleum services

General and cold oil

1. Nature and complete description of liquid to be handled. Specific gravity over range of pumping temperatures.
2. Capacity. Maximum, minimum, and normal.
3. Suction conditions with special consideration to NPSH.
4. Discharge conditions.
5. Viscosity and variations if any.
6. Corrosive and abrasive properties of liquid.
7. Type of service relative to pump control.

Hot oil

The information listed under *General and cold oil,* plus the following:

8. Range of pumping temperatures.
9. Vapor pressure at pumping temperature.
10. Correction of capacity on hot-to-cold operating range. Resulting effect on power consumption with reference to possible driver overload.
11. Special stuffing box requirements. Local practice with respect to mechanical seals, pressure-sealing arrangements, and smothering glands.

Volatile fractions

The information needed for ethanes, propanes, butanes, gasolines, kerosenes, and other light fractions is the same as listed under *General and cold oil,* plus the following:

8. Range of pumping temperatures.
9. Vapor pressure at pumping temperature.
10. Special stuffing box requirements as described above.
11. Local practice with reference to protection against fire hazards (such as firewalls and special driver enclosures).

Pipeline service

The information listed under *General and cold oil,* plus the following:

8. Graphical representation of required pressure gradients for entire pipeline.
9. Possible variations in capacity or in the nature of products handled.
10. Method of operation—manual or automatic control.
11. Special stuffing box requirements as described above.
12. Local practice with reference to fire hazard.
13. Local preference for series, single, or multiple units.

Process and paper pulp service

The information needed for sugar, brine, brewery, and paper stock service is as follows:

1. The nature of the liquid.
2. Acid or alkaline.
3. Amount of solids in suspension carried, if any. How solids are carried—floating or through agitation and flow. Characteristics of the solids—inert or subject to dewatering or swelling. Consistency of the solids and past experience in pumping the liquid.
4. Pumping temperatures.
5. Specific gravity and viscosity at pumping temperature.
6. Capacity. Maximum, minimum, and normal.
7. Discharge conditions.
8. Suction conditions.

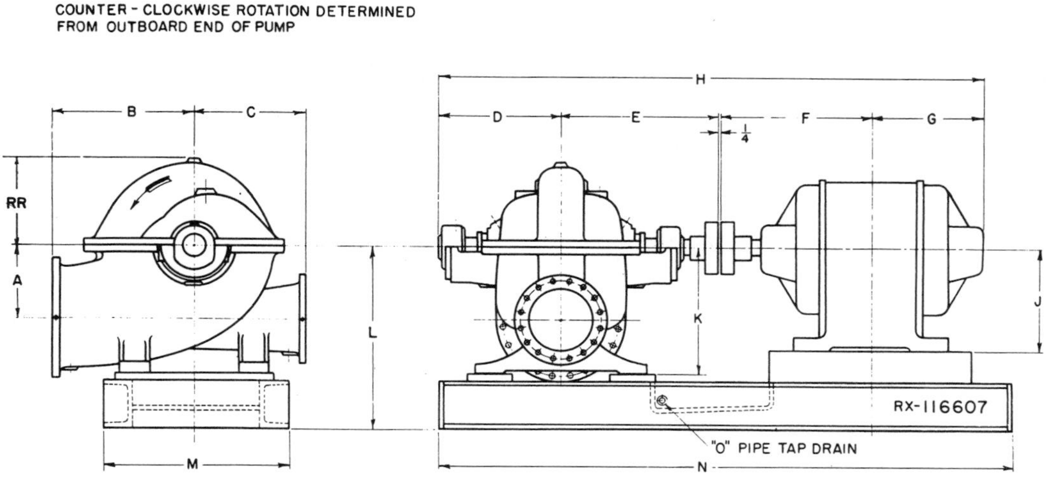
COUNTER - CLOCKWISE ROTATION DETERMINED
FROM OUTBOARD END OF PUMP
B
C
RR
A
L
M
H
D
E
F
G
1/4
K
J
RX-116607
"O" PIPE TAP DRAIN
N

SIZE & TYPE	DISCHARGE DIAMETER, INS.	SUCTION DIAMETER, INS.	A	B	C	D	E	*F	*G	*H	*J	K	*L	*M	*N	O	RR
10 LN-18	10	12	$12\frac{1}{2}$	21	19	$20\frac{13}{16}$	$26\frac{3}{8}$	$25\frac{3}{4}$	$19\frac{5}{8}$	$92\frac{13}{16}$	$14\frac{1}{2}$	$19\frac{1}{2}$	$26\frac{1}{2}$	$32\frac{1}{4}$	$81\frac{1}{4}$	$\frac{3}{4}$	$13\frac{15}{16}$
10 LN-22	10	14	14	25	$19\frac{1}{2}$	$22\frac{3}{8}$	$29\frac{1}{4}$	$29\frac{1}{4}$	$22\frac{1}{8}$	$103\frac{1}{4}$	17	$22\frac{1}{2}$	30	$37\frac{1}{2}$	$86\frac{1}{4}$	$\frac{3}{4}$	$15\frac{3}{4}$
12 LN-17	12	16	14	24	21	$22\frac{7}{8}$	$28\frac{7}{8}$	$26\frac{3}{4}$	$20\frac{5}{8}$	$99\frac{3}{8}$	$14\frac{1}{2}$	23	30	33	$83\frac{1}{4}$	$\frac{3}{4}$	$14\frac{1}{4}$
12 LN-21	12	18	15	$28\frac{1}{2}$	$22\frac{1}{2}$	$24\frac{3}{4}$	$31\frac{3}{4}$	$32\frac{1}{4}$	$25\frac{1}{8}$	$114\frac{1}{8}$	17	$24\frac{1}{2}$	$31\frac{1}{2}$	36	$94\frac{3}{4}$	$\frac{3}{4}$	$16\frac{3}{4}$
12 LN-32	12	20	18	32	30	$29\frac{1}{2}$	$38\frac{1}{2}$	$44\frac{5}{8}$	30	$148\frac{3}{4}$	25	30	40	40	130	$\frac{3}{4}$	23

*APPROXIMATE ONLY, DEPENDING ON SIZE OF MOTOR.

FLANGE DIMENSIONS.					
DIAMETER OF OPENING	DIAMETER OF FLANGE	SIZE OF BOLT	DIAMETER OF BOLT CIRCLE	NUMBER OF BOLTS	FLANGE THICKNESS
10	16	$\frac{7}{8}$	$14\frac{1}{4}$	12	$1\frac{3}{16}$
12	19	$\frac{7}{8}$	17	12	$1\frac{1}{4}$
14	21	1	$18\frac{3}{4}$	12	$1\frac{3}{8}$
16	$23\frac{1}{2}$	1	$21\frac{1}{4}$	16	$1\frac{7}{16}$
18	25	$1\frac{1}{8}$	$22\frac{3}{4}$	16	$1\frac{9}{16}$
20	$27\frac{1}{2}$	$1\frac{1}{8}$	25	20	$1\frac{11}{16}$

Fig. 23.4 Typical elevation drawing

Used in proposals for double-suction, single-stage pumps with axially split casings.

9. Sanitary precautions to be observed. Local preferences for specific details or materials of construction.

Mining service

1. Nature of mine water. Chemical analysis, presence or absence of suspended materials, abrasive quality of the solids in suspension.
2. Capacity requirements.
3. Discharge pressure. Static head and friction losses.
4. Suction conditions.
5. Location of installation. Importance of space factor. Size limitations imposed by transportation through the mine shaft.
6. Special considerations for the drivers. Description of special insulation or enclosure necessary for the electric motors, if any.
7. Previous experience with centrifugal pumps handling this mine water. What materials stood up best, which ones proved unsuitable?

Hydraulic pressure service

1. Liquid used. Water or oil.

2. Capacity. Maximum, minimum, and normal. Rate of capacity change. Rapid fluctuations may require special precautions.

3. Total head conditions. Static head, friction losses, pressures to be maintained.

4. Suction conditions.

5. Type of pressure system. Method of pump discharge—directly to press or nozzles, or through an accumulator. A sketch of the installation is very helpful.

6. Type of control. Characteristics of accumulator, if used. Maintains a constant pressure automatically and pump is shut down during the no-load operation, or pump operates at no flow. If the latter, arrangements used to protect pump against overheating, that is, control of bypass recirculation.

ORDERING EQUIPMENT

Normally, the customer will select and order the equipment after careful study of several proposed installations submitted to him by manufacturers in response to his original inquiry, which described his specific requirements and operating conditions.

Promises of shipment made by pump manufacturers date from the receipt of complete manufacturing information; therefore, to avoid delays in the preparation of drawings and in the issuance of specifications to the shop for manufacturing, it is necessary that the purchaser's order contain all required information. In addition to the complete conditions of service for the pump as well as the driver, the order should specify the desired direction of rotation, a statement as to whether manufacturing should proceed without awaiting approval of elevation drawings, and other pertinent information. For example, if the pump is to be driven by a steam turbine

equipped with an excess-pressure regulator, and if the turbine is to be ordered by the pump manufacturer, the order should specify the location of the pilot lines from the regulator and the value of the excess pressure to be maintained.

If the order for the pump requires the pump manufacturer to furnish the driver, he will order it and obtain a certified elevation drawing from the driver manufacturer before proceeding with the combined elevation drawing. If the driver is ordered separately, the purchaser must expedite the receipt of the driver print necessary for the preparation of the combined drawing. The proposal submitted by the manufacturer usually includes a bulletin describing the construction features of the equipment as well as an outline drawing of the pump in question, as shown in Fig. 23.4. Whereas the dimensions given on that drawing are not to be used for final construction purposes, they are sufficiently detailed, not only to determine the desired direction of rotation, but also to prepare a preliminary layout of the installation.

Since drivers can generally be obtained with either direction of rotation, the choice of rotation is mainly directed towards locating the suction and discharge openings of the pump in proper relation to the proposed piping. Manufacturers of standardized lines of single-stage end-suction pumps generally limit their designs to a single direction of rotation, whether these pumps are close-coupled mounted on a motor (Fig. 1.14), or driven by a separate driver. This decision reduces the cost of pattern development and increases the effectiveness of stock inventory without complicating the user's problems, as it does not materially affect nozzle location.

In addition to showing the general dimensions of the equipment and the location of the baseplate foundation bolts, an outline elevation drawing may also depict a great number of details such as the location and size of all auxiliary cooling or lubricating piping, and space requirements for dismantling. The complexity of the information that will be contained on an outline drawing reflects the severity of the conditions of service involved and, therefore, the complexity of construction of the pump itself.

The construction of equipment for small- or medium-size pumps intended for general service is usually standard, and unless some special features have been discussed with the pump manufacturer before placement of the order, it is practical to release the pump for manufacturing without formal approval by the customer.

Frequently, the construction of the pump itself may be of a standard nature, but approval may be required for the general arrangement, including the baseplate construction. In such an event, the order should authorize the manufacturer to proceed with construction of the pump but withhold approval for the manufacture of the baseplate.

V
INSTALLATION,
OPERATION,
and
MAINTENANCE

24 *Installation*

The general information contained in this chapter and in Chap. 25 should be supplemented by the specific instructions prepared by the manufacturer for the centrifugal pump in question. These instructions are contained in the instruction booklets that manufacturers pack in the box of fittings accompanying the pump or send to the customer when the pump is shipped.

Instruction books

An instruction book contains direction for the following phases of pump use:

1. Installation for maximum service with minimum wear and expense
2. Adjustment and operation for optimum performance
3. Maintenance and repair when necessary.

Because instruction books are intended to help keep the machinery efficient and reliable at all times, they should always be available to the following personnel: (1) Construction men responsible for installation, (2) operators who use the equipment and make periodic checks or tests, (3) maintenance men who repair and service the equipment, and (4) engineers who determine the proper use of the apparatus. If the books contain a parts list and instructions for ordering repair parts, a copy should be available to personnel responsible for ordering repair parts.

The number of instruction books needed, therefore, depends upon the extent to which these functions are departmentalized. If one man is in charge of installing, operating, and maintaining a centrifugal pump, he only needs one copy. If each function is performed by a separate department, however, each department should have at least one copy of the instruction book.

Too often, sufficient instruction books are requested, but are never transmitted to the proper personnel once they have been received. Later, when a department needs the books, it may be difficult to locate them in a short time.

A request for abnormal quantities of instruction books adds to the cost of providing this material, costs which are passed on to the customer. On the other hand, books requested after the pump is delivered may be more expensive, particularly if they are prepared for equipment deviating from standard. It is wise, therefore, to order the exact number of books required when the pump is ordered. It is equally important that the books be given to the personnel who will need them and not be kept locked away.

Preparation for shipment

The general procedure in preparing pumps for shipment after manufacture is practically identical with all manufacturers. All flanges and exposed finished metal parts are cleaned of foreign matter and treated with an anticorrosion compound, such as grease, vaseline, or heavy oil. For protection during shipment and erection, all pipe flanges, pipe openings and nozzles are protected by wooden service flanges or by screwed-in metal plugs, which prevent the entrance of dirt, dust, moisture, or foreign matter. All small piping is cleaned, and protective guards are installed, if necessary.

Usually the driver is delivered to the pump manufacturer, where it is assembled and aligned with the pump on a common baseplate. The baseplate is drilled for driver-mounting, but the final dowelling is left for the field, after final alignment. When size and weight permits, the unit is shipped in the assembled state, that is, pump and driver on the baseplate. Sometimes, however, large drivers are shipped directly to be mounted in the field. Then the baseplate should be drilled at the job site.

Sometimes it is impractical to ship the pump mounted on its baseplate, as when the suction and discharge nozzles are located vertically downwards and extend for an appreciable distance below the baseplate, or if size or weight limitations exist. These pumps must be shipped separately or turned upside down and temporarily secured to the baseplate.

Care of equipment in the field before use

The equipment should be inspected and checked against the shipping manifest immediately on receipt of the shipment. Any damage or shortage should be reported to the transportation company's local agent.

If the pumping equipment is received at the site before it can be used, it should be immediately stored in a dry location. The protective flanges and finishes should remain on the pumps. The bearings and couplings must be carefully protected against sand, grit, or other foreign matter, and the pump rotor should be turned over by hand at frequent intervals to prevent rusting or binding. Sometimes a rotor will become slightly bound in storage. To free it, the thrust bearing should be removed or dismantled, and oil throwers, guards, or other parts which may restrict end movement should be loosened. Moving the rotor a few times will free it at the wearing rings or wherever it has become bound at the internal clearances.

If a pump must be stored for an extended period of time, more thorough precautions are required. The pump should be carefully dried internally with hot air or a vacuum-producing device to avoid rusting of internal parts. Once free of moisture, the pump should be filled with a protective fluid, such as light oil, kerosene, or antifreeze. All accessible parts, such as bearings and couplings, should be dismantled, dried, and coated with vaseline or acid-free heavy oil, after which they should be properly tagged, wrapped, and boxed to avoid metal-to-metal contact, and stored.

If rust preventative has been used on stored parts, they should be thoroughly cleaned before final installation. Extreme care must be taken to assure that all traces of the protective coating are removed and the bearings are relubricated.

General rules for pump location

Pumps installed indoors, in poorly lighted and cramped locations, or where dirt and moisture accumulate, are improperly placed for dismantling and repair; they will be neglected and both pump and driver may become damaged. Pumps should be placed in light, dry, and clean locations whenever possible.

If a motor-driven unit will be operated in a damp, moist, or dusty location, the proper motor must be selected. Pumps and drivers designed for outdoor installation are specially constructed to withstand exposure to weather and usually are readily available for overhaul.

Sufficient room must always be provided for dismantling the pump; that is, enough headroom must be allowed for horizontal pumps with axially split casings, so that the upper half of the casing may be lifted free of the rotor. Some high-pressure pumps are radially split, and their rotor is removed longitudinally (Fig. 3.16). Space must be provided so that the rotor can be pulled out without canting it. For large pumps with heavy casings and rotors, a travelling crane or facilities for attaching a hoist should be provided over the pump location.

When pumping equipment must be used at levels where flooding is possible, two courses of action may be followed: (1) A vertical wet-pit pump may be used, or (2) auxiliary wet-pit drainage pumps must be provided as an insurance against damage to the main equipment.

In normal installations, pumps should be located as close as possible to the source of liquid supply. When convenient, the pump centerline should be placed below the level of the liquid in the suction reservoir. The manufacturer's recommendations for suction conditions should always be followed.

For most pumping units, more satisfactory service is obtained when rigid foundations are provided.

Foundations

The foundations are any structure heavy enough to afford permanent rigid support to the full area of the baseplate and to absorb any normal strains or shocks.

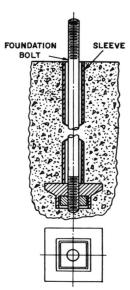

Fig. 24.1   Foundation bolt

Concrete foundations built up from solid ground are the most satisfactory. In building the foundation, allowance should be made between the rough surface of the concrete and the underside of the baseplate for grouting. Whereas most pumping units are mounted on baseplates, very large equipment may be mounted directly on the foundations. In these cases, soleplates should be provided under the pump and driver feet. In this manner, the alignment can be corrected with shims and the unit can be removed and replaced without difficulty, if necessary.

The space required by the pumping unit and the location of the foundation bolts are determined from the drawings supplied by the manufacturer. Each foundation bolt (Fig. 24.1) should be surrounded by a pipe sleeve, three or four diameters larger than the bolt. After the concrete foundations are poured, the pipe is held solidly in place while the bolt may be moved to conform to the corresponding hole in the baseplate.

When a unit is mounted on steelwork or other structure, it should be placed as near as possible to the main members, beams, and walls, and be supported so that the

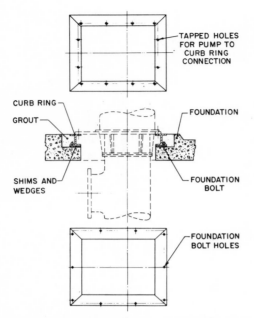

Fig. 24.2 Curb ring details, rectangular type for below-ground discharge vertical pump

baseplate cannot be distorted or the alignment disturbed by any yielding or springing of the structure or of the baseplate.

Mounting of vertical wet-pit pumps

A curb ring or soleplate must be used as a bearing surface for the support flange of a vertical wet-pit pump. The mounting face must be machined to an even flat surface, since the curb ring or soleplate will be used in aligning the pump.

If the discharge pipe is located below the support flange of the pump (below ground discharge), the curb ring or soleplate must be large enough to allow passage of the discharge elbow during assembly. A rectangular ring (Fig. 24.2) should be used.

If the discharge pipe is located above the support flange of the pump (above-ground discharge), a round curb ring or soleplate should be provided. Sufficient clearance should be left on the inner diameter to allow passage of all sections of the pump located below the support flange, such as

manhole openings on the side of the casing (Fig. 24.3).

A typical method of arranging a grouted soleplate for a vertical pump is shown in Fig. 24.4.

If the discharge is below ground (see Fig. 14.18), and a Dresser type coupling is used, it is necessary to watch for the moment which may be imposed on the structure. Generally, the pump casing must be securely attached to some rigid structural members with tie rods. If vertical wet-pit pumps are very long, some steadying device is required, regardless of the location of the discharge or the type of pipe connection. Here again, tie rods can be used to connect the unit to a wall, or a small clearance can be provided around a flange to prevent excessive displacement of the wet-pit pump in the horizontal plane.

Alignment

When a unit consisting of pump, base, coupling, and driver is assembled at the factory, the baseplate is placed on a flat, even surface, the pump and driver are mounted on the plate, and the coupling halves are accurately aligned with shims

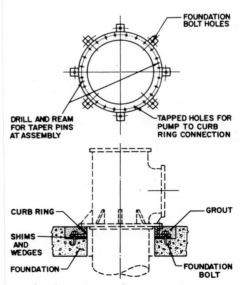

Fig. 24.3 Curb ring details, round type for above-ground discharge vertical pump

under the pump and driver mounting surfaces where necessary. The pump is usually dowelled to the baseplate at the factory, but, as described previously, the driver is dowelled after installation at the site because factory alignment cannot be maintained with sufficient accuracy to start and operate the unit without realignment in the field.

Baseplates of sufficient strength and rigidity and mounting feet and bolting of sufficient size to permit operation without field realignment would be prohibitive in size, weight, and cost. During shipping, furthermore, the pump, its driver, and the baseplate are subjected to such forces that serious misalignment is introduced. When the baseplate is mounted in place and the piping is connected, further misalignment can occur, often serious enough to cause coupling and bearing failures and, in some cases, shaft breakage. If the pump handles high-temperature liquids, the final field alignment should be made with both pump and driver brought up to normal operating temperature.

With units of moderate size it is usually unnecessary to remove the pump or driver from its baseplate when leveling. The unit should be placed over the foundation and supported by short strips of steel plate or shim stock close to the foundation bolts, allowing a space from ¾ in. to 2 in. between the bottom of the baseplate and the top of the foundation for grouting. The shim stock should extend fully across the supporting edge of the baseplate. The coupling bolts should be removed before the unit is leveled and the coupling halves are aligned. Wedges are sometimes used instead of flat strips, although they are less satisfactory. Opinions differ with respect to removal of shims or wedges after grouting, as discussed later in this chapter. The projecting edges of the pads supporting the pump and driver feet, when scraped clean, can be used with a small spirit level for leveling the baseplate. Where possible, it is preferable to place the level on some exposed part of the pump shaft, sleeve, or planed surface of the casing. The steel supporting strips or shim stock under the baseplate should be adjusted until the pump shaft is level, the flanges of suction and discharge nozzles are vertical or horizontal, as required, and the pump is at the specified height and location. When the baseplate has been leveled, the nuts on the foundation bolts should be made hand-tight.

During pump and base leveling, the accurate alignment of the unbolted coupling halves between pump and driver shafts must be maintained. Before alignment, both pump and driver rotors should be revolved by hand to insure that they move freely. A straight-edge should be placed across the top and side of the coupling, and, at the same time, the faces of the coupling halves should be checked with a tapered thickness gauge or feeler gauges (Fig. 24.5) to see that they are parallel. For all alignment checks, including parallelism of coupling faces, both shafts should be pressed hard over to one side when taking the readings.

When the peripheries of the coupling halves are true circles of the same diameter and the faces are flat, exact alignment exists when the distance between the faces is the same at all points and a straight-edge will lie squarely across the rims at any point. If the faces are not parallel, the thickness gauge or feelers will show a variation at different points. If one coupling is higher than the other, the amount may be determined by the straight-edge and feeler gauges.

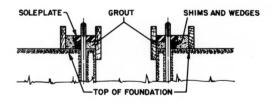

Fig. 24.4 Grouting form for vertical pump soleplate

Sometimes coupling halves are not true circles and are not identically the same diameter, because of manufacturing tolerances. The trueness of either coupling half is checked by revolving it, holding the other coupling half stationary, and checking the alignment at each quarter turn of the half being rotated. Then the half previously held stationary should be revolved and the alignment checked. A variation within manufacturing limits may be found in either of the half couplings and proper allowance for this must be made when aligning the unit.

A dial indicator bolted to the pump half of the coupling may be used to check both radial and axial alignment instead of a straight-edge and feeler gauge (Fig. 24.6). With the button resting on the other coupling periphery, the dial should be set at zero and a mark chalked on the coupling half at the point where the button rests. For any check (top, bottom, or sides) both shafts should be rotated by the same amount, that is, all readings on the dial must be made with its button on the chalk mark. The dial readings will indicate whether the driver must be raised, lowered, or moved to either side. After any movement, it is necessary to check that the coupling faces remain parallel to one another.

For example, if the dial reading at the starting point is set to zero and the diametrically opposite reading at the bottom or sides shows ±0.020 in., the driver must be raised or lowered by shimming, or moved to one side or the other by half of this reading.

When an extension coupling connects the pump to its driver, a dial indicator should be used to check the alignment (Fig. 24.7). The extension piece between the coupling halves should be removed, exposing the coupling hubs. The coupling nut on the end of the shaft should be used to clamp a suitable extension arm or bracket long enough to extend across the space between the coupling hubs. The dial test in-

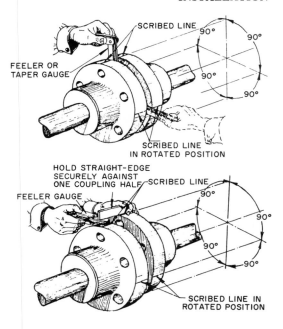

Fig. 24.5 Coupling alignment, using feeler gages

dicator is mounted on this arm and alignment is checked both for concentricity of the hub diameters and parallelism of the hub faces. Changing the arm from one hub to the other provides an additional check.

The clearance between the faces of couplings of the pin and buffer type and the ends of shafts in other types should be set so that the faces cannot touch, rub, or exert a pull on either pump or driver. The amount of this clearance will vary with the size and type of coupling used. Sufficient clearance will allow unhampered endwise movement of the shaft of the driving element to the limit of its bearing clearance. On motor-driven units, the magnetic center of the motor will determine the running position of the motor half coupling. This position should be checked by operating the motor while it is disconnected; the direction of rotation of the motor should also be checked. If current is not available, the motor shaft should be moved in both directions as far as the bearings will permit, then the shaft should be adjusted centrally between these limits. The unit should be assembled with the correct gap between coupling halves.

Large horizontal sleeve-bearing electric motors are not generally equipped with thrust bearings. The rotor is permitted to float, and, whereas it will seek its magnetic center, a force of rather small magnitude can cause it to move off this center. Sometimes it will move enough to cause the shaft collar to contact the bearing, generally leading to serious motor bearing difficulties. To avoid this, pump and motor manufacturers have agreed to use a limited float coupling between the pump and motor on all large size units, taking advantage of the fact that all pumps are equipped with adequate thrust bearings. In this manner, the motor rotor is kept within a restricted position.

The motor manufacturers mark the two extremes of motor shaft travel and make the total axial clearance between the shaft collars and the bearing shoulders not less than $\frac{1}{2}$ in. In turn, the pump manufacturers furnish flexible couplings that will limit the end float of the motor rotor to $\frac{3}{16}$ in.

When the pumps are driven by steam turbines, the final alignment should be made with the driver heated to its operating temperature. Where this is not possible at the time of alignment, suitable allowance must be made for the shaft being elevated when the unit expands. The alignment should always be checked when the unit is at operating temperature and adjusted as required before placing the pump in service. Heat applied to steam inlet and exhaust

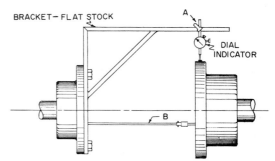

Fig. 24.7 Use of dial indicator for alignment of extension type coupling

(*Courtesy John Waldron Corp.*)

piping causes expansion, and the installation must be so made that the turbine nozzles are not subjected to piping strains.

When the unit is accurately leveled and aligned, the hold-down bolts should be gently and evenly tightened before grouting.

The alignment must be rechecked after suction and discharge piping have been bolted to the pump, to test the effect of piping strains.

The pump and driver alignment should be occasionally rechecked, because piping strains may develop after a unit has been operating for some time, causing misalignment. This is especially true when the pump handles hot liquids, as there is a likelihood of growth or change in the shape of the piping. Pipe flanges at the pump should be disconnected after a period of operation to check the effect of the expansion of the piping, and adjustments should be made to compensate for this.

Grouting

Ordinarily, the baseplate is grouted before the piping connections are made and before the alignment of the coupling halves is finally rechecked. The purpose of baseplate grouting is to prevent lateral shifting of the baseplate and increase its mass to reduce vibration, as well as smoothing irregularities in the foundation.

The usual mixture for grouting a pump base is composed of one part pure portland

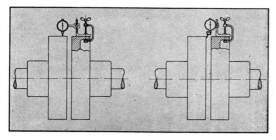

Fig. 24.6 Use of dial indicator for alignment of standard coupling

cement and two parts building sand, with sufficient water to cause the mixture to flow freely under the base (heavy cream consistency). In order to reduce settlement, it is best to mix the grout and let it stand for a couple of hours, remixing it thoroughly before use, without adding any water.

The top of the rough concrete foundation should be well saturated with water before grouting. A wooden form is built around the outside of the baseplate to contain the grout, either tight against the lower outer edge of the base or a slight distance removed from the edge as convenient. For convenience in getting the grout under the base, tin plate funnels are used at several points around the edge. Grout is added until the entire space under the base is filled (Fig. 24.8). The grout holes in the base serve as vents to allow the air to escape. A stiff wire should be used through the grout holes to work the grout and release any air pockets.

The exposed surfaces of the grout should be covered with wet burlap to prevent cracking from drying too rapidly. When the grout is sufficiently set (about 48 hours) so that the forms may be removed, the exposed surfaces of the grout and foundation are finished to a smooth surface. When the grout is hard (72 hours or more), the hold-down bolts should be finally tightened and

the coupling halves rechecked for alignment.

Erectors and experts on the use of concrete differ among themselves on whether or not leveling strips or wedges should be removed after grouting. An article published a few years ago contained the following two statements in two portions of the text:

Shims or wedges and screws are usually removed after the grout has hardened.

and

Wedges may be of steel or hardwood, but wood wedges are suitable only for small machines. Wood wedges should be removed. Removal of steel wedges is left to the discretion of the erectors. If they are to be removed, however, care should be taken to see that *all* are removed, as a single wedge left in place may cause damage.

This apparent vacillation of various authorities and erectors is influenced by the type of experience they have had early in their career, that is, whether they have mainly dealt with reciprocating or rotary machinery. Removal of shims or wedges is mandatory with reciprocating machinery or any other installation where impact, pounding action, or appreciable vibration are to be expected in normal operation. The following quotation is from a Worthington Corporation instruction book for vertical diesel engines:

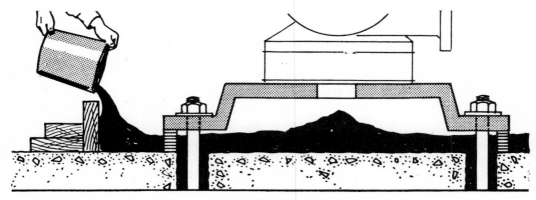

Fig. 24.8 Application of grouting

Before the grout has entirely set but after it will no longer flow, remove the excess grout between the engine holding-down flanges and the dam and trim the grout in line with the vertical edge of the flanges. At the same time, dig around the shims and wedges sufficiently to facilitate their removal later. The grout should set at least 48 hours and longer if possible, to properly harden, after which all shims and wedges must be removed and the space they occupied regrouted. CAUTION: THE REMOVAL OF THE SHIMS AND WEDGES, USED IN THE LEVELING OF THE UNIT, IS ABSOLUTELY ESSENTIAL. If not done, the unit will be suspended, with metal to metal contact, on the wedges and shims and in time, due to reciprocating forces, become loose on the foundation.

With rotary machinery, such as centrifugal pumps and centrifugal compressors, however, the danger of loosening is almost nonexistent, and most erectors who specialize in this type of machinery leave the shims in place. The loosening problem is affected by the care used in choosing the grout material and in mixing it. Excess water in the grouting mix is almost certain to result in shrinking and damage to the grout. A plain cement-sand mortar shrinks considerably when it dries and is, therefore, a poor choice for grouting material. A number of grout mixes guaranteed not to shrink when properly used are available and cost little more than a plain cement-sand mixture. A bulletin issued by the Master Builders Co. describing their Embeco Pre-Mixed Grout carries the statement: "It is not necessary to remove shims or loosen leveling screws with this Grout."

No harm will result from removing the shims, and erectors can safely follow their own preference in this matter, as long as they are dealing with centrifugal pumps and are using suitable grouting material.

Whatever the practice with regard to the shims, it is imperative that pump and driver alignment be rechecked thoroughly after the grout has hardened permanently and at reasonable intervals later. The fact that a baseplate has been leveled before and during grouting should not be considered sufficient evidence of the permanent evenness of pump and driver mounting surfaces. Proper alignment and shimming between baseplate mounting surfaces and machine feet is still necessary.

Dowelling of pump and driver on the baseplate

When a pumping unit handles hot liquids, dowelling of both the pump and its driver should be delayed until the unit has been operated. A final recheck of alignment with coupling bolts removed and with pump and driver at operating temperature is advisable before dowelling.

Some special types of multistage pumps for handling very hot liquids are constructed with a key and keyway in the casing foot and base (Fig. 24.9). One end of the pump is securely bolted, whereas the other is bolted with spring washers under the nuts on the casing feet, allowing one end to move laterally when expanded. Generally, all hot liquid multistage pumps should be dowelled at the thrust bearing end, either in the ordinary manner or with the dowels crosswise. Dowels at the other end, if used, are fitted in a similar manner to the key and keyway, that is, parallel to the pump shaft, to allow the casing to expand when heated.

Piping

Piping strains

Satisfactory operation cannot be maintained when piping imposes forces and torques on the pump. A pump can be easily sprung and pulled out of position by drawing up on the bolts in the piping flanges. Piping flanges must be brought squarely together before the bolts are tightened. The suction and discharge piping and all associated valves and similar equipment should be supported and anchored near the pump, but independent of it, so that no strain is transmitted to the pump casing.

When large pumps are involved, or when major temperature changes are expected, the pump manufacturer supplies the user with information on maximum permissible piping stresses. Typical diagrams are illustrated in Fig. 24.10 for a multistage pump with side suction and side discharge nozzles and an axially split casing and in Fig. 24.11 for multistage pump with top suction and discharge and a radially split double casing.

Suction piping

The major source of trouble in centrifugal pump installations, other than misalignment, is a faulty suction line. The suction piping should be as short and as direct as possible. If a long suction line is required, the pipe size should be increased to reduce friction losses. (The only exceptions are boiler feed pumps, where difficulties may arise during transient conditions of load change if the suction piping volume is excessive. This is a complex and very special subject; therefore the pump manufacturer should be consulted on this matter.) The suction piping should be laid out with a continual rise towards the pump, without any high spots, to prevent the formation of air pockets, which invariably cause troubles. Only eccentric reducers, installed straight side up (Fig. 24.12), should be used between the suction piping and the pump suction nozzle.

Elbows and other fittings next to the pump suction should be selected and arranged carefully, or the flow into the impeller will be unfavorably disturbed. Long-radius elbows are generally preferred for suction lines, as they create less friction and provide a more uniform flow distribution than standard elbows.

On vertical suction piping, the inlet should preferably be submerged in the liquid to four times the piping diameter.

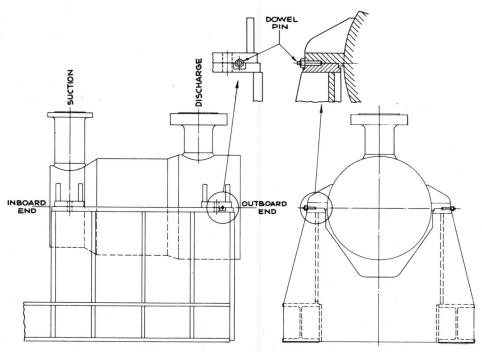

Fig. 24.9 Dowelling of a high-temperature multistage pump

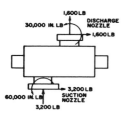

Fig. 24.10 Maximum permissible pipe stresses and torques for multistage pump with side suction and discharge and an axially split casing

After installation, the suction piping should be blanked off and hydrostatically tested for air leaks before the initial start-up.

Discharge piping

Generally, both a check valve and a gate valve are installed in the discharge line. The check valve is placed between the pump and the gate valve and protects the pump against reverse flow in the event of unexpected driver failure. The gate valve is used when priming the pump or when shutting it down for inspection and repairs.

Manually operated valves that are difficult to reach should be fitted with a sprocket rim wheel and a chain. These valves are sometimes motorized and operated by remote control.

Expansion joints

Expansion joints are sometimes used in the discharge and suction lines of centrifugal pumps, to avoid transmitting any piping strains to the pump, whether these strains are from expansion during handling of hot liquids, misalignment of the piping, or any other cause. On occasion, expansion joints are formed by looping the pipe, as is customary with steam piping. More often, expansion joints are of the slip joint or corrugated diaphragm type. They eliminate

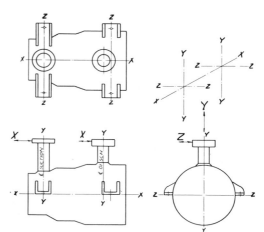

Fig. 24.11 Permissible pipe stresses and torques for multistage pump with top suction and discharge and a radially split double casing

Allowable forces, in lb

	X	Y	Z
Suction	15,800	15,800	11,200
Discharge	19,300	15,800	11,200
Total	19,300	15,800	11,200

Allowable moments, in ft-lb

	X-X axis	Y-Y axis	Z-Z axis
Suction	22,900	45,400	22,900
Discharge	21,600	43,200	21,600
Total	22,900	45,400	22,900

piping stresses, but introduce an entirely new problem, namely, a reaction and a torque on the pump and its foundations. Unless they are applied correctly, therefore, expansion joints can cause greater problems than they alleviate. Precautions must be observed in the design of the piping and

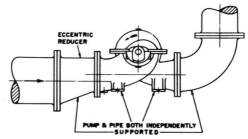

Fig. 24.12 Correct method of installing suction piping

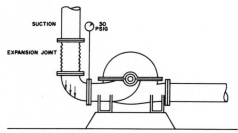

Fig. 24.13 Expansion joint in a suction line

positioning of the expansion joints so that reactions due to flow and pressure conditions are absorbed by the strategic placing of anchors, hangers, and bolts controlling movement. Normally, pumps, their drivers (steam turbines), and baseplates will withstand certain limited reactions; however, special study and calculation of these reactions is important.

When analyzing expansion joints, it is necessary to remember that action equals reaction. For example, if an expansion joint is located vertically ahead of the suction elbow (Fig. 24.13), the downward force will be the product of the pressure (30 psig) and of the area on which this pressure acts in the expansion joint. This area can generally be approximated by assuming that the pressure is applied over an area corresponding to the mean diameter of the expansion joint element or corrugated bellows. In the case illustrated in Fig. 24.13, if the pipe is 8 in., the expansion joint would have $13\frac{1}{4}$-in. OD and $8\frac{5}{8}$-in. ID in the corrugations. The effective area would be approximately 93 sq in. and the downward force 30×93 or 2,790 lb. Some engineers feel that this force should be further corrected by the spring constant of the expansion joint, which is 800 pounds for $\frac{1}{2}$-in. expansion of each corrugation. However, the displacements are never of the order of magnitude where this factor would be significant, and, therefore, it can be neglected. This 2,790-lb force, acting downward, applies a couple to the pump and may twist the pump on its foundation if no provision has been made to counteract the force.

Similarly, when an expansion joint is located in the 6-in. discharge pipe in Fig. 24.14, the 100-psig pressure is acting over the approximately 53-sq in. effective area of the expansion joint in the discharge pipe, and the horizontal reaction against the pump is 5,300 pounds. If there is no expansion joint, the pump and the pipe will pull against each other, and if the flange bolts hold, there is no stress on the pump, its foundation bolts, or the foundation. But with an expansion joint interposed between the pump and the piping, the 5,300 lb act on the pump and tend to pull it off the foundation bolts. Depending upon the size and number of these bolts, the stress on the foundation bolts may become excessive.

Expansion joints are not generally recommended for high-head pump installations because the reaction forces rise rapidly with increased sizes and pressures. With a 24-in. pipe, an effective corrugation area of 600 sq in., and a 160-psi pressure, the reaction force would reach 96,000 lb or 48 tons. The casing would have to be almost entirely sunk and anchored in concrete to hold it in place.

A vertical pump with side discharge (Fig. 24.15) will have a side force equal to the discharge pressure times the effective expansion joint area, or

$$20 \text{ psi} \times 900 \text{ sq in.} = 18,000 \text{ lb}$$

The vertical force upwards on the elbow is balanced by the hydraulic thrust, as far as the plate support is concerned. Therefore, the only downward force on the supports is the dead weight of the unit. On the other hand, the thrust on the motor bear-

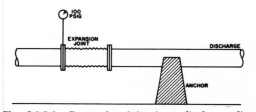

Fig. 24.14 Expansion joint in a discharge line

ings is the dead weight plus the hydraulic thrust.

Figure 24.16 illustrates some problems introduced by the use of expansion joints. The high-head (220 to 325 ft) vertical pumps in this installation were provided with a sleeve-type flexible pipe connection on the downstream side of the cone-type discharge valves. The valves, the pumps, and the pipe fittings between them were subjected to a high hydraulic reaction. To transmit this reaction to the foundation, the 45-deg elbows were substantially anchored and the pump casings were securely tied into the foundations by casting the concrete up around part of the pump casing as illustrated. On the largest, highest head pump this was still insufficient anchorage, and the flexible pipe connection had to be replaced with flanged fittings.

Fig. 24.16 Expansion joints in an installation of vertical pumps

Suction strainers

Another common source of troubles in centrifugal pump installations is the entrance of foreign matter of various sizes into the pump. This foreign matter, if sufficiently large, can clog up the pump and reduce its capacity or completely stop it from pumping. Small particles of foreign matter can cause serious damage by lodging between close running clearances. A typical problem concerns boiler feed pumps in a steam power plant. The life of these pumps and of their component parts is unbeliev-

ably short when dirt, welding beads, mill scale, and similar foreign matter is permitted to enter the feedwater circuit. The abrasive effects of mill scale or hard brittle oxides on close clearance parts of the pump will require endless cleaning, inspection, and hours of hard work to keep the pump serviceable, if certain basic preventive measures are ignored.

Whereas the description of the precautions to be taken during the initial start-up would normally be included with Chap. 25, Operation of Centrifugal Pumps, strainers must be installed in the suction lines and are, therefore, discussed in this chapter.

Boiler feed pumps have internal running clearances from 0.020 in. to 0.012 in. on the diameter, that is, 0.010 in. to 0.006 in. radially; small particles of foreign matter, such as mill scale left in the piping or brittle oxides, can cause severe damage if they get into these clearances. Generally, a seizure does not occur while the pump is running but when it is brought down to rest. Since boiler feed pumps are frequently started and stopped during the initial plant start-up period, seizures are very likely to

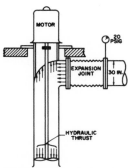

Fig. 24.15 Expansion joint in the side discharge piping of a vertical pump

occur at this time if foreign matter is present.

The actual method used in cleaning the condensate lines and the boiler feed pump suction piping varies considerably in different central stations. The essential ingredient in all cases, however, is the use of a temporary strainer located at a strategic point. Generally, the cleaning out starts with a thorough flushing of the condenser and deaerating heater, if one is used in the feedwater cycle. It is preferable to flush all the piping to waste with hot water, if possible, before finally connecting the boiler feed pumps, as more dirt and mill scale can be loosened at higher temperatures. Some central stations use a hot phosphate and caustic solution for this purpose.

Temporary screens or appropriate size strainers must be installed in the suction line as close to the pump as possible. The choice of the size of the openings is difficult to make. If 8-mesh screening with 0.025-in. wire is used, the openings are 0.1 in., too coarse to remove particles large enough to cause difficulties at the pump clearances, which may be from 0.006 in. to 0.01 in. radially. If a large amount of finely divided solids is present, and if the pump is stationary during flushing, some solids will probably pack into the clearances and cause damage when the pump is started.

The safest solution consists of using a strainer with 40 to 60 mesh and flushing with the pump stationary, until the strainer remains essentially clean for a half day or longer. After that, a coarser mesh can be used if a higher rate of circulation is necessary. It is very important to turn the pumps by hand both before and after flushing to check whether any foreign matter has washed into the clearances. If the pump drags after flushing, it must be cleared before operation.

Unless the system is thoroughly flushed before starting the pump, the use of a fine mesh screen may cause trouble. For example, 40-mesh screening with 0.015-in.

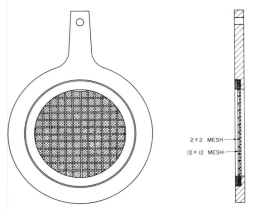

Fig. 24.17   Flat suction strainer
(*Courtesy Leslie Co.*)

wire leaves only 0.010-in. openings, and these would clog up instantly, unless a very thorough cleaning job was done initially.

Pressure gauges must be installed both upstream and downstream of the screen and the pressure drop across it watched very carefully. As soon as dirt begins to build up on the screen and the pressure drop increases, the pump should be stopped and the screen cleaned out.

Types of strainers

Although conventional flat screen strainers (Fig. 24.17) placed between two flanges are used very frequently to catch foreign matter and are inexpensive, they are not really economical. Considerable expense is involved in removing, cleaning, and replacing such strainers, and pipe fitters must be on duty to break and make the joints.

One of the most satisfactory strainers is the type illustrated in Fig. 24.18. It has a sloping screen, which affords a lower pressure drop and, therefore, longer periods between cleaning than the conventional flat screen strainer. This strainer is provided with a draw-off cock for removing accumulated foreign matter. Thus, in many cases, dirt accumulation can be dropped out by opening this draw-off cock without actually shutting down the system. The strainer is also provided with a valved opening on its downstream side, permitting back-washing of the screen if necessary. It is possible to clean this strainer in ten minutes of complete outage time as against approximately four hours required for the conventional flat strainers.

If the size of the free straining area is properly selected to minimize the pressure drop, the start-up strainer may be left in the line for a considerable period before the internal screen is removed. Alternately, the entire unit may be removed and replaced with a spool piece. The strainer is then available for the next initial start-up.

Venting and drainage

Vent valves are generally installed at one or more high points of the pump casing waterways to allow air or vapor trapped in the casing to escape. These valves are used during the priming of the pump or later, during its operation, if the pump becomes air- or vapor-bound. In most cases, these valves need not be piped up to any special place, since their use is infrequent and the vented air or vapors can be allowed to escape into the surrounding atmosphere. Vents from pumps handling flammable or toxic and corrosive liquids, however, must be connected so that they endanger neither the operating personnel nor the installation itself. Also, the suction vents of steam power plant pumps handling condensate or feedwater are generally piped up to the source of the pump suction.

All drain and drip connections should be piped up to a point where the leakage can be disposed of or collected for re-use, depending whether or not the drains are worth reclaiming.

Warmup piping

When a pump must come up to operating temperature before starting, or be kept ready to start at full temperature, warmup flow should be circulated through the pump. A number of arrangements are possible to accomplish this. If the pump oper-

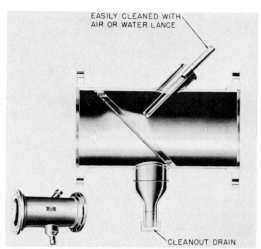

Fig. 24.18 Improved suction strainer
(*Courtesy Leslie Co.*)

ates under positive pressure on suction, the pumped liquid can be permitted to drain out through the pump and a casing connection to some point in the cycle at a pressure lower than the suction pressure. Alternately, some liquid can be made to flow back from the discharge header through a jumper line around the check valve into the pump and out into the suction header. An orifice to regulate the amount of warmup flow is generally provided in this jumper line. The pump manufacturer's recommendations should be followed in all cases with regard to the best means to provide for an adequate warmup procedure.

Balancing device leakoff

Whenever a multistage pump is equipped with a balancing device to counteract the axial thrust generated by a group of single suction impellers all facing in the same direction, it is necessary to return the balancing device leakoff into the cycle at a point where the pressure is equivalent to the suction pressure. In this case, the pumped liquid is subject to a definite temperature rise at light flow operation, and the balancing device leakoff itself undergoes a further temperature rise as it flows

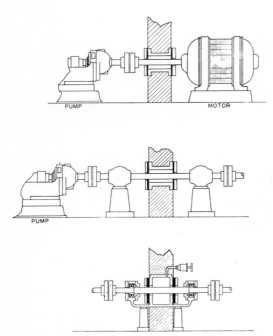

Fig. 24.19   Firewall extensions

through the balancing device because of the degradation of pressure energy. Therefore, if the leakoff is returned directly to the pump suction, it will mix with the incoming liquid and raise the temperature of the mixture. If the pump is handling liquid at or near the boiling point, it is safer to return the leakoff to a point in the cycle where the pressure is equal to the suction pressure, but where the added heat can be dissipated. For example, in the case of a boiler feed pump taking its suction from a deaerating heater, the balancing device leakoff should be returned directly to the heater rather than to the pump suction proper.

The return line should be arranged so that the back pressure on the balancing device cannot inadvertently be increased to more than the value of the suction pressure. Therefore, whereas this return piping needs to be valved off to permit the dismantling of the pump when necessary, any valve interposed between the balancing device relief chamber and the final return point should be locked in the open position. The operators should be warned that if this

valve is ever closed during inspection or overhaul, it must be opened again before the pump is started.

Instrumentation

A number of instruments are essential to maintain a close check on the performance and condition of the installed centrifugal pumps. A compound-pressure gauge should be connected to the suction of the pump and a pressure gauge connected to the discharge. Pressure taps are generally provided in the suction and discharge flanges for this purpose. The gauges should be mounted in a convenient location, so that they can be easily observed.

It is also advisable to provide a flow-metering device, as it is impractical to determine the capacity delivered by the pump with any degree of accuracy without one. Depending upon the importance of the installation, indicating meters may be supplemented by recording attachments.

Whenever pumps incorporate various leakoff arrangements, such as a balancing device or pressure-reducing labyrinths, the quantity of these leakoffs should be checked by installing measuring orifices and differential gauges in the leakoff lines.

Firewall extensions

When a pump handles a flammable volatile liquid such as gasoline, there is danger of an explosion unless an explosion-proof driver is used. To avoid this risk it is frequently the practice to locate the pump and its driver in separate rooms, with a firewall between them. This requires the use of a shaft extension with a stuffing box so that the driver can be connected to the pump through the wall. Various types of firewall extensions have been used, some of which are shown in Fig. 24.19. These firewall extensions have a seal at each end, made of a close fitting felt gasket or of one or more rings of packing, followed by a gland. The space between the two seals is usually filled with a light grease to form a vapor-tight seal.

25 *Operation*

Centrifugal pumps are generally selected for a given capacity and total head when operating at rated speed. These characteristics are referred to as rated conditions of service and, with few exceptions, represent the conditions at which the pump will operate most of the time. Pump efficiency should be at its maximum for these conditions of service, and pumps are so selected wherever possible.

Often, however, pumps are required to operate at capacities and heads differing considerably from the rated conditions. Examples are applications for steam power plant services, where boiler feed, condensate, and heater drain pumps may be called upon to deliver any flow to the boiler from full capacity to zero, depending upon the load carried at the moment by the turbogenerator. Condenser circulating pumps are subject to somewhat lesser variations, but nevertheless these pumps may have to run against widely varying total heads and, therefore, at varying capacities. General service pumps in a variety of applications may also be called upon to operate over a very wide range of flows. It is very important, therefore, that the user of centrifugal pumps acquaint himself with the effects of operating pumps at capaci-

ties and heads other than the rated conditions and with the limitations imposed upon such operation by hydraulic, mechanical, or thermodynamic considerations.

This chapter will discuss, therefore, not only the normal recommended procedures of starting, operating, and stopping centrifugal pumps, but also the various problems introduced by operation under abnormal conditions.

Operation of centrifugal pumps at reduced flows

The subject of radial thrust in volute pumps is discussed in Chap. 2. As described, even dual-volute pumps are not always suitable for operation at all flows down to zero. Therefore, it is imperative to adhere to the limitations on the minimum recommended flow for sustained operation given by the manufacturer.

Another problem arises from the operation of a centrifugal pump at extremely reduced flows—the thermodynamic problem caused by the heating up of the liquid handled by the pump. The difference between the brake horsepower consumed and the water horsepower developed represents

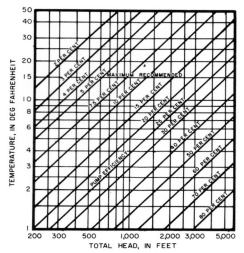

Fig. 25.1   Temperature rise in centrifugal pumps in terms of total head and pump efficiency

the power losses within the pump itself, except for a small amount lost in the pump bearings. These power losses are converted into heat and transferred to the liquid passing through the pump.

If the pump is operating against a completely closed valve, the power losses are equal to the shutoff brake horsepower and, since no flow takes place through the pump, all this power goes into heating the small quantity of liquid contained within the pump casing.

As this process occurs, the pump casing itself heats up and a certain amount of heat is dissipated by radiation and convection to the surrounding atmosphere. If the amount of heat added to the liquid is small, it can be transmitted through the casing with a low differential in temperature between the liquid in the casing and the outside air. If the power loss is very high, however, the liquid temperature might have to reach an exceedingly high value, far in excess of the boiling temperature at suction pressure, before the amount of dissipated heat equals that generated in the pump proper. Operation of the pump under such conditions would have disastrous effects.

Very obviously the rate of heating the liquid in the pump casing for a given power loss depends both upon the volume of water contained in the casing and upon the surface of casing that can dissipate heat. For practical reasons, dissipation of heat by radiation can be ignored, and the temperature rise can be determined from the formula:

$$T_r = \frac{42.4 P_{so}}{W_p C_p + W_w C_w}$$

in which:

T_r = temperature rise, in degrees Fahrenheit per minute

P_{so} = brake horsepower at shutoff

42.4 = conversion from brake horsepower into British thermal units per minute

W_p = net weight of pump, in pounds

C_p = specific heat of pump metal (approximately 0.13)

W_w = net weight of liquid in pump, in pounds

C_w = specific heat of liquid (1.0 if liquid is water).

Because the temperature rise may be so rapid that there is no time to transmit the heat from the liquid to the pump casing, it is generally safer to neglect the casing altogether. The formula is then simplified to:

$$T_r = \frac{42.4 P_{so}}{W_w C_w}$$

For example, if the pump handles water ($C_w = 1.0$) and contains 100 lb of liquid, and if the brake horsepower at shutoff is 100, the water temperature will increase at the rate of 42.4°F per minute. Operation at shutoff under these conditions is very dangerous. But with a low-head, high-capacity pump that contains 5,000 lb of water and that takes the same amount of power at shutoff, the rate of temperature increase will be only 0.85°F per minute—hardly serious if the operation against a closed discharge valve is not prolonged.

If liquid is flowing through the pump, conditions become stabilized and the amount by which the temperature at the discharge will exceed the suction temperature can be calculated for any given flow. This rise is determined by multiplying the difference between the brake horsepower and the water horsepower by 2,545 (Btu equivalent of 1 hp-hr) and thus determining the total heat imparted to the liquid in British thermal units. Assuming that the liquid is water (specific heat = 1 Btu per lb per deg) this total heat is divided by the flow in pounds per hour and yields the temperature rise in degrees Fahrenheit. A more readily used formula is:

$$T_r = \frac{H}{778}\left(\frac{1}{E} - 1\right)$$

in which:

T_r = temperature rise, in degrees Fahrenheit per minute

H = total head, in feet

E = pump efficiency at the capacity involved.

The chart in Fig. 25.1 gives a graphical solution of this formula and allows determination of the minimum permissible operating capacity, once the maximum permissible temperature rise has been selected. When liquids other than water are handled by the pump, it becomes necessary to correct the resulting answer for the difference in the specific heats of the liquids.

The chart can be used to plot a temperature rise curve directly on the performance curve of a centrifugal pump (Fig. 25.2), which represents the characteristics of a boiler feed pump designed to handle 550 gpm of 250°F feedwater against a total head of 1,800 ft. As shown, the temperature rise increases very rapidly with a reduction in flow. This is caused by the fact that the losses at low deliveries are greater when the flow of liquid that must absorb the heat developed in the pump is low. For example, Fig. 25.2 shows that for a capacity of 50 gpm, the temperature rise is 17°F, whereas

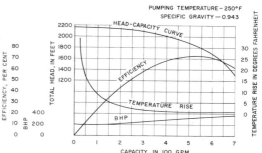

Fig. 25.2 Centrifugal pump performance curve
Temperature rise is also indicated.

at the full capacity of 550 gpm, it is less than 1°F.

If the pump is fitted with a balancing device, a certain portion of the suction capacity is returned either to the pump suction or to the suction supply vessel. Then, the discharge capacity does not represent the true flow through the pump. The formula for the temperature rise and the chart in Fig. 25.1 can still be used, provided a correction is made to take care of the increase in pump flow representing the balancing device leakoff.

As the temperature rise varies with the pump capacity, the minimum permissible capacity from the thermodynamic point of view will depend on the maximum permissible temperature rise, which varies over a wide range, depending upon the type of installation. With hot-water pumps, for example those on boiler feed service, the temperature rise should generally be limited to 15°F. Generally, the minimum permissible flow required to hold the temperature rise in boiler feed pumps to 15°F, is 30 gpm for each 100 bhp at shutoff.

On general-service pumps, where the permissible temperature rise will vary with the service, the operating temperature, and the size of the pump, a convenient approximation of the minimum permissible flow is given by the relation: Minimum flow, in gallons per minute = 6.0 (brake horsepower at shutoff)/permissible temperature rise, in degrees Fahrenheit. This formula includes a safety factor of approximately 20 per cent.

When the centrifugal pump handles cold water, the temperature rise may be permitted to reach 50° or 100°F. The minimum capacity is then established as that capacity at which the temperature rise corresponds to the maximum permitted.

If the pump is required to operate at shutoff or at extremely low flows, some means must be provided to prevent operation at flows below the minimum permissible, regardless of whether the discharge valve or the check valve is closed. This is accomplished by installing a bypass in the discharge line from the pump, located on the pump side of the check and gate valves and leading to some lower pressure point in the installation where the excess heat absorbed through operation at light flows may be dissipated. Under no circumstances should this bypass be led directly back to the pump suction, as there would be no means to dissipate the excess heat, and the entire purpose of the bypass arrangement would be defeated.

In boiler feed pump installations where the pumps take their suction from a de-aerating heater, the logical place to return the recirculation flow is the heater itself (Fig. 25.3). Since the water in the bypass line is under a pressure equal to the pump discharge pressure, an orifice must be located in this line to limit the flow through the line to the desired value. When the differential pressure broken down by the orifice is low, a single drilled orifice in a stainless steel rod of 3 to 6 in. long can be used. Figure 25.4 shows flows through orifices of different diameters and with various head differentials.

Elbows should not be located too close after an orifice to avoid damage to the piping. If the piping must bend, the orifice should be followed by a 12 to 18 in. length of straight pipe and T-joint (Fig. 25.5) leading back to the heater. The dead end of the T-joint should then be followed by another length of straight pipe, terminating in a pipe coupling. The end of the pipe cou-

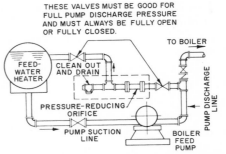

Fig. 25.3 Arrangement of bypass recirculation for boiler feed service

pling should be fitted with a stainless steel plug, which will take the brunt of the high velocity stream coming through the orifice. The plug will be cheaper and simpler to replace, if necessary, than an elbow in the piping.

When higher pressures than those indicated in Fig. 25.4 are encountered, or when the noise incidental to the breaking down of the discharge pressure through a single orifice must be reduced, special multiple pressure-reducing orifices can be obtained and installed.

The arrangement illustrated in Fig. 25.3 is based on a manual control of the bypass. Two valves are used, one on each side of the orifice. One of these remains locked in the open position and is closed only to isolate the orifice for inspection or replacement. The second is the control valve and remains either open or closed. This valve

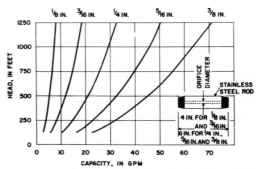

Fig. 25.4 Flows through single orifices drilled in a stainless-steel rod

When the centrifugal pump handles cold water, the temperature rise may be permitted to reach 50° or 100°F. The minimum capacity is then established as that capacity at which the temperature rise corresponds to the maximum permitted.

If the pump is required to operate at shutoff or at extremely low flows, some means must be provided to prevent operation at flows below the minimum permissible, regardless of whether the discharge valve or the check valve is closed. This is accomplished by installing a bypass in the discharge line from the pump, located on the pump side of the check and gate valves and leading to some lower pressure point in the installation where the excess heat absorbed through operation at light flows may be dissipated. Under no circumstances should this bypass be led directly back to the pump suction, as there would be no means to dissipate the excess heat, and the entire purpose of the bypass arrangement would be defeated.

In boiler feed pump installations where the pumps take their suction from a de-aerating heater, the logical place to return the recirculation flow is the heater itself (Fig. 25.3). Since the water in the bypass line is under a pressure equal to the pump discharge pressure, an orifice must be located in this line to limit the flow through the line to the desired value. When the differential pressure broken down by the orifice is low, a single drilled orifice in a stainless steel rod of 3 to 6 in. long can be used. Figure 25.4 shows flows through orifices of different diameters and with various head differentials.

Elbows should not be located too close after an orifice to avoid damage to the piping. If the piping must bend, the orifice should be followed by a 12 to 18 in. length of straight pipe and T-joint (Fig. 25.5) leading back to the heater. The dead end of the T-joint should then be followed by another length of straight pipe, terminating in a pipe coupling. The end of the pipe cou-

THESE VALVES MUST BE GOOD FOR FULL PUMP DISCHARGE PRESSURE AND MUST ALWAYS BE FULLY OPEN OR FULLY CLOSED.

Fig. 25.3 Arrangement of bypass recirculation for boiler feed service

pling should be fitted with a stainless steel plug, which will take the brunt of the high velocity stream coming through the orifice. The plug will be cheaper and simpler to replace, if necessary, than an elbow in the piping.

When higher pressures than those indicated in Fig. 25.4 are encountered, or when the noise incidental to the breaking down of the discharge pressure through a single orifice must be reduced, special multiple pressure-reducing orifices can be obtained and installed.

The arrangement illustrated in Fig. 25.3 is based on a manual control of the bypass. Two valves are used, one on each side of the orifice. One of these remains locked in the open position and is closed only to isolate the orifice for inspection or replacement. The second is the control valve and remains either open or closed. This valve

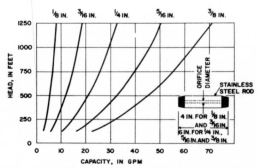

Fig. 25.4 Flows through single orifices drilled in a stainless-steel rod

is often automatically controlled and is responsive to the metering of the flow through the pump. For small installations, automatic control would be too expensive. If absolute assurance that the pump will never operate against complete shutoff is necessary, the control valve can be locked in the open position.

Regardless of the centrifugal pump installation, it must not operate against complete shutoff for a long enough period to cause a dangerous rise in temperature.

Priming

Centrifugal pumps should almost never be started until they are fully primed, that is, until they have been filled with the liquid pumped and all the air has escaped. The exceptions involve self-priming pumps and some special large-capacity, low-head, and low-speed installations where pump priming before starting is not practical and the priming is almost simultaneous with the starting. Priming of centrifugal pumps is described in Chap. 21.

Warmup

Pumps that handle hot liquids should be maintained at approximately operating temperature when idle. A constant small flow through the pump will provide for this. Many arrangements are available for this warmup procedure. In some cases, flow goes from the open suction, through the pump, and out through a warmup valve on the pump side of the discharge valve. The drains from the warmup valve are returned to the pumping cycle at some lower pressure point than the pump suction. In other cases, flow goes through a jumper line around the discharge check valve, through the pump, and into the common suction header (Fig. 25.6). The exact arrangement to be used should be recommended by the pump manufacturer.

Some pump designs are capable of starting up cold in an emergency, whereas others should never be exposed to this sud-

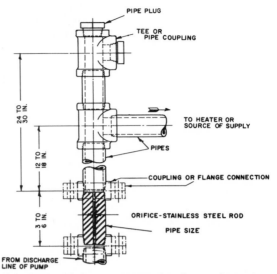

Fig. 25.5 Piping arrangement for a bypass orifice

den shock; the pump manufacturer should be consulted in each particular case. Some general considerations are as follows:

Impeller mounting

If pump impellers are mounted with a slight shrink fit, starting the pump cold will have no injurious effects. If, however, the impellers are mounted with a no-interference fit on a shaft, the material of which will expand more rapidly than the impeller material, the shrink fit is effected when the pump comes up to operating temperature.

A cold start in such a case may lead to operation with slightly loose impellers.

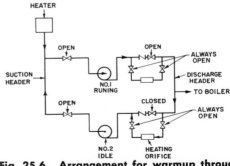

Fig. 25.6 Arrangement for warmup through a jumper line around discharge check valve

Double-casing pumps

With double-casing pumps, such as used for high-pressure boiler feed service (see Fig. 3.14 and 3.15), the relative position of the inner assembly and of the outer casing must be examined carefully to see whether or not cold starts will lead to difficulties. When hot water is suddenly admitted to a cold double-casing pump, the relative expansion of the outer casing barrel and of the inner element goes through two separate and distinct phases. At first, the inner element, which is much lighter than the barrel and which is in more intimate contact with the hot water, expands at a considerably faster rate than the outer casing itself. To simplify the analysis, it can be assumed that the inner element reaches its final temperature before any appreciable temperature change has taken place in the outer casing. Then, as the pump continues to operate, the outer casing heats up and reaches its own final temperature at some later time. If the casing barrel is not lagged, the temperature on its external surface may be somewhat lower than the internal temperature, but this is negligible.

A particular example concerns an 8-stage boiler feed pump, designed to operate at 320°F, with an inner element of 5 per cent chrome stainless steel, and an outer barrel of forged SAE-1020 steel. The coefficients of expansion of the two metals in question can be considered to be the same and equal to 0.0000065 in. per in. per deg F. The length of the inner element within the barrel (dimension A in Fig. 25.7) is 42 in.

If, when the pump is at rest on standby service, the metal temperature is permitted to fall to 120°F, the sudden admission of 320°F feedwater will cause the inner element to expand 0.055 in. with relation to the casing barrel when it reaches the final temperature. Ultimately, the outer casing will also come up to temperature and will expand by an equal amount, nullifying the initial expansion of the inner element.

The effect of this initial relative expansion, followed by a return to the initial

relative position of inner element and outer casing, will have different effects on different designs, depending on whether or not the construction of the double-casing pump permits free movement of the inner element within the barrel. If it does, the events which take place will have but little effect on the unit. If, however, the inner element is constrained within fixed limits, it becomes necessary to interpose some form of compressible gasket between the inner element and either the barrel or the discharge head. (Sometimes these gaskets are incorporated at both points.) These gaskets must absorb the difference in expansion just calculated. The reliability and life of the compressible gaskets determines whether or not the unit can safely be started cold.

Some consideration should also be given to the effect of the warmup method selected. It is true that, in some cases, the pump will be subjected to a certain amount of distortion because heat may flow unevenly to various parts of the pump. When warmup has been properly completed, however, this distortion should disappear; the distortion will not have affected the pump while it was at rest. A careful analysis is necessary to check that a pump started up cold does not undergo this type of distortion as it is coming up to temperature, since interference at the running joints or misalignment at the bearings will be a possible cause of trouble. Although some pumps can be started up cold, whereas others should not, all pumps handling hot liquids will profit from starting warmed up if the warmup operation insures a thorough and even distribution of heat to all parts of the pump.

Final checks before startup

After a centrifugal pump has been properly installed and all necessary precautions have been taken in aligning it with its driver, it is ready for service on its initial start. A few last-minute checks are recommended. The bearings and the lubrication system must be clean. Before putting the

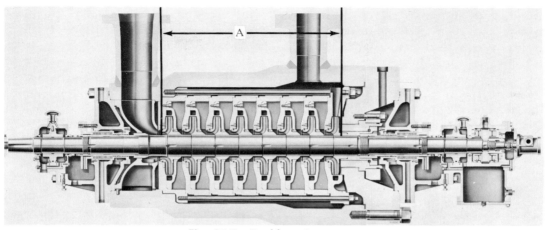

Fig. 25.7 Double-casing pump

pump in service, the bearing covers should be removed, the bearings flushed with kerosene, and thoroughly cleaned. Waste should not be used to clean bearings as lint can find its way into the lubricant; clean rags are superior for this purpose. All oil or grease used in the lubrication system must be free from water, grit, or other contaminant. The bearings should be filled with new lubricant in accordance with the manufacturer's recommendations.

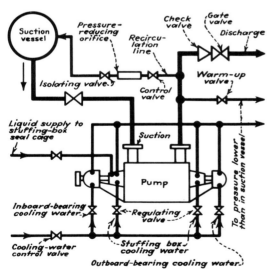

Fig. 25.8 Connections and auxiliary services to a centrifugal pump

Of importance in establishing starting and stopping procedures.

With the coupling disconnected, the driver should be tested again for correct direction of rotation. Generally, an arrow is inscribed on the pump casing to show the correct rotation.

All parts should be finally inspected before starting. It must be possible to revolve the pump rotor by hand, and in the case of a pump handling hot liquids, the rotor must be free to rotate with the pump cold or heated. If the rotor is bound or even drags slightly, the pump must not be operated until the cause of the trouble is ascertained and has been corrected.

Starting and stopping procedures

The steps necessary to start a centrifugal pump will depend upon its type and the service on which it is installed (Fig. 25.8). Many installations require steps that are unnecessary in other installations. For example, standby pumps are often held ready for immediate starting, particularly centrifugal boiler feed pumps. The suction and discharge gate valves are held open, and reverse flow through the pump is prevented by the check valve in the discharge valve.

The methods used to start pumps are greatly influenced by the performance characteristics of the pump in question, that is, by the shape of its power-capacity curve. High- and medium-head pump (low and medium specific speeds) curves rise from the shutoff condition to the normal operating capacity condition (see Fig. 17.32 and 17.33); therefore, these pumps should be started against a closed discharge valve in order to reduce the starting load on the driver. The use of a check valve in the discharge line is equivalent to a closed valve for this purpose, as long as another pump is already on the line. The check valve will not lift until the pump being started has come up to a speed sufficient to generate a head high enough to lift the check valve from its seat. Precautionary measures against pump overheating from operation at shutoff may have to be employed in certain cases by installing a recirculation bypass.

The power consumption curve of low-head pumps (high specific speed) of the mixed flow and propeller type has the opposite characteristic, rising sharply with a reduction in capacity, as shown in Fig. 17.35. Such pumps, therefore, should be started with the discharge valve wide open, against a check valve if required, to prevent back flow.

Assuming that the pump in question is motor driven, that its shutoff horsepower does not exceed the safe motor horsepower, and that it is to be started against a closed gate valve, the starting procedure is as follows:

1. Prime the pump, opening the suction valve and closing the drains to prepare the pump for operation.

2. Open the valve in the cooling water supply to the bearings.

3. Open the valve in the cooling water supply if the stuffing boxes are water cooled.

4. Open the valve in the sealing liquid supply if the pump is so fitted.

5. Open the warmup valve of a pump handling hot liquids if the pump is not normally kept at operating temperature. When the pump is warmed up, close the valve.

6. Open the valve in the recirculating

line if the pump should not be operated against dead shutoff.

7. Start the motor.

8. Open the discharge valve slowly.

9. Observe the leakage from the stuffing boxes and adjust the sealing liquid valve for proper flow to insure the lubrication of the packing. If the packing is new, do not tighten up on the gland immediately, but let the packing run in before reducing the leakage through the stuffing boxes.

10. Check the general mechanical operation of the pump and motor.

11. Close the valve in the recirculating line once there is sufficient flow through the pump to prevent overheating.

If the unit is to be started against a closed check valve with the discharge gate valve open, the steps would be the same, except that the discharge gate valve would be opened some time before the motor is started.

In some cases, the cooling water to the bearings and the sealing water to the seal cages is provided by the pump itself. This eliminates the need for steps (2) through (4) in the starting procedure.

The procedure for stopping a pump also depends upon the type and service of the pump. Generally, the procedure to stop a pump that can operate against a closed gate valve is as follows:

1. Open the valve in the recirculating line.

2. Close the gate valve.

3. Stop the motor.

4. Open the warmup valve if the pump is to be kept up to operating temperature.

5. Close the valve in the cooling water supply to the bearings and to water-cooled stuffing boxes.

6. If the sealing liquid supply is not required when the pump is idle, close the valve in this supply line.

7. Close the suction valve and open the drain valves, as required by the particular installation, or if the pump is to be opened up for inspection.

If the pump does not permit operation against a closed gate valve, steps (2) and (3) are reversed. Many installations permit stopping the motor before closing the discharge gate valve.

Generally, the starting and stopping of steam-turbine-driven pumps require the same steps and the same sequence as for motor-driven pumps. As a rule, steam turbines require warming up before starting and have drains and seals that must be opened and closed before and after operation. The operator should, therefore, conscientiously follow the steps outlined by the turbine manufacturer in starting and stopping the turbine. This is also true in the case of internal combustion engines used to drive pumps.

Auxiliary services on standby pumps

Standby pumps are frequently started up from a remote location, and several methods of operation are available for the auxiliary services, such as the cooling water supply to the bearings or to water-cooled stuffing boxes. The choice among these methods must be dictated by the specific circumstances surrounding each case. The most logical methods are the following:

1. A constant flow can be maintained through the bearing jackets or oil coolers and through the stuffing box lantern rings, whether the pump is running or standing still on standby service.

2. The service connections may be opened automatically whenever the pump is started up.

3. The service connections may be kept closed while the pump is idle and the operator instructed to open them within a short interval after the pump has been put on the line automatically.

Method (1) wastes the cooling water and can be harmful. The necessity of regulating the amount of cooling water to the pump bearings is frequently overlooked, and, generally, the error is to overcool

rather than to supply insufficient cooling water. Many ball bearing failures are due to the bearing being almost refrigerated so that the resulting condensation on the cold inside walls of the bearing housing mixes with the lubricating oil or grease. Rusting and pitting of the balls leads to obvious trouble. The outflowing cooling water temperature should not fall much below 105° to 115°F.

Cooling water is frequently available at temperatures as low as 60° to 70°F, and if allowed to flow through the bearing housing of an idle pump installed in a warm or moist atmosphere, it may lead to bearing troubles. While the pump is not running, no heat is generated at the bearings and the bearing housings will be maintained at exactly the temperature of the cooling water.

Sometimes cooling or sealing water to the pump stuffing boxes must be maintained whether or not the pump is running. Some examples are pumps that handle liquids corrosive to the packing or liquids that may crystallize and deposit on the shaft sleeves.

If the stuffing boxes are equipped with water-cooled jackets, leaving the connections open at all times may be wasteful, but presents no particular danger.

In method (2), individual water supply lines can be equipped with spring-loaded pressure control valves as illustrated in Fig. 25.9. The pressure side of the diaphragm is connected to the pump discharge by means of a pilot line, so that the valves will open as soon as the pump starts and develops pressure.

If the standby pump is motor driven, solenoid-operated control valves can be installed in the water cooling supply lines. The valves remain closed under spring action as long as the solenoid is de-energized and open as soon as the motor is put on the line, energizing the solenoid.

Valves, whether controlled by pressure or by solenoids, should be supplied with locking devices. The operators can lock them in the open position, when they have time to attend to a pump that has been started automatically.

If operators are available near the pump location, and if the pump is of such a design and on such a service that it can be operated a few minutes without cooling or sealing water supply, method (3) may be the most suitable.

General rules on pump operation

Running a pump dry

Only a centrifugal pump with excessive clearances between stationary and rotating parts could be run dry for an indefinite period of time. Most centrifugal pumps have close clearance leakage joints and cannot run dry at all, or in some cases for longer than a few seconds, without being seriously damaged.

The one exception to this rule is a particular design of automatic priming used with large, low-head pumps. The pump is started dry when the vacuum pump is started and runs dry for not more than two minutes, at which time the priming has been completed and the pump goes into normal operation. To insure successful operation under these conditions, the clearances at the wearing rings are made slightly larger than in the normal design.

Throttling at the pump suction

Throttling the suction of a centrifugal pump causes a reduction of the absolute pressure at the inlet to the impeller. This can be made to result in a reduction of capacity by forcing the pump to operate "in the break," and reducing the delivered capacity by altering the shape of the head-capacity curve. Such operation is harmful to the pump, unless, as in the case of a condensate pump, it is specifically designed for it. Pump efficiency is reduced when operated "in the break," but most important,

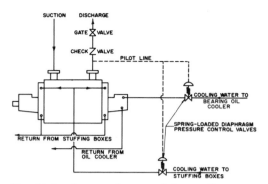

Fig. 25.9   Automatic control for cooling water supply

erosion and premature destruction is caused by the cavitation induced when the suction is throttled.

Pump capacity can be reduced simply and safely by throttling the discharge. In this manner, artificial friction losses are introduced by throttling, and a new system-head curve is obtained which intersects the head-capacity curve at the desired flow.

Throttling at the suction is permissible only when the suction pressure exceeds the minimum requirements by a large margin, such as the case of the second pump of a series unit. The effect, however, is not to reduce capacity by operation in the break, but rather by the reduction of the total net head generated by the series unit. This causes the head-capacity characteristics and the system-head curve to intersect at a lower rate of flow.

Restarting motor-driven pumps stopped by power failure

If a check valve protects a pump against reverse flow after a power failure, there is generally no reason why the pump should not be restarted once current has been re-established. The type of motor control used will determine whether or not the pump will start again automatically once power is restored. Starters are made with low-voltage protection, with low-voltage release, or without either. Starters with low-voltage protection will de-energize under low-voltage conditions, or following power failure, and the units they control must be restarted manually. Starters with low-voltage protection can only be used with momentary-contact pilot devices and cannot be used with maintained-contact pilot devices, such as float switches, unless auxiliary relays are incorporated in the controls.

If the starter does not incorporate low-voltage protection, resumption of power will always cause the unit to start again automatically. Because pumps operating on a suction lift may lose their prime during the period when power is off, starters should be provided with low-load protection for such installations. This does not apply, of course, if the pumps are automatically primed, or if some protection device is incorporated so that the pump cannot run unless it is primed.

Operation of centrifugal and reciprocating pumps in parallel

Whereas often centrifugal pumps can be operated in parallel with reciprocating pumps, the general performance of the centrifugal pump will be affected both mechanically and hydraulically by the pulsations of the reciprocating pump. A triplex pump would have the least effect on the operation of the centrifugal unit whereas actual troubles can be experienced if attempts are made to parallel a centrifugal pump with a single-acting, single-cylinder pump. It is important not to use a common suction line for a centrifugal and a reciprocating pump, especially if they operate under a high-suction lift.

A large number of problems encountered in the operation of centrifugal pumps cannot be discussed here. The pump manufacturer should be consulted if difficulties are encountered or if the choice between two different solutions of an operating problem seems difficult.

26 *Maintenance*

Because of the wide variation in types, sizes, parts, and design of centrifugal pumps, any description of maintenance must be restricted to the most common types of centrifugal pumps. The manufacturer's instruction books should be carefully studied before any attempt is made to service any particular pump.

The extent of the knowledge the maintenance personnel should have about the pumps in their care depends upon the demands and complexities of the system in which the pumps are installed. In most cases, complete information of the mechanical construction as given in the instruction books is sufficient. Generally, the maintenance personnel need only know the rated conditions of service, which are usually given on the pump nameplate. Occasionally, they also need more complete information on the pump characteristics to provide adequate inspection and maintenance. In such cases a performance curve of the pump is required and, if the curve is not included in the instruction book, it should be obtained from the pump manufacturer.

Considerable difference of opinion exists concerning good practice in rebuilding worn parts such as wearing rings or shaft sleeves. As far as possible, successful procedures have been described in the chapters dealing with individual pump parts, even though some of these procedures are not fully recognized as good practice.

Centrifugal pumping equipment maintenance problems vary from simple to complex ones. The type of service for which the pump is intended, the general construction of the pump, the relative complexity of the required repairs, the facilities available at the site, and other factors enter into the decision whether the necessary repairs will be carried out at the installation or at the pump manufacturer's plant. Sometimes, especially when sufficient spare equipment is available, a pump needing repairs is sent to the manufacturer's plant for complete overhauling. Otherwise, repairs or overhauls are made locally by the mechanics servicing the installation. When expedient, service engineers from the pump manufacturer's plant perform repairs at the site.

Daily observation

Pump installations that are constantly attended should be inspected hourly and daily. A card record system is unnecessary for these inspections, but the operator should immediately report any irregularity in pump operation. A change in the sound of a running pump should be investigated immediately. Bearing temperatures should be observed hourly. An abrupt change in bearing temperature is much more indicative of trouble than a constant high temperature. If the pump is fitted with ring-oiled bearings, the proper functioning of the oil rings should be observed.

Stuffing box operation should also be observed hourly. The stuffing box leakage should be checked to see whether it is sufficient to provide cooling and lubrication of the packing but not excessive and wasteful.

The pressure gauges and flow indicator, if these are installed, should also be checked hourly for proper operation. Recording instruments, if available, should be checked daily to insure that the capacity output, pressure, or power consumption do not indicate that something needs attention.

Semiannual inspection

The stuffing box gland should be checked semiannually for free movement. The gland bolts and nuts should be cleaned and oiled and inspected to see if the packing needs replacement.

The pump and driver alignment should be checked and corrected if necessary. Oil-lubricated bearings should be drained and refilled with fresh oil. Grease-lubricated bearings should be checked to see if the correct amount of grease is provided and if it is still of suitable consistency.

Annual inspection

Centrifugal pumps should be very thoroughly inspected once a year. In addition to the semiannual maintenance procedure, the bearings should be removed, cleaned, and examined for flaws. The bearing housings should be carefully cleaned. Antifriction bearings should be examined for scratches and wear after cleaning. Immediately after inspection, antifriction

bearings should be coated with oil or grease to prevent dirt or moisture from getting into them.

The packing should be removed and the shaft sleeves, or shaft, if sleeves are not used, should be examined for wear.

The coupling halves should be disconnected and alignment checked. In the case of horizontal pumps with babbitt bearings, the vertical shaft movement for both ends should be checked with the packing out and the coupling disconnected. Any vertical movement more than 150 per cent of the original play requires an investigation to determine the cause. The end play allowed by the bearings should also be checked, and if more than recommended by the manufacturer, the cause should be determined and corrected.

Drains, sealing water piping, cooling water piping, and other piping should be checked and flushed. If an oil cooler is used, it should be flushed and cleaned.

The pump stuffing boxes should be repacked and the coupling reconnected.

If instrument and metering devices are available, these should be recalibrated and a test made to determine whether proper performance is obtained. If internal repairs are made, the pump should again be tested after completion of the repairs.

Complete overhaul

General rules cannot easily be made to determine the proper frequency and regularity of complete overhauls of centrifugal pumps. The type of service for which the pump is intended, the general construction of the pump, the liquid handled, the materials used, the average operating time of the pump, and the evaluation of overhaul costs against possible power savings from renewed clearances, all enter into the decision on the frequency of complete overhauls. Some pumps on severe service may need a complete overhaul monthly, whereas other applications only require overhaul every two to four years or even less frequently.

Most pump designers and specialists consider that a centrifugal pump need not be opened for inspection unless either factual or circumstantial evidence indicates that overhaul is necessary.

Factual evidence

Some types of factual evidence are a decline in pump performance, excessive noise or bearing temperatures, driver overload, or similar troubles. Proper instrumentation is of paramount importance to the satisfactory operation and life of centrifugal pumping equipment. Equipment for determining pump capacities and pressures is as important as any other maintenance tools and should always be available.

A schedule should be established for frequent complete tests of the pumping unit and the results of these tests compared with the performance of the pump in its initial condition. Any sudden decline in performance can then be detected immediately. This performance comparison, and not the passing of a fixed period of time, should be relied upon to establish whether or not sufficient internal wear has occurred to require a complete overhaul. Running a complete test is less expensive than opening up a pump for inspection and does not require taking the unit out of service.

The life of a centrifugal pump is determined by the extent of the internal wear and the effect of this wear on the pump performance. Two separate causes lead to increased internal clearances: (1) The erosive action of the liquid flowing past the wearing rings and various other internal running clearances, and (2) the infrequent momentary contacts that sometimes occur during pump operation.

No matter how rigidly a pump is designed, certain conditions can arise during abnormal operation (such as cavitation, loss of prime, or sudden hydraulic shocks) that impose a momentary vibration on the rotating element sufficient to cause a slight contact at the running joints. Wear will occur through the rubbing action of two

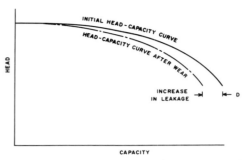

Fig. 26.1 Effect of internal wear on a centrifugal pump head-capacity curve

metal parts, even though the contact may remain too slight to cause galling and seizure of the rotor in the stationary parts. A similar condition can exist if the shaft deflection under stationary conditions exceeds the internal clearances. This can occur even though the deflection is reduced to less than the running clearance by the supportive action of internal clearance joints acting as additional steadying bearings, lubricated by the liquid pumped. In these cases, a slight amount of wear will occur every time the pump starts.

The exact amount of wear caused by erosion and momentary contacts cannot be determined exactly. The more rigid the pump construction, however, the less cause there will be to expect wear through contact at the running clearances.

As running clearances increase with wear, a greater portion of the gross capacity of the pump is short-circuited through the clearances and has to be repumped again. The effective or net capacity delivered by the pump against a given head is reduced by an amount equal to the increase in leakage. While in theory the leakage varies with the square root of the differential pressure across a running joint, and therefore with the square root of the total head, it is accurate enough to assume that the increase in leakage remains constant at all heads. Figure 26.1 shows the effect of increased leakage on the shape of the head-capacity curve. Subtracting the additional internal leakage from the initial capacity at each head gives a new head-capacity curve as shown.

A certain amount of capacity margin should always be included in the design capacity to allow for the expected reduction in capacity caused by wear of internal clearances. The decision on when to overhaul a pump depends, therefore, on the amount of margin initially included with the pump selection and on the evaluation of the reduced operating economy against the cost of the overhaul.

Circumstantial evidence

Circumstantial evidence refers to data accumulated through past experience, either with the pump in question or with similar equipment on similar service. For example, if a group of boiler feed pumps built of chrome stainless steel alloys has run continuously for 80,000 hours without the need for a complete overhaul, a duplicate unit will not require inspection before it has operated for 80,000 hours.

Pumps on severe service that have required overhaul at three-month intervals may be replaced by better built or sturdier units. Nevertheless, until the new equipment has proved itself and a new experience pattern has been established, the pump should be opened at the end of three months to evaluate the effect of the better construction or better materials.

Exceptions

Corrosion-erosion troubles, which cannot be classified with either of the preceding types of evidence, will not necessarily be obvious in the performance characteristics of the pump obtained by means of routine tests. If these troubles are permitted to continue unattended, however, they can easily result in the destruction of the pump, beyond any possibility of repair. Corrosion-erosion troubles are generally foreseeable, however. For example, a pump handling corrosive chemicals that is built of ordinary materials or of materials untested in that particular application may become

rapidly and severely damaged. In such a case, the pump should be opened for inspection soon after initial installation and at frequent intervals thereafter, until the life of the pump materials under the actual operating conditions has been determined.

Another exception is in the case of an operator who prefers to rely on periodic visual examinations and actual measurements of clearances. If he cannot be convinced that this procedure is unnecessary, the exception is fully warranted because the operator's confidence in his equipment is much more important than the fact that a piece of equipment has been dismantled one or two extra times during its useful life.

An adequate store of spare parts should be maintained at all times to insure rapid restoration to service in the event of an unexpected overhaul and to avoid any delay in obtaining special repair parts from the manufacturer.

Complete dismantling of a centrifugal pump

Centrifugal pumps should be dismantled with great care. The suction and discharge valves should be closed and the pump casing drained. All necessary piping and parts that would interfere with the disassembly of the pump, such as bearing covers, should be taken apart as required by the manufacturer's instructions.

The upper half of pumps with axially split casings should be lifted straight up after the dowels and the nuts of the casing bolts have been removed, to prevent damage to internal parts (Fig. 26.2). The rotor should also be removed vertically to prevent injury to the impellers, wearing rings, and other parts.

During the dismantling procedure, the various parts removed must be marked to insure proper reassembly. All individual parts and all important joints should be carefully examined. If the pump has been operating satisfactorily with only a slight

Fig. 26.2   Pump with axially split casing

reduction in head and capacity due to increased leakage, a decision on reconditioning will depend on several factors:

1. Availability of spare parts
2. Length of time the pump can be left out of service
3. Economic considerations and importance of getting the greatest service from the unit without overhauls.

Generally, worn parts should be renewed if the pump is not to be examined until the next routine period, regardless of the performance of the unit, because when parts with metal seats in new or good condition are assembled in contact with dirty or worn parts, the new parts are very likely to wear out rapidly.

Maintenance of specific pump parts

Maintenance of the following specific pump parts is described in the chapters listed:

1. Casings—Chap. 2
2. Impellers and wearing rings—Chap. 4
3. Shafts and shaft sleeves—Chap. 7
4. Stuffing boxes and packing—Chap. 8
5. Bearings—Chap. 11.

Rotor reassembly

Impellers must be remounted on the pump shaft so that they will rotate in the proper direction, always away from the curvature of their vanes (Fig. 26.3).

Special care is required in the reassembly of multistage pump rotors with axially split casings. These casings are made from castings and, when the pump is built, it is sometimes necessary to allow variations in longitudinal dimensions on the casings. This is done by making assembly floor adjustments to the rotor, in order to preserve the designed lateral clearances and to place the individual impellers in their correct positions with respect to their volutes or diffusers. When making field renewals of rotors or of stationary parts, all lateral distances should be compared with those on the old parts and where lateral end movement is affected, these distances should be duplicated. The assembled rotor and stationary parts (such as casing wearing rings, stage-pieces, diffusers) should be placed in the lower half of the casing and the total lateral clearance checked. When the thrust bearing is assembled and the shaft is in its proper position, this total clearance should be suitably divided and the impellers centrally located in their volutes or diffusers. The shaft nuts can be manipulated for final adjustments.

To avoid shaft distortion, all abutting joints (Fig. 26.4) must be square with the shaft axis and with each other, and the impeller and shaft sleeve nuts must not be tightened with excessive force. Otherwise, the metal may be crushed at these joints, exerting severe moments on the shaft. The shaft may become bowed under the influence of these moments and develop a marked vibration, in addition to the possibility of rubbing and binding at the internal running joints. When using locking screws of the safety type, the assembly should be checked, using a dial test indicator, to make sure that the shaft is not bent. If possible, after assembly, the rotor should be supported in its bearings or on

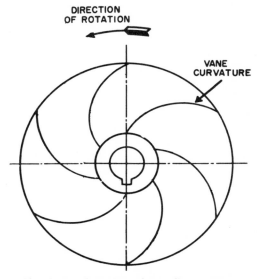

Fig. 26.3   Direction of impeller rotation

centers to check for concentricity and any eccentricity corrected.

Pump reassembly

If the pump casing is split axially, great care must be exercised in replacing the upper casing half and tightening up the casing bolts. If more than one row of bolts is used, the row nearest to the pump central axis should be tightened up first. After all the bolts have been tightened once, they should be tightened again to insure the tightness of the casing joint. They should then be tightened once more when the pump has been brought up to operating temperature.

Spare and repair parts

The service for which a centrifugal pump is used will determine, to a great extent, the minimum number of spare parts that should be carried in stock at the installation site. The minimum for any centrifugal pump should include a set of wearing rings, a set of shaft sleeves (or a shaft if no sleeves are used) and a set of bearings. It is often advisable to carry a complete spare rotor for installation in the pump when examination shows that the pump rotor has become excessively worn, or if it becomes accidentally damaged. Sufficient stock of spare packing for the stuffing boxes and material for the gasket of axially split casing pumps should always be in stock.

Spare parts should be purchased at the time the order for the complete unit is placed. Depending upon the contemplated method for wearing ring overhaul, the spare wearing rings are ordered either same size as the wearing rings used in the assembly of the new pump or bored undersize.

The pump serial number and size as stamped on the manufacturer's nameplate should always be given when ordering spare and replacement parts after the pump has been received in the field, so that the manufacturer may identify the pump and furnish repair parts of correct size and materials. Most centrifugal pumps are of standard de-

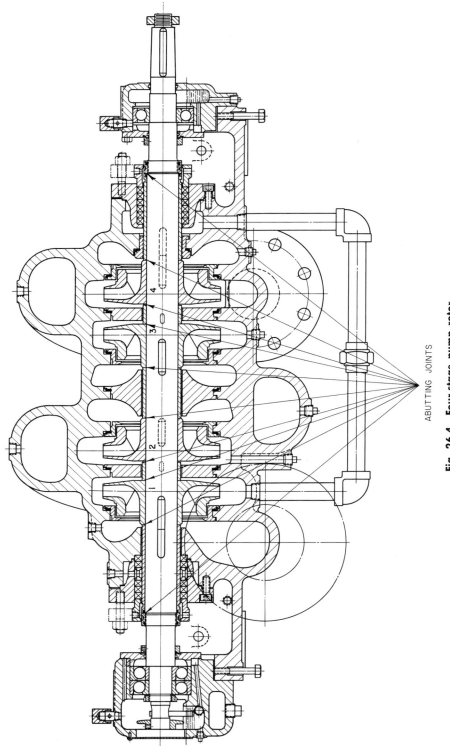

ABUTTING JOINTS

Fig. 26.4 Four-stage pump rotor

reduce pump capacity and possibly restore pump operation to a range in which sufficient NPSH is available at the pump suction. If this step eliminates the crackling noise, the diagnosis is correct, and corrective measures will consist either in increasing the NPSH for the normal range of operating capacities or in replacing the existing impeller with one that can operate with an existing NPSH that cannot be altered.

A rumbling noise in the discharge waterways of the casing is usually caused by operation of the pump at part-load capacities when the pump is not hydraulically suitable for such operation, or by operation of the pump at capacities well in excess of those for which it was designed.

Water hammer is caused by a sudden change in the velocity of flow of a column of liquid and, in general, is serious only when long pipe lines are involved. Water hammer can be avoided by starting a pump against a closed gate valve and then opening the valve slowly afterward. This procedure causes a very gradual increase in the velocity of the liquid in the pipe line. During the stopping cycle, it is necessary to shut the gate valve slowly and stop the pump after the valve has been fully closed. This method fails when a unit or units are stopped suddenly by a control or power failure. Additional provision for the control of water hammer must be provided in installations in which the resulting pressure due to water hammer can reach a dangerous level.

In such installations, slow opening and closing check valves are generally employed to increase the flow gradually when the pump is started and to decrease the flow gradually when the pump is stopped. In other installations, a special protective valve is employed. This valve opens wide quickly with the drop of pressure that is part of a water hammer cycle, and then closes slowly to throttle the resulting back flow. Sometimes air-charged surge tanks or air chambers have been used satisfactorily.

Because a motor-driven pump will stop delivering water almost instantaneously when the power is cut off, water hammer is a more serious problem in this type of installation. In a few very important high-head installations, the stopping time has been lengthened by adding a flywheel to the unit. Although the actual time it takes for the unit to stop pumping is lengthened very little, this slower stopping cycle helps greatly in minimizing the water hammer.

Air leakage into a pump

If a pump operates with a suction lift, air will sometimes leak into the pump through the stuffing boxes unless lantern rings are installed in the boxes and liquid under positive pressure is piped to the lantern rings. Air may also leak under the shaft sleeves. Sometimes the suction piping itself is not quite air-tight and air enters it to collect at the top of the casing; at other times, the water handled by the pump is saturated with air and this air may be liberated in the pump.

If a limited amount of air is allowed to leak into the suction side of a centrifugal pump, the air will be carried through the pump. This ability of a centrifugal pump to carry a small amount of air through its passages is utilized in some installations involving a pneumatic tank that adds air to the tank. If air leaks into the suction passages and passes through the pump, there is a reduction in the amount of water handled and the power required. If the amount of air allowed to leak into the suction side is increased, a point at which the pump will be unable to carry the air through with the water is reached, and the pump will lose its prime.

Sometimes a pump that operates satisfactorily under full discharge will lose its suction when throttled to a lower rate of flow. This condition may be caused by the presence of air leakage into the pump. At full discharge capacity, the velocities in the

TABLE 27.1　CHECK CHART FOR CENTRIFUGAL PUMP TROUBLES

Symptom	Possible causes	Key	
Pump does not deliver water	1, 2, 3, 4, 6, 11, 14, 16, 17, 22, 23	1. Pump not primed	⎫
Insufficient capacity delivered	2, 3, 4, 5, 6, 7, 8, 9, 10, 11, 14, 17, 20, 22, 23, 29, 30, 31	2. Pump or suction pipe not completely filled with liquid 3. Suction lift too high 4. Insufficient margin between suction pressure and vapor pressure 5. Excessive amount of air or gas in liquid 6. Air pocket in suction line	⎬ Suction troubles
Insufficient pressure developed	5, 14, 16, 17, 20, 22, 29, 30, 31	7. Air leakage into suction line 8. Air leakage into pump through stuffing boxes 9. Foot valve too small	⎭
Pump loses prime after starting	2, 3, 5, 6, 7, 8, 11, 12, 13	10. Foot valve partially clogged 11. Inlet of suction pipe insufficiently submerged 12. Water seal pipe plugged	
Pump requires excessive power	15, 16, 17, 18, 19, 20, 23, 24, 26, 27, 29, 33, 34, 37	13. Seal cage improperly located in stuffing box, preventing 　　　sealing fluid entering space to form the seal 14. Speed too low 15. Speed too high 16. Wrong direction of rotation	
Stuffing box leaks excessively	13, 24, 26, 32, 33, 34, 35, 36, 38, 39, 40	17. Total head of system higher than pump design head 18. Total head of system lower than pump design head 19. Specific gravity of liquid different than design	⎫
Packing has short life	12, 13, 24, 26, 28, 32, 33, 34, 35, 36, 37, 38, 39, 40	20. Viscosity of liquid differs from that for which designed 21. Operation at very low capacity 22. Parallel operation of pumps unsuitable for such operation	⎬ System troubles

Symptom	Possible causes
Pump vibrates or is noisy	2, 3, 4, 9, 10, 11, 21, 23, 24, 25, 26, 27, 28, 30, 35, 36, 41, 42, 43, 44, 45, 46, 47
Bearings have short life	24, 26, 27, 28, 35, 36, 41, 42, 43, 44, 45, 46, 47
Pump overheats and seizes	1, 4, 21, 22, 24, 27, 28, 35, 36, 41

Mechanical troubles

23. Foreign matter in impeller
24. Misalignment
25. Foundations not rigid
26. Shaft bent
27. Rotating part rubbing on stationary part
28. Bearings worn
29. Wearing rings worn
30. Impeller damaged
31. Casing gasket defective, permitting internal leakage
32. Shaft or shaft sleeves worn or scored at the packing
33. Packing improperly installed
34. Incorrect type of packing for operating conditions
35. Shaft running off-center due to worn bearings or misalignment
36. Rotor out of balance resulting in vibration
37. Gland too tight, resulting in no flow of liquid to lubricate packing
38. Failure to provide cooling liquid to water-cooled stuffing boxes
39. Excessive clearance at bottom of stuffing box between shaft and casing, causing packing to be forced into pump interior
40. Dirt or grit in sealing liquid, leading to scoring of shaft or shaft sleeve
41. Excessive thrust caused by a mechanical failure inside the pump or by the failure of the hydraulic balancing device, if any
42. Excessive amount of grease or oil in the housing of an anti-friction bearing or lack of cooling, causing excessive bearing temperature
43. Lack of lubrication
44. Improper installation of antifriction bearings (damage during assembly, incorrect assembly of stacked bearings, use of unmatched bearings as a pair)
45. Dirt getting into bearings
46. Rusting of bearings due to water getting into housing
47. Excessive cooling of water-cooled bearing resulting in condensation in the bearing housing of moisture from the atmosphere

pump casing are sufficiently high to wash the air into the discharge pipe, thus constantly purging the air from the pump casing. When the capacity is reduced, the lower velocities are insufficient to wash out the air that accumulates in the pump. This air accumulates within the central portion of the impeller and is prevented from being vented out by the ring of water that is thrown off by the impeller. The ability of the impeller to generate total head is reduced by the presence of the air in its central portion. This air eventually prevents all further pumping action and the pump loses its prime.

A pump may deliver its normal rated capacity when it is started but gradually slacken off until it handles a fraction of its normal flow. This behavior indicates an air leak. Air is accumulating within the pump and reducing the effective capacity. It is possible to check the accuracy of this diagnosis after stopping the pump and letting the air rise to the top of the casing and of the suction volutes. If the pump operates under a suction head, opening the casing and suction vents to the atmosphere will clear the air out and the pump capacity should be restored after the pump is restarted. If the pump operates under a suction lift, opening the vents will obviously not reprime the pump, and it should be reprimed by whatever means are normally used for this purpose.

Some noisy pumps operating with a suction lift can be quieted by allowing a small leakage of air into the suction. This practice is considered poor. The admission of air quiets the pump only because the pump is operating with an excessive suction lift and the air cushions the cavitation phenomena. The amount of air necessary to quiet the pump varies with the capacity, and a variation in capacity due to variation in operating head will require a different amount of air. Furthermore, the presence of air will generally decrease the efficiency of the pump and may accelerate corrosion.

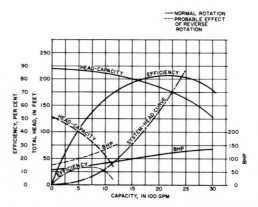

Fig. 27.1   Effect of incorrect rotation on performance of a 6-in. double-suction, single-stage pump at 1,750 rpm

In addition, the presence of air in the pumped water is frequently undesirable. The fact that a small amount of air leaking into the suction line will quiet certain noisy pumps should be used as a test to determine if the cause of the noise is on the suction side, but the admission of air is not a remedy.

System troubles

Some system troubles are relatively easy to correct. For instance, if the motor leads are improperly connected, the pump will rotate in the wrong direction. As soon as this condition is diagnosed, the motor leads can be reversed and the installation made ready for proper operation. When a centrifugal pump is operated in a reverse direction, its performance characteristic is completely abnormal and highly inefficient (Fig. 27.1). The motor of a newly installed pump should be tested for correct rotation when the motor is disconnected from the pump. This is particularly important for vertical turbine pumps and for other pumps that have shafting sections joined by screwed couplings. With this type of shafting, incorrect rotation, even for testing, would result in one or more couplings becoming unscrewed and would require drawing the pump out to reconnect the joint. Damage to some of the pump parts might even result.

Another cause of major loss of capacity and increase in power consumption is the incorrect reassembly of a pump. As mentioned in previous chapters, centrifugal pump impellers have backward curved vanes, that is, they revolve *away* from the curvature of their vanes (Fig. 27.2A). If a pump is dismantled completely and then reassembled, an error in mounting a double-suction impeller is especially easy to overlook. Such an impeller is symmetrical about its central axis and will fit on the shaft even if it is turned in the wrong direction (Fig. 27.2B). The hydraulic performance is the same as would be obtained in a normal casing with correct rotation, but with *forward curved vanes*. If an impeller is reversed on the shaft, the effect is not the same as running the pump in reverse rotation (Fig. 27.2C).

The probable performance of a pump rotating in the proper direction both with a properly mounted impeller and with a reversed impeller is shown in Fig. 27.3. The loss in head and capacity and the increase in power consumption when the impeller is mounted in reverse is much less severe than with a correctly mounted impeller that is running in reverse.

Occasionally, the nonreturn or check valve used to prevent reverse flow through a pump when the unit is shut down, fails to function, and remains open. If this fault occurs, there is a back flow of water through the pump causing the pump to become a

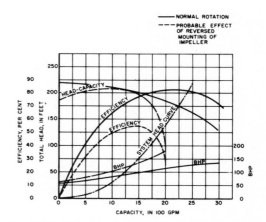

Fig. 27.3 Effect of reversed mounting of impeller on performance of a 6-in. double-suction, single-stage pump at 1,750 rpm

water turbine and, if the driver does not act as a brake, the unit rotates in the reverse direction at the runaway speed of the pump. This runaway speed depends on two factors: The type of the pump and the net head available. With an effective net head equal to the design pumping head, the runaway speed varies from a maximum of approximately 175 per cent of rated speed with low-head, high specific speed types of propeller pumps, to even less than rated speed for high-head, low specific speed types. Most pumping systems involve both a static and frictional component. Thus, when reverse flow occurs, the net head, which is now the static head minus friction head, is usually much less than the normal pumping head (which is static head plus friction head). This lower net head available at the time of reverse flow causes the runaway speed to have a lower value than would result with a net head under reverse flow equal to the design head.

An operating speed that is too low or too high is generally the cause of faulty operation only in turbine- or engine-driven units. Correction of this trouble may sometimes be more involved than the correction of wrong direction of rotation for a motor, but it is not too difficult to effect. It is necessary to remember, however, that if the

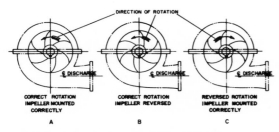

Fig. 27.2 Pump assembly and rotation

pump driver is subject to speed variation, the first step in diagnosing a variation of operating conditions from expected value is to determine the operating speed and see if it corresponds to the design.

The problem of parallel operation of centrifugal pumps has been discussed in Chap. 18. If it is attempted to operate centrifugal pumps that, by virtue of their characteristics, are unsuitable for·parallel operation, the choice of corrective measures is limited. In some conditions, it may be found that as long as parallel operation is restricted to the higher capacity range of the various pumps, no difficulties arise. It becomes necessary, then, to maintain only enough pumps on the line to insure their individual operation in this higher range of capacities. If high-head pumps are involved and, especially, if these pumps handle liquid at or near the boiling point, it is absolutely essential to keep the protective recirculation bypass lines open at all times in order to eliminate the hazard of a check valve slamming on one of the pumps and causing the unit to burn when an unforeseen reduction in total capacity occurs. Manual control of all the units provides a second solution. This procedure is both tedious and uneconomical. The third and last solution consists in revamping the pumps in question and replacing the hydraulic element with parts that will correct the shape of the characteristic curves and thus make parallel operation feasible.

A break in the discharge line of a centrifugal pump is not a very frequent source of trouble. If the discharge line break occurs near the pump, the pump will operate against a very low head and, as a result, will deliver nearly its maximum capacity. In some installations, this capacity will produce a dangerous overload on the driver. Should the break occur at some distance from the pump, a larger frictional component may be involved—especially if the static portion of the total head is low —and the capacity of the pump will be limited.

Foreign matter in the pump

The presence of foreign matter in the impeller or casing of centrifugal pumps can lead to serious trouble and should, unless the pumps are specially designed to handle such material, be carefully prevented. Strainers or screens should always be located in the suction line if foreign matter may be expected in the liquid handled by the pump. Even under normal operation with the presence of foreign matter unlikely —as in boiler feed pumps—welding beads, bolts, nuts, and other objects may find their way into the pump suction in the early stages of operation because the pump is piped without precautionary flushing of the suction piping. Temporary screens or strainers should, therefore, be used in all pump installations. After the pump has been in successful operation for a while, the screens may be removed.

If the suction eye of one side of a double-suction impeller becomes completely clogged, the pump capacity is reduced by approximately one-half. A complete blocking of one side does not impose an excessive load on the thrust bearing because there is little difference in pressure on the two eye areas. A partial blockage of one or both sides, however, often results in throwing the rotor out of balance and may impose an additional load on the bearings.

Misalignment and damage to internal parts

Misalignment difficulties should never be permitted to arise. There is absolutely no excuse for not following properly the instructions provided by the pump manufacturer—instructions that are generally explicit and not difficult to follow. Aside from obvious effects on the pump bearings, serious misalignment will frequently interfere with the proper functioning of the stuffing boxes and impose a severe stress on the shaft, occasionally leading to failure of the shaft due to fatigue.

Rarely, a pump may develop mechanical troubles, when it is first being started, because of rubbing of rotating parts, a bent shaft, damaged internal parts, defective casing gaskets, a rotor out of balance, and the like. Most of these defects should be discovered prior to shipment during the test at the manufacturer's plant; they can, however, develop after a pump is dismantled and reconditioned in the field. No criticism of maintenance personnel is meant here; but it must be recognized that personnel who engage in such maintenance work at infrequent intervals cannot be expected to be as familiar with the necessary operations and precautions as personnel regularly employed in such work.

First on the list of the difficulties that may occur in field reassembly of rotors is contact between rotating and stationary parts of the rotor. Precautions must be taken to eliminate the possibility of the occurrence of such contact. If a pump is started before it has been thoroughly checked for the concentricity of the rotating element with the stationary parts the possibility of a damaged or even seized rotor is increased, particularly if the pump is stainless steel fitted, because such materials gall easily on making contact.

Other practices that may, under some conditions, cause the binding of a rotor in its internal clearances are starting a pump that handles high temperature liquids before warming it up thoroughly and, also, running it at a much reduced speed (say one-half or one-fourth of normal) a number of times and then, by cutting the power, letting it run down to a stop several times. The pump may seize, because if any of its internal parts are rubbing even very slightly, heat is generated, and if the pump is not up to speed, there is not enough liquid flowing past the internal clearances to keep the contacting elements cooled. If the power is cut off when the pump seizes, the rotor will lock solidly in the wearing rings while rolling to a stop. It is therefore recommended that when a pump is started up cold or only partly warmed up, the unit should be brought up to full operating speed and kept going at all times. If, through faulty internal alignment, a slight momentary contact does take place, the chances are that the pump will not seize.

If the axial contacting faces of impeller hubs, shaft sleeves, and shaft nuts are not 100 per cent perpendicular to the bores of these parts—such alignment is difficult to attain in the field without very accurate machining facilities—excessive tightening of the shaft nuts will lead to bending of the shaft and to potential contact of the rotating parts. For this reason, a reassembled rotor should always be checked for concentricity and true running before being inserted into a pump casing.

Stuffing box packing

Stuffing box packing troubles are one of the commonest causes of centrifugal pump failure. The conditions that contribute to stuffing box difficulties are:

1. Shaft running off center because of excessive wear in the bearings, a bent shaft, or misalignment. This condition can be readily checked—first by disconnecting the coupling and rechecking the alignment, and secondly by mounting an indicator on the pump casing in the vicinity of the stuffing box to determine whether the shaft revolves concentrically.

2. Shaft or shaft sleeves worn and scored at the packing. A routine examination of these parts will reveal whether they must be renewed or repaired.

3. Shaft vibration due to unbalance in the rotor, cavitation, operation at extremely light flows or beyond the recommended maximum capacity, or instability in parallel operation.

4. Plugging of the water seal connection or improper location of the seal so that no sealing liquid can enter the stuffing box. The presence of dirt or grit in the sealing liquid will similarly cause stuffing box difficulties by scoring the shaft or shaft sleeves.

5. Excessive tightening of the gland with resulting absence of the leakage that lubri-

cates the packing. Hourly and daily observation of the pump operation, together with the knowledge that some leakage is necessary for proper stuffing box operation, will prevent troubles from this cause.

6. Failure to provide suitable cooling through water-cooled stuffing boxes if the pump is so fitted.

7. Excessive clearance between the bottom of the stuffing box and the shaft or shaft sleeve, which causes the packing to be gradually pushed into the pump interior. This condition can arise when the shaft or shaft sleeves are repaired by grinding them down excessively instead of replacing them or building them up to original dimensions.

8. Packing not properly selected for pressure, temperature, or rubbing speed conditions.

9. Packing not properly inserted into the stuffing box because the individual rings are too short and the gap between ring ends is excessive, or because the ring joints are not staggered.

Bearings

Bearing troubles probably occur in centrifugal pump installations as frequently as packing and shaft sleeve difficulties. The following list of contributory causes is relatively lengthy and should be thoroughly analyzed if bearing replacement becomes a consistent problem:

1. Excessive radial or thrust load caused by some mechanical failure inside the pump or by failure of a hydraulic thrust balancing device if one is used.

2. Excessive grease or oil in the housing of an antifriction bearing. This defect will cause overheating and failure of the bearing.

3. Excessive heating of the bearing due to improper dissipation of heat or a failure of the cooling medium supply.

4. Lack of lubrication.

5. Dirt or foreign matter in the lubricant.

6. Improper installation of antifriction bearings (damage during assembly, incorrect assembly of stacked bearings, use of unmatched bearings as pairs, and so forth).

7. Rusting of antifriction bearings because of water in the bearing housing.

8. Excessive cooling of water-cooled bearings, resulting in condensation of atmospheric moisture in the bearing housing.

9. Misalignment of the unit, bent shaft, or severe vibration, which impose excessive loads on the bearings.

It should be noted that most antifriction bearing manufacturers issue excellent treatises on their equipment, in which the subject of bearing failures is thoroughly discussed. They provide photographs of the various types of failures, thus enabling the user to analyze his own specific problem and to diagnose his troubles by visual comparison. Because antifriction bearings are widely used in centrifugal pumps as well as in other equipment, it is recommended that the maintenance personnel have such literature available and become familiar with the various symptoms of antifriction bearing failure.

Corrosion and erosion

Among the most insidious and most costly of the troubles that may affect centrifugal pumps is the failure of the materials used in the construction of the pump to withstand the corrosive or erosive properties of the liquids handled. Unfortunately, this problem is extremely involved and cannot be treated extensively in a book of this character. Some general information on the subject is presented in Chap. 16. For more detailed information, the reader may consult the extensive literature available. It should be noted, in passing, that the present trend in material selection is to be conservative—that is, to choose materials more resistant to corrosion and erosion than may be necessary, in order to avoid expensive breakdowns and interruption of service.

Data Section

Index of figures and tables in the text that are of special interest

TABLE A.1 CONVERSIONS, CONSTANTS, AND FORMULAS

Volume and Weight

1 U. S. gallon = 8.34 lbs × Sp Gr
1 U. S. gallon = 0.84 Imperial gallon
1 cu ft of liquid = 7.48 gal
1 cu ft of liquid = 62.32 lbs × Sp Gr
Specific gravity of sea water = 1.025 to 1.03
1 cu meter = 264.5 gal
1 barrel (oil) = 42 gal

Capacity and Velocity

gpm = cu ft per sec × 449

$$\text{gpm} = \frac{\text{lbs per hour}}{500 \times \text{Sp. Gr.}}$$

gpm = 0.069 × boiler Hp
gpm = 0.7 × bbl/hour = 0.0292 bbl/day
gpm = 0.227 metric tons per hour

1 mgd = 694.5 gpm

$$V = \frac{\text{gpm} \times 0.321}{\text{area in sq in}} = \frac{\text{gpm} \times 0.409}{D^2}$$

$$V = \sqrt{2gH}$$

gpm = gallons per minute
Sp Gr = specific gravity based on water at 62°F

Hp = horsepower
bbl = barrel (oil) = 42 gal
mgd = million gallons per day of 24 hours
V = velocity in ft/sec
D = diameter in inches
g = 32.16 ft/sec/sec
H = head in feet

Head

$$\text{Head in feet} = \frac{\text{Head in psi} \times 2.31}{\text{Sp Gr}}$$

1 foot water (cold, fresh) = 1.133 inches of mercury
1 psi = 0.0703 kilograms per sq centimeter
1 psi = 0.068 atmosphere

$$H = \frac{V^2}{2g}$$

psi = pounds per square inch

Power and Torque

1 horsepower = 550 ft-lb per sec
= 33,000 ft-lb per min
= 2545 btu per hr
= 745.7 watts
= 0.7457 kilowatts

Power and Torque—(*Continued*)

$$\text{bhp} = \frac{\text{gpm} \times \text{Head in feet} \times \text{Sp Gr}}{3960 \times \text{efficiency}}$$

$$\text{bhp} = \frac{\text{gpm} \times \text{Head in psi}}{1714 \times \text{efficiency}}$$

Navy formula to determine Hp rating of motor:

$$\text{Hp} = \frac{\text{ohp}\left(1.05 + \dfrac{1.35}{\text{ohp} + 3}\right)}{\text{efficiency of pump}}$$

where ohp is the output horsepower or water horsepower work done by the pump which is determined by:

$$\text{ohp} = \frac{\text{gpm} \times \text{Head in feet} \times \text{Sp Gr}}{3960}$$

$$\text{or ohp} = \frac{\text{gpm} \times \text{Head in psi}}{1714}$$

$$\text{Torque in lbs feet} = \frac{\text{Hp} \times 5252}{\text{rpm}}$$

bhp = brake horsepower
rpm = revolutions per minute

Miscellaneous Centrifugal Pump Formulas

$$\text{Specific speed} = N_s = \frac{\sqrt{\text{gpm}} \times \text{rpm}}{H^{3/4}}$$

where H = head per stage in feet

$$\text{Diameter of impeller in inches} = d = \frac{1840 \, Ku\sqrt{H}}{\text{rpm}}$$

where Ku is a constant varying with impeller type and design. Use H at shut-off (zero capacity) and Ku is approx. 1.0

$$\text{At constant speed:} \quad \frac{d_1}{d_2} = \frac{\text{gpm}_1}{\text{gpm}_2} = \frac{\sqrt{H_1}}{\sqrt{H_2}} = \frac{\sqrt[3]{\text{bhp}_1}}{\sqrt[3]{\text{bhp}_2}}$$

At constant impeller diameter

$$\frac{\text{rpm}_1}{\text{rpm}_2} = \frac{\text{gpm}_1}{\text{gpm}_2} = \frac{\sqrt{H_1}}{\sqrt{H_2}} = \frac{\sqrt[3]{\text{Bhp}_1}}{\sqrt[3]{\text{Bhp}_2}}$$

TABLE A.2   MEASUREMENT CONVERSION

Angstrom Unit . . . . . 10^{-10} meter
1/10,000 micron
0.003937 millionths of an
inch

Atmosphere . . . . 14.7 pounds (English)
14.223 pounds (Russian)

Btu (*British Thermal Unit*) 778 foot pounds
0.2930 watt hour
0.252 calorie

Calorie . . . . 1 kilogram of water raised
1 degree Centigrade
3.97 Btu

Centare . (*square meter*) 10.764 square feet

Centimeter . . . . 0.3937 inch

Cheval (*French hp.*) . . . 0.986 horsepower

Circular Mil . . . . Area of circle whose
diameter is one mil or
1/1,000 inch
0.000000785 square inch

Cubic Centimeter
(*milliliter*)   0.061 cubic inch

Cubic Foot . . . . 1,728 cubic inches
7.48 gallons
60 pints
8/10 bushel
62.32 lbs water (62°F)
1,000 ounces of water,
approx.
0.028 cubic meter
28.32 liters

Cubic Inch . . . . 16.39 cubic centimeters

Cubic Meter . . . . 35.315 cubic feet
1.308 cubic yards

Cubic Yard . . . . 27 cubic feet
0.765 cubic meter

Decimeter . . . . 3.937 inches

Foot . . . . 12 inches
0.385 meter

Foot Pound . . . . 0.1364 kilogrammeter

Gallon . . . . 231 cubic inches
4 quarts
8 pints
3.785 liters
128 fluid ounces
8.33 pounds of water

Gallon per Minute 449 cubic feet per second
0.227 metric tons per hour

Gallon (*British Imperial*) 277.3 cubic inches
1.201 U. S. gallons
10 lbs water at 15°C.
4.546 liters

Grain . . . . 1/7000 lb avoirdupois
0.0584 gram

Gram . . . . 15.43 grains
0.0353 ounce
0.0022 pound

Horsepower . . . . 33,000 ft lb per minute
42.41 Btu per minute
1.014 cheval
746 watts

Hundredweight
(*British*) . . . . 112 pounds
50.80 kilograms

Inch . . . . 39,540 ½ wave lengths of
red ray of cadmium
25.4 millimeters

Joule . . . . 1 watt second

Kilogram . . . . 2.2046 pounds
35.274 ounces
15432.36 grains
0.0011 short ton
0.00098 long ton

**Kilogram per Cubic
Meter** . . . . 0.0624 lbs per cu ft

**Kilogram
per Square
Centimeter** . . . . . . 14.225 lbs per sq in

**Kilogram
per Square
Meter** . . . . 0.205 lbs per sq ft

Kilometer . . . . 1,000 meters
0.621 mile

Kilowatt . . . . 1.34 horsepower
44,257 ft lb per minute
56.87 Btu per minute

Liter . . . . 1.000027 cubic decimeter
1.057 quart
0.264 gallon
61.02 cubic inches
.035 cubic feet
33.8147 fluid ounces
270.518 fluid drams

Liter per Second . . . 2.12 cu ft per minute
0.474 U. S. Gal per min

Meter . . . . 39.37 inches
3.28 feet
1.09 yards

Metric Ton . . . . 2204.6 pounds
1.1023 short tons

Micron . . . . 0.001 millimeter
10,000 Angstrom units
39.37 millionths of an in

Microgram . . . . 1/1000 milligram

Mil . . . . 0.001 inch
25.4 microns
0.0254 millimeter

Mile . . . . 1,760 yards
5,280 feet
1.61 kilometers

Milligram . . . . 0.0154 grain

Milliliter . . . . 1.000027 cu centimeter
0.0610 cubic inch

Myriameter . . . . 10,000 meters
6.2137 miles

Ounce . . . . 437.5 grains
0.911 troy ounces
28.35 gram

Ounce (*Fluid*) . . . . . . 1.805 cubic inches
29.573 milliliters

Ounce (*Fine*) . . . . . . Troy ounce
480 grains
31.104 grams

Pied (*French foot*) . . . . . 12 Paris inches
0.325 meter

Pint . . . . 0.4732 liter
16 fluid ounces

Pound Avoirdupois . 16 ounces
7,000 grains
454 grams
0.454 kilogram
14.58 troy ounces

**Pound per Cubic
Foot** . . . . 16.02 kilogram per cubic
meter

Pound per Sq In . . . 2.31 feet head of water at
1.00 sp gr
0.0703 kilogram per sq
centimeter

Pound per Sq Ft . . . 4.88 kilogram per square
meter

Quart . . . . 2 pints
¼ gallon
0.946 liter

Quart (*British quarter
hundredweight*) . . . . . . 28 pounds
12.70 kilograms

Quintal . . . . 100 kilograms
220.46 pounds

Stere . . . . 1 cubic meter

Square Centimeter 0.155 square inch

Square Foot . . . . . . 0.093 square meter
144 square inches

Square Inch . . . . . . 6.452 square centimeters

Square Kilometer . . 0.386 square mile

Square Meter (*centare*) 10.764 square feet
1.196 square yard

Square Mil . . . . 0.000001 square inch
0.000645 sq centimeter

Square Mile . . . . . . 640 acres
3,097,600 square yards
2.59 square kilometers

Square Millimeter . 0.00155 square inch

Square Yard . . . . . . 0.835 square meter

Stone (*British*) . . . . . 14 pounds
6.35 kilograms

Ton (*short*) . . . . 2,000 pounds
907 kilograms

Ton (*long*) . . . . 2,240 pounds
1,016 kilograms
270 gallons

Ton per Hour (*metric*) 4.4 gallons per minute

Tonne (*metric*) . . . . . 1,000 kilograms
2204.62 pounds

Yard . . . . 3 feet
36 inches
0.914 meter

Use of Table A.4

For laminar flow, the pressure loss is directly proportional to the viscosity and the velocity of flow. Therefore, for intermediate values of viscosity and rate of flow (gpm), the pressure loss can be obtained by direct interpolation of Table A.4. For pipe sizes not shown, the pressure loss will vary inversely as the fourth power of the inside diameters for the same discharge rate.

The values of pressure loss, which will be found to the left of the heavy line on the first two pages of Table A.4, fall within the turbulent flow region rather than in the laminar or viscous flow region. For interpolation of rate of flow and pipe size in this region of turbulent flow, use the following methods. For intermediate capacities select the friction for the next lower capacity value and multiply by the square of the ratio of the capacities. For intermediate diameters the friction per 100 ft will vary inversely as the fifth power of the inside diameter for the same discharge rate. For intermediate values of viscosity in this region, the "Pipe Friction Manual" of the Hydraulic Institute should be consulted.

Example

Find the pressure loss for 50 gpm of oil in 200 ft of 2-in. Schedule 40 pipe. The oil has a viscosity of 2,000 ssu and a specific gravity of 0.9. Solution: From Table A.4, the loss for 100 ft of pipe is 32.1 psi for a specific gravity of 1.00. For a specific gravity of 0.9 and for 200 ft of pipe, therefore:

$$\text{Pressure loss} = 32.1 \times \frac{200}{100} \times 0.9$$

$$= 57.8 \text{ psi}$$

Friction loss in fittings with viscous flow

When the piping system includes valves and fittings, the following must be considered:

1. For turbulent flow the values of the equivalent lengths of straight pipe for valves and fittings, as given in Fig. 18.7 on page 233, should be used.

2. For laminar flow the losses in valves and fittings can only be approximated. For liquids of relatively low viscosity, where the flow is adjacent to the turbulent region, the values of the equivalent straight pipe for valves and fittings, given in Fig. 18.7, can be used.

For the very highest viscosities, the effects of the valves or fittings are small, and it is probably only necessary to include its actual length as part of the pipe length. For the intermediate viscosities the approximate equivalent length can be estimated by interpolation, using Table A.3 as a guide.

TABLE A.3   FRICTION LOSS IN FITTINGS FOR INTERPOLATED INTERMEDIATE VISCOSITIES

It should be noted that this table is only an approximation. Very little reliable test data on losses in valves and fittings for laminar flow are available. Figure 18.7 is on page 233.

	3–30 gpm ssu	30–50 gpm ssu	50–100 gpm ssu	100–250 gpm ssu	250–1,000 gpm ssu
Use full value from Fig. 18.7 when viscosity is	100	200	300	400	500
Use ¾ value from Fig. 18.7 when viscosity is	1,000	2,000	3,000	4,000	5,000
Use ½ value from Fig. 18.7 when viscosity is	10,000	20,000	30,000	40,000	50,000
Use ¼ value from Fig. 18.7 when viscosity is	100,000	200,000	300,000	400,000	500,000
Use actual length of valves and fittings when the viscosity exceeds	500,000	500,000			

TABLE A.4   FRICTION LOSS FOR VISCOUS LIQUIDS

Loss in pounds per square inch per 100 feet of new Schedule 40 steel pipe based on specific gravity of 1.00. (For a liquid having a specific gravity other than 1.00, multiply the value from the table by the specific gravity of that liquid. For commercial installations, it is recommended that 15 per cent be added to the values in this table. No allowance for aging of pipe is included.)

GPM	Pipe Size	VISCOSITY—SAYBOLT SECONDS UNIVERSAL																	
		100	200	300	400	500	1000	1500	2000	2500	3000	4000	5000	6000	7000	8000	9000	10,000	15,000
3	½	11.2	23.6	35.3	47.1	59	118	177	236	294	353	471	589	706	824	942	.....	.....	.....
	¾	3.7	7.6	11.5	15.3	19.1	38.2	57	76	96	115	153	191	229	268	306	344	382	573
	1	1.4	2.9	4.4	5.8	7.3	14.5	21.8	29.1	36.3	43.6	58	73	87	101	116	131	145	218
5	¾	6.1	12.7	19.1	25.5	31.9	64	96	127	159	191	255	319	382	446	510	573	637	956
	1	2.3	4.9	7.3	9.7	12.1	24.2	36.3	48.5	61	73	97	121	145	170	194	218	242	363
	1¼	0.77	1.6	2.4	3.3	4.1	8.1	12.2	16.2	20.3	24.3	32.5	40.6	48.7	57	65	73	81	122
7	¾	8.5	17.9	26.8	35.7	44.6	89	134	178	223	268	357	446	535	624	713	803	892	.....
	1	3.2	6.8	10.2	13.6	17	33.9	51	68	85	102	136	170	203	237	271	305	339	509
	1¼	1.1	2.3	3.4	4.5	5.7	11.4	17	22.7	28.4	34.1	45.4	57	68	80	91	102	114	170
10	1	4.9	9.7	14.5	19.4	24.2	48.5	73	97	121	145	194	242	291	339	388	436	485	727
	1¼	1.6	3.3	4.9	6.5	8.1	16.2	24.3	32.5	40.6	48.7	65	81	97	114	130	146	162	243
	1½	0.84	1.8	2.6	3.5	4.4	8.8	13.1	17.5	21.9	26.3	35	43.8	53	61	70	79	88	131
15	1	11	14.5	21.8	29.1	36.3	73	109	145	182	218	291	363	436	509	581	654	727	.....
	1¼	2.8	4.9	7.3	9.7	12.2	24.3	36.5	48.7	61	73	97	122	146	170	195	219	243	365
	1½	1.3	2.6	3.9	5.3	6.6	13.1	197	26.3	32.8	39.4	53	66	79	92	105	118	131	197
20	1	18	18	29.1	38.8	48.5	97	145	194	242	291	388	485	581	678	775	872	.....	.....
	1¼	4.9	6.4	9.7	13	16.2	32.5	48.7	65	81	97	130	162	195	227	260	292	325	487
	1½	2.3	3.5	5.3	7	8.8	17.5	26.3	35	43.8	53	70	88	105	123	140	158	175	263
	2	0.64	1.3	1.9	2.6	3.2	6.4	9.6	12.9	16.1	19.3	25.7	32.1	38.5	45	51	58	64	96
25	1½	3.5	4.4	6.6	8.8	11	21.9	32.8	43.8	55	66	88	110	131	153	176	197	219	328
	2	1	1.6	2.4	3.2	4	8	12.1	16.1	20.1	24.1	32.1	40.2	48.2	56	64	72	80	121
	2½	0.4	0.79	1.2	1.6	2	4	5.9	7.9	9.9	11.8	15.8	19.7	23.7	27.6	31.6	35.5	39.5	59
30	1½	5	5.3	7.9	10.5	13.1	26.3	39.4	53	66	79	105	131	158	184	210	237	263	394
	2	1.4	1.9	2.9	3.9	4.8	9.6	14.5	19.3	24.1	28.9	38.5	48.2	58	67	77	87	96	145
	2½	0.6	0.95	1.4	1.9	2.4	4.7	7.1	9.5	11.8	14.2	19	23.7	28.4	33.2	37.9	42.6	47.4	71
40	1½	8.5	9	10.5	14	17.5	35	53	70	88	105	140	175	210	245	280	315	350	526
	2	2.5	2.5	3.9	5.1	6.4	12.9	19.3	25.7	32.1	38.5	51	64	77	90	103	116	129	193
	2½	1.1	1.3	1.9	2.5	3.2	6.3	9.5	12.6	15.8	19	25.3	31.6	37.9	44.2	51	57	63	95
50	1½	12.5	14	14	17.5	21.9	43.8	66	88	110	131	175	219	263	307	350	394	438	657
	2	3.7	4	4.8	6.4	8	16.1	24.1	32.1	40.2	48.2	64	80	96	112	129	145	161	241
	2½	1.6	1.7	2.4	3.2	4	7.9	11.8	15.8	19.7	23.7	31.6	39.5	47.4	55	63	71	79	118
60	2	5	5.8	5.8	7.7	9.6	19.3	28.9	38.5	48.2	58	77	96	116	135	154	173	193	289
	2½	2.2	2.4	2.8	3.8	4.7	9.5	14.2	19	23.7	28.4	37.9	47.4	57	66	76	85	95	142
	3	0.8	0.8	1.2	1.6	2	4	6	8	9.9	11.9	15.9	19.9	23.9	27.9	31.8	35.8	39.8	60
70	2½	2.8	3.2	3.4	4.4	5.5	11.1	16.6	22.1	27.6	33.2	44.2	55	66	77	88	100	111	166
	3	1	1.1	1.4	1.9	2.3	4.6	7	9.3	11.6	13.9	18.6	23.2	27.8	32.5	37.1	41.7	46.4	70
	4	0.27	0.31	0.47	0.63	0.78	1.6	2.4	3.1	3.9	4.7	6.3	7.8	9.4	11	12.5	14.1	15.6	23.5
80	2½	3.6	4.2	4.2	5.1	6.3	12.6	19	25.3	31.6	37.9	51	63	76	88	101	114	126	190
	3	1.3	1.4	1.6	2.1	2.7	5.3	8	10.6	13.3	15.9	21.2	26.5	31.8	37.1	42.4	47.7	53	80
	4	0.36	0.36	0.54	0.72	0.89	1.8	2.7	3.6	4.5	5.4	7.2	8.9	10.7	12.5	14.3	16.1	17.9	26.8
100	2½	5.3	6.1	6.4	6.4	8	15.8	23.7	31.6	39.5	47.4	63	79	95	111	127	142	158	237
	3	1.9	2.2	2.2	2.7	3.3	6.6	9.9	13.3	16.6	19.9	26.5	33.1	39.8	46.4	53	60	66	99
	4	0.52	0.57	0.67	0.89	1.1	2.2	3.4	4.5	5.6	6.7	8.9	11.2	13.4	15.6	17.9	20.1	22.3	33.5

◄———— **TURBULENT FLOW** ————►◄———————— **LAMINAR FLOW** ————————►

TABLE A.4 (CONT.)   FRICTION LOSS FOR VISCOUS LIQUIDS

Loss in pounds per square inch per 100 feet of new Schedule 40 steel pipe based on specific gravity of 1.00. (For a liquid having a specific gravity other than 1.00, multiply the value from the table by the specific gravity of that liquid. For commercial installations, it is recommended that 15 per cent be added to the values in this table. No allowance for aging of pipe is included.)

GPM	Pipe Size	VISCOSITY—SAYBOLT SECONDS UNIVERSAL																	
		100	200	300	400	500	1000	1500	2000	2500	3000	4000	5000	6000	7000	8000	9000	10,000	15,000
120	3	2.7	3.1	3.2	3.2	4	8	11.9	15.9	19.9	23.9	31.8	39.8	47.7	56	64	72	80	119
	4	0.73	0.81	0.81	1.1	1.3	2.7	4	5.4	6.7	8	10.7	13.4	16.1	18.8	21.4	24.1	26.8	40.2
	6	.098	0.11	0.16	0.21	0.26	0.52	0.78	1.0	1.3	1.6	2.1	2.6	3.1	3.6	4.2	4.7	5.2	7.8
140	3	3.4	4	4.3	4.3	4.6	9.3	13.9	18.6	23.2	27.8	37.1	46.4	56	65	74	84	93	139
	4	0.95	1.1	1.1	1.3	1.6	3.1	4.7	6.3	7.8	9.4	12.5	15.6	18.8	21.9	25	28.2	31.3	46.9
	6	0.13	0.15	0.18	0.24	0.30	0.61	0.91	1.2	1.5	1.8	2.4	3.0	3.6	4.2	4.9	5.5	6.1	9.1
160	3	4.4	5	5.7	5.7	5.7	10.6	15.9	21.2	26.5	31.8	42.4	53	64	74	85	95	106	159
	4	1.2	1.4	1.4	1.4	1.8	3.6	5.4	7.2	8.9	10.7	14.3	17.9	21.5	25	28.6	32.2	35.7	54
	6	0.17	0.18	0.21	0.28	0.35	0.69	1.0	1.4	1.7	2.1	2.8	3.5	4.2	4.9	5.5	6.2	6.9	10.4
180	3	5.3	6.3	7	7	7	11.9	17.9	23.9	29.8	35.8	47.7	60	72	84	95	107	119	179
	4	1.5	1.8	1.8	1.8	2	4	6	8	10.1	12.1	16.1	20.1	24.1	28.1	32.2	36.2	40.2	60
	6	0.2	0.24	0.24	0.31	0.39	0.78	1.2	1.6	2	2.3	3.1	3.9	4.7	5.5	6.2	7	7.8	11.7
200	3	6.5	7.7	8.8	8.8	8.8	13.3	19.9	26.5	33.1	39.8	53	66	80	93	106	119	133	199
	4	1.8	2.2	2.2	2.2	2.2	4.5	6.7	8.9	11.2	13.4	17.9	22.3	26.8	31.3	35.7	40.2	44.7	67
	6	0.25	0.3	0.3	0.35	0.43	0.87	1.3	1.7	2.2	2.6	3.5	4.3	5.2	6.1	6.9	7.8	8.7	13
250	4	2.6	3.2	3.5	3.5	3.5	5.6	8.4	11.2	14	16.8	22.3	27.9	33.5	39.1	44.7	50	56	84
	6	0.36	0.43	0.45	0.45	0.54	1.1	1.6	2.2	2.7	3.3	4.3	5.4	6.5	7.6	8.7	9.8	10.8	16.3
	8	.095	0.12	0.12	0.15	0.18	0.36	0.54	0.72	0.9	1.1	1.5	1.8	2.2	2.5	2.9	3.3	3.6	5.4
300	4	3.7	4.3	5	5	5	6.7	10.1	13.4	16.8	20.1	26.8	33.5	40.2	47	54	60	67	101
	6	0.5	0.6	0.65	0.65	0.65	1.3	2	2.6	3.3	3.9	5.2	6.5	7.8	9.1	10.4	11.7	13	19.5
	8	0.13	0.17	0.17	0.18	0.22	0.43	0.65	0.87	1.1	1.3	1.7	2.2	2.6	3	3.5	3.9	4.3	6.5
400	6	0.82	1	1.1	1.2	1.2	1.7	2.6	3.5	4.3	5.2	6.9	8.7	10.4	12.1	13.9	15.6	17.3	26
	8	0.23	0.27	0.29	0.29	0.29	0.58	0.87	1.2	1.5	1.7	2.3	2.9	3.5	4.1	4.6	5.2	5.8	8.7
	10	0.08	0.09	0.1	0.1	0.12	0.23	0.35	0.47	0.58	0.7	0.93	1.2	1.4	1.6	1.9	2.1	2.3	3.5
500	6	1.2	1.5	1.6	1.8	1.8	2.2	3.2	4.3	5.4	6.5	8.7	10.8	13	15.2	17.3	19.5	21.7	32.5
	8	0.33	0.39	0.44	0.47	0.47	0.72	1.1	1.5	1.8	2.2	2.9	3.6	4.3	5.1	5.8	6.5	7.2	10.8
	10	0.11	0.14	0.15	0.15	0.15	0.29	0.44	0.58	0.73	0.87	1.2	1.5	1.8	2	2.3	2.6	2.9	4.4
600	6	1.8	2.2	2.3	2.4	2.6	2.7	3.9	5.2	6.5	7.8	10.4	13	16	18.2	20.8	23.4	26	39
	8	0.47	0.57	0.62	0.67	0.67	0.87	1.3	1.7	2.2	2.6	3.5	4.3	5.2	6.1	6.9	7.8	8.7	13
	10	0.16	0.18	0.2	0.22	0.22	0.35	0.52	0.7	0.87	1.1	1.4	1.8	2.1	2.4	2.8	3.3	3.5	5.2
700	6	2.3	2.7	3	3.2	3.5	3.6	4.6	6.1	7.6	9.1	12.1	15.2	18.4	21.2	24.3	27.3	30.3	45.5
	8	0.6	0.74	0.82	0.89	0.93	1	1.5	2	2.5	3	4.1	5.1	6.1	7.1	8.1	9.1	10.1	15.2
	10	0.2	0.25	0.27	0.3	0.3	0.41	0.61	0.82	1	1.2	1.6	2	2.4	2.9	3.3	3.7	4.1	6.1
800	6	2.8	3.5	3.7	4	4.2	4.8	5.2	6.9	8.7	10.4	13.9	17.3	20.8	24.3	27.7	31.2	34.7	52
	8	0.78	0.94	1	1.1	1.2	1.2	1.7	2.3	2.9	3.5	4.6	5.8	6.9	8.1	9.3	10.4	11.6	17.3
	10	0.26	0.3	0.34	0.38	0.4	0.47	0.7	0.92	1.2	1.4	1.9	2.3	2.8	3.3	3.7	4.2	4.7	7
900	6	3.5	4.3	4.6	5.0	5.2	6	6	7.8	9.8	11.7	15.6	19.5	23.4	27.3	31.2	35.1	39	58.5
	8	0.95	1.1	1.3	1.4	1.5	1.5	2	2.6	3.3	3.9	5.2	6.5	7.8	9.1	10.4	11.7	13	19.5
	10	0.32	0.37	0.43	0.46	0.5	0.52	0.79	1.1	1.3	1.6	2.1	2.6	3.1	3.7	4.2	4.7	5.2	7.9
1000	8	1.1	1.4	1.5	1.6	1.8	1.9	2.2	2.9	3.6	4.3	5.8	7.2	8.7	10.1	11.6	13	14.5	21.7
	10	0.38	0.45	0.5	0.55	0.6	0.6	0.87	1.2	1.5	1.8	2.3	2.9	3.5	4.1	4.7	5.2	5.8	8.7
	12	0.17	0.2	0.22	0.24	0.25	0.29	0.43	0.58	0.72	0.87	1.2	1.5	1.7	2	2.3	2.6	2.9	4.3

TURBULENT FLOW ◄───►◄───LAMINAR FLOW───►

TABLE A.4 (CONT.)   FRICTION LOSS FOR VISCOUS LIQUIDS

Loss in pounds per square inch per 100 feet of new Schedule 40 steel pipe based on specific gravity of 1.00. (For a liquid having a specific gravity other than 1.00, multiply the value from the table by the specific gravity of that liquid. For commercial installations, it is recommended that 15 per cent be added to the values in this table. No allowance for aging of pipe is included.)

GPM	Pipe Size	VISCOSITY—SAYBOLT SECONDS UNIVERSAL														
		20,000	25,000	30,000	40,000	50,000	60,000	70,000	80,000	90,000	100,000	125,000	150,000	175,000	200,000	500,000
3	2	19.3	24.1	28.9	38.5	48.2	58	67	77	87	96	120	145	169	193	482
	2½	9.5	11.8	14.2	19	23.7	28.4	332	37.9	42.6	47.4	59	71	83	95	237
	3	4	5	6	8	9.9	11.9	13.9	15.9	17.9	19.9	24.9	29.8	34.8	39.8	99
5	2	32	40	48.2	64	80	96	112	129	145	161	201	241	281	321	803
	2½	15.8	19.7	23.7	31.6	39.5	47.4	55	63	71	79	99	118	138	158	395
	3	6.6	8.3	9.9	13.3	16.6	9.9	23.2	26.5	29.8	33	41.4	49.7	58	66	166
7	2	45	56	67	90	112	135	157	180	202	225	281	337	393	450	......
	2½	22.1	27.6	33.2	44.2	55	66	77	88	100	111	138	166	194	221	553
	3	9.3	11.6	13.9	18.6	23.2	27.8	32.5	37.1	41.7	46.4	58	70	81	93	232
10	2½	31.6	39.5	47.4	63	79	95	111	126	142	158	197	237	276	316	790
	3	13.3	16.6	19.9	26.5	33.1	39.8	46.4	53	60	66	83	99	116	133	331
	4	4.5	5.6	6.7	8.9	11.2	13.4	15.6	17.9	20.1	22.3	27.9	33.5	39.1	44.7	112
15	2½	47.4	59	71	95	118	142	166	190	213	237	296	355	415	474	......
	3	19.9	24.9	29.8	39.8	49.7	60	70	80	89	99	124	149	174	199	497
	4	6.7	8.4	10.1	13.4	16.8	20.1	23.5	26.8	30.2	33.5	41.9	50	59	67	168
20	3	26.5	33.1	39.8	53	66	80	93	106	119	133	166	199	232	265	663
	4	8.9	11.2	13.4	17.9	22.3	26.8	31.3	35.7	40.2	44.7	56	67	78	89	223
	6	1.7	2.2	2.6	3.5	4.3	5.2	6.1	6.9	7.8	8.7	10.8	13	15.2	17.3	43.3
25	3	33.1	41.4	49.7	66	83	99	116	133	149	166	207	249	290	331	828
	4	11.2	14	16.8	22.3	27.9	33.5	39.1	44.7	50	56	70	84	98	112	279
	6	2.2	2.7	3.3	4.3	5.4	6.5	7.6	8.7	9.8	10.8	13.5	16.3	19	21.7	54
30	3	39.8	49.7	60	80	99	119	139	159	179	199	249	298	348	398	......
	4	13.4	16.8	20.1	26.8	33.5	40.2	46.9	54	60	67	84	101	117	134	335
	6	2.6	3.3	3.9	5.2	6.5	7.8	9.1	10.4	11.7	13	16.3	19.5	22.7	26	65
40	3	53	66	80	106	133	160	186	212	239	265	331	398	464	532	......
	4	17.9	22.3	26.8	35.7	44.7	54	63	72	80	89	112	134	156	179	447
	6	3.5	4.3	5.2	6.9	8.7	10.4	12.1	13.9	15.6	17.3	26	30.3	34.7		87
50	4	22.3	27.9	33.5	44.7	56	67	78	89	101	112	140	168	196	223	559
	6	4.3	5.4	6.5	8.7	10.8	13	15.2	17.3	19.5	21.7	27.1	32.5	37.9	43.3	108
	8	1.5	1.8	2.7	2.9	3.6	4.3	5.1	5.8	6.5	7.2	9	10.8	12.6	14.5	36.1
60	4	26.8	33.5	40.2	54	67	80	94	107	121	134	168	201	235	268	670
	6	5.2	6.5	7.8	10.4	13	16	18.2	20.8	23.4	26	32.5	39	45.5	52	130
	8	1.7	2.2	2.6	3.5	4.3	5.2	6.1	6.9	7.8	8.7	10.8	13	15.2	17.3	43.4
70	4	31.3	39.1	46.9	63	78	94	110	125	141	156	196	235	274	313	782
	6	6.1	7.6	9.1	12.1	15.2	18.4	21.2	24.3	27.3	30.3	37.9	45.5	53	61	152
	8	2	2.5	3	4.1	5.1	6.1	7.1	8.1	9.1	10.1	12.6	15.2	17.7	20.2	51
80	6	6.9	8.7	10.4	13.9	17.3	20.8	24.3	27.7	31.2	34.7	43.3	52	61	69	173
	8	2.3	2.9	3.5	4.6	5.8	6.9	8.1	9.3	10.4	11.6	14.5	17.3	20.2	23.1	58
	10	0.93	1.2	1.4	1.9	2.3	2.8	3.3	3.7	4.2	4.7	5.8	7	8.2	9.3	23.3
90	6	7.8	9.8	11.7	15.6	19.5	23.4	27.3	31.2	35.1	39	48.7	59	68	78	195
	8	2.6	3.3	3.9	5.2	6.5	7.8	9.1	10.4	11.7	13	16.3	19.5	22.8	26	65
	10	1.1	1.3	1.6	2.1	2.6	3.1	3.7	4.2	4.7	5.2	6.6	7.9	9.2	10.5	26.2
100	6	8.7	10.8	13	17.3	21.7	26	30.3	34.7	39	43.3	54	65	76	87	217
	8	2.9	3.6	4.3	5.8	7.2	8.7	10.1	11.6	13	14.5	18.1	21.7	25.3	28.9	72
	10	1.2	1.5	1.8	2.3	2.9	3.5	4.2	4.7	5.2	5.8	7.3	8.7	10.2	11.6	29.1

◄———————————————————————————LAMINAR FLOW———————————————————————————►

TABLE A.4 (CONT.)   FRICTION LOSS FOR VISCOUS LIQUIDS

Loss in pounds per square inch per 100 feet of new Schedule 40 steel pipe based on specific gravity of 1.00. (For a liquid having a specific gravity other than 1.00, multiply the value from the table by the specific gravity of that liquid. For commercial installations, it is recommended that 15 per cent be added to the values in this table. No allowance for aging of pipe is included.)

GPM	Pipe Size	VISCOSITY—SAYBOLT SECONDS UNIVERSAL														
		20,000	25,000	30,000	40,000	50,000	60,000	70,000	80,000	90,000	100,000	125,000	150,000	175,000	200,000	500,000
120	6	10.4	13	15.6	20.8	26	31.2	36.4	41.6	46.8	52	65	78	91	104	260
	8	3.5	4.3	5.2	6.9	8.7	10.4	12.1	13.9	15.6	17.3	21.7	26	30.4	34.7	87
	10	1.4	1.8	2.1	2.8	3.5	4.2	4.9	5.6	6.3	7	8.7	10.5	12.2	14	34.9
140	6	12.1	15.2	18.2	24.3	30.3	36.4	42.5	48.5	55	61	76	91	106	121	303
	8	4	5.1	6.1	8.1	10.1	12.1	14.2	16.2	18.2	20.2	25.3	30.4	35.4	40.5	101
	10	1.7	2	2.4	3.3	4.1	4.9	5.7	6.5	7.3	8.1	10.2	12.2	14.3	16.3	40.7
160	6	13.9	17.3	20.8	27.7	34.7	41.6	48.5	56	62	69	87	104	121	139	347
	8	4.6	5.8	6.9	9.3	11.6	13.8	16.2	18.5	20.8	23.1	28.9	34.7	40.5	46.2	116
	10	1.9	2.3	2.8	3.7	4.7	5.6	6.5	7.5	8.4	9.3	11.6	14	16.3	18.6	46.6
180	6	15.6	19.5	23.4	31.2	39	46.8	55	62	70	78	98	117	137	156	390
	8	5.2	6.5	7.8	10.4	13	15.6	18.2	20.8	23.4	26	32.5	39	45.5	52	130
	10	2.1	2.6	3.1	4.2	5.2	6.3	7.3	8.4	9.4	10.5	13.1	15.7	18.3	21	52
200	8	5.8	7.2	8.7	11.6	14.5	17.3	20.2	23.1	26	28.9	36.1	43.4	51	58	145
	10	2.3	2.9	3.5	4.7	5.8	7	8.2	9.3	10.5	11.6	14.6	17.5	20.4	23.3	58
	12	1.2	1.5	1.7	2.3	2.9	3.5	4.1	4.6	5.2	5.8	7.2	8.7	10.1	11.6	28.9
250	8	7.2	9	10.8	14.5	18.1	21.7	25.3	28.9	32.5	36.1	45.2	54	63	72	181
	10	2.9	3.6	4.4	5.8	7.3	8.7	10.2	11.6	13.1	14.6	18.2	21.8	25.5	29.1	73
	12	1.5	1.8	2.2	2.9	3.6	4.3	5.1	5.8	6.5	7.2	9	10.9	12.7	14.5	36.2
300	8	8.7	10.8	13	17.3	21.7	26	30.4	34.7	39	43.4	54	65	76	87	217
	10	3.5	4.4	5.2	7	8.7	10.5	12.2	14	15.7	17.5	21.8	26.2	30.6	34.9	87
	12	1.7	2.2	2.6	3.5	4.3	5.2	6.1	7	7.8	8.7	10.9	13	15.2	17.4	43.4
400	8	11.6	14.5	17.3	23	28.9	34.7	40.5	46.2	52	58	72	87	101	116	289
	10	4.7	5.8	7	9.3	11.6	14	16.3	18.6	21	23.3	29.6	34.9	40.7	46.6	116
	12	2.3	2.9	3.5	4.6	5.8	7	8.1	9.3	10.4	11.6	14.5	17.4	20.3	23.2	58
500	8	14.5	18.1	21.7	28.9	36.1	43.4	51	58	65	72	90	108	126	145	361
	10	5.8	7.3	8.7	11.6	14.6	17.5	20.4	23.3	26.2	29.1	36.4	43.7	51	58	146
	12	2.9	3.6	4.3	5.8	7.2	8.7	10.1	11.6	13	14.5	18.1	21.7	25.3	28.9	72
600	8	17.3	21.7	26	34.7	43.4	52	61	69	78	87	108	130	152	173	434
	10	7	8.7	10.5	14	17.5	21	24.4	27.9	31.4	34.9	43.7	52	61	70	175
	12	3.5	4.3	5.2	7	8.7	10.4	12.2	13.9	15.6	17.4	21.7	26.1	30.4	34.7	87
700	8	20.2	25.3	30.3	40.5	51	61	71	81	91	101	126	152	177	202	506
	10	8.2	10.2	12.2	16.3	20.4	24.4	28.5	32.6	36.7	40.7	51	61	71	82	204
	12	4.1	5.1	6.1	8.1	10.1	12.2	14.2	16.2	18.2	20.3	25.3	30.4	35.5	40.5	101
800	8	23.1	28.9	34.7	46.2	58	69	81	93	104	116	145	173	202	231	578
	10	9.3	11.6	14	18.6	23.3	27.9	32.6	37.3	41.9	46.6	58	70	82	93	233
	12	4.6	5.8	7	9.3	11.6	13.9	16.2	18.5	20.8	23.1	28.9	34.7	40.5	46.3	116
900	8	26	32.5	39	52	65	78	91	104	117	130	163	195	228	260	650
	10	10.5	13.1	15.7	21	26.2	31.4	36.7	41.9	47.1	52	66	79	92	105	262
	12	5.2	6.5	7.8	10.4	13	15.6	18.2	20.8	23.4	26.1	32.6	39.1	45.6	52	130
1000	8	28.9	36.1	43.4	58	72	87	101	116	130	145	181	217	253	289	723
	10	11.6	14.6	17.5	23.3	29.1	34.9	40.7	46.6	52	58	73	87	102	116	291
	12	5.8	7.2	8.7	11.6	14.5	17.4	20.3	23.2	26.1	28.9	36.2	43.4	51	58	145

←————————————————————LAMINAR FLOW————————————————————→

TABLE A.5   VISCOSITY OF COMMON LIQUIDS

Liquid	*Sp Gr at 60 F	VISCOSITY		
		SSU	Centistokes	At F
Freon	1.37 to 1.49 @ 70 F		.27–.32	70
Glycerine (100%)	1.26 @ 68 F	2,950 813	648 176	68.6 100

Glycol:
  Propylene
  Triethylene
  Diethylene
  Ethylene
	1.038 @ 68 F
	1.125 @ 68 F
	1.12
	1.125
		240.6
		185.7
		149.7
		88.4
			52
			40
			32
			17.8
				70
				70
				70
				70

Hydrochloric Acid (31.5%)	1.05 @ 68 F		1.9	68
Mercury	13.6		.118 .11	70 100
Phenol (Carbolic Acid)	.95 to 1.08	65	11.7	65
Silicate of Soda	40 Baumé 42 Baumé	365 637.6	79 138	100 100
Sulfuric Acid (100%)	1.83	75.7	14.6	68
FISH AND ANIMAL OILS: Bone Oil	.918	220 65	47.5 11.6	130 212
Cod Oil	.928	150 95	32.1 19.4	100 130
Lard	.96	287 160	62.1 34.3	100 130
Lard Oil	.912 to .925	190 to 220 112 to 128	41 to 47.5 23.4 to 27.1	100 130
Menhadden Oil	.933	140 90	29.8 18.2	100 130
Neatsfoot Oil	.917	230 130	49.7 27.5	100 130
Sperm Oil	.883	110 78	23.0 15.2	100 130
Whale Oil	.925	163 to 184 97 to 112	35 to 39.6 19.9 to 23.4	100 130
MINERAL OILS: Automobile Crankcase Oils (Average Midcontinent Paraffin Base): SAE 10	**.880 to .935	165 to 240 90 to 120	35.4 to 51.9 18.2 to 25.3	100 130
SAE 20	**.880 to .935	240 to 400 120 to 185	51.9 to 86.6 25.3 to 39.9	100 130
SAE 30	**.880 to .935	400 to 580 185 to 255	86.6 to 125.5 39.9 to 55.1	100 130

*Unless otherwise noted.
**Depends on origin or percent and type of solvent.

Revised 1958. Copyright 1955 by the Hydraulic Institute.

TABLE A.5 (CONT.)   VISCOSITY OF COMMON LIQUIDS

Liquid	*Sp Gr at 60 F	VISCOSITY		
		SSU	Centistokes	At F
SAE 40	**.880 to .935	580 to 950 255 to 80	125.5 to 205.6 55.1 to 15.6	100 130 210
SAE 50	**.880 to .935	950 to 1,600 80 to 105	205.6 to 352 15.6 to 21.6	100 210
SAE 60	**.880 to .935	1,600 to 2,300 105 to 125	352 to 507 21.6 to 26.2	100 210
SAE 70	**.880 to .935	2,300 to 3,100 125 to 150	507 to 682 26.2 to 31.8	100 210
SAE 10W	**.880 to .935	5,000 to 10,000	1,100 to 2,200	0
SAE 20W	**.880 to .935	10,000 to 40,000	2,200 to 8,800	0
Automobile Transmission Lubricants: SAE 80	**.880 to .935	100,000 max	22,000 max	0
SAE 90	**.880 to .935	800 to 1,500 300 to 500	173.2 to 324.7 64.5 to 108.2	100 130
SAE 140	**.880 to .935	950 to 2,300 120 to 200	205.6 to 507 25.1 to 42.9	130 210
SAE 250	**.880 to .935	Over 2,300 Over 200	Over 507 Over 42.9	130 210
Crude Oils: Texas, Oklahoma	.81 to .916	40 to 783 34.2 to 210	4.28 to 169.5 2.45 to 45.3	60 100
Wyoming, Montana	.86 to .88	74 to 1,215 46 to 320	14.1 to 263 6.16 to 69.3	60 100
California	.78 to .92	40 to 4,840 34 to 700	4.28 to 1,063 2.4 to 151.5	60 100
Pennsylvania	.8 to .85	46 to 216 38 to 86	6.16 to 46.7 3.64 to 17.2	60 100
Diesel Engine Lubricating Oils (Based on Average Midcontinent Paraffin Base): Federal Specification No. 9110	**.880 to .935	165 to 240 90 to 120	35.4 to 51.9 18.2 to 25.3	100 130
Federal Specification No. 9170	**.880 to .935	300 to 410 140 to 180	64.5 to 88.8 29.8 to 38.8	100 130
Federal Specification No. 9250	**.880 to .935	470 to 590 200 to 255	101.8 to 127.8 43.2 to 55.1	100 130
Federal Specification No. 9370	**.880 to .935	800 to 1,100 320 to 430	173.2 to 238.1 69.3 to 93.1	100 130
Federal Specification No. 9500	**.880 to .935	490 to 600 92 to 105	106.1 to 129.9 18.54 to 21.6	130 210

*Unless otherwise noted.
**Depends on origin or percent and type of solvent.

Revised 1958. Copyright 1955 by the Hydraulic Institute.

TABLE A.5 (CONT.)   VISCOSITY OF COMMON LIQUIDS

Liquid	*Sp Gr at 60 F	VISCOSITY		
		SSU	Centistokes	At F
Diesel Fuel Oils:				
No. 2 D	**.82 to .95	32.6 to 45.5	2 to 6	100
		39	1 to 3.97	130
No. 3 D	**.82 to .95	45.5 to 65	6 to 11.75	100
		39 to 48	3.97 to 6.78	130
No. 4 D	**.82 to .95	140 max	29.8 max	100
		70 max	13.1 max	130
No. 5D	**.82 to .95	400 max	86.6 max	122
		165 max	35.2 max	160
Fuel Oils:				
No. 1	**.82 to .95	34 to 40	2.39 to 4.28	70
		32 to 35	2.69	100
No. 2	**.82 to .95	36 to 50	3.0 to 7.4	70
		33 to 40	2.11 to 4.28	100
No. 3	**.82 to .95	35 to 45	2.69 to .584	100
		32.8 to 39	2.06 to 3.97	130
No. 5A	**.82 to .95	50 to 125	7.4 to 26.4	100
		42 to 72	4.91 to 13.73	130
No. 5B	**.82 to .95	125 to	26.4 to	100
		400	86.6	122
		72 to 310	13.63 to 67.1	130
No. 6	**.82 to .95	450 to 3,000	97.4 to 660	122
		175 to 780	37.5 to 172	160
Fuel Oil — Navy Specification	**.989 max	110 to 225	23 to 48.6	122
		63 to 115	11.08 to 23.9	160
Fuel Oil — Navy II	1.0 max	1,500 max	324.7 max	122
		480 max	104 max	160
Gasoline	.68 to .74		.46 to .88	60
			.40 to .71	100
Gasoline (Natural)	76.5 degrees API		.41	68
Gas Oil	28 degrees API	73	13.9	70
		50	7.4	100
Insulating Oil:				
Transformer, switches and circuit breakers		115 max	24.1 max	70
		65 max	11.75 max	100
Kerosene	.78 to .82	35	2.69	68
		32.6	2	100
Machine Lubricating Oil (Average Pennsylvania Paraffin Base):				
Federal Specification No. 8	**.880 to .935	112 to 160	23.4 to 34.3	100
		70 to 90	13.1 to 18.2	130

*Unless otherwise noted.
**Depends on origin or percent and type of solvent.

Revised 1958. Copyright 1955 by the Hydraulic Institute.

TABLE A.5 (CONT.)   VISCOSITY OF COMMON LIQUIDS

Liquid	*Sp Gr at 60 F	VISCOSITY		
		SSU	Centistokes	At F
Federal Specification No. 10	**.880 to .935	160 to 235 90 to 120	34.3 to 50.8 18.2 to 25.3	100 130
Federal Specification No. 20	**.880 to .935	235 to 385 120 to 185	50.8 to 83.4 25.3 to 39.9	100 130
Federal Specification No. 30	**.880 to .935	385 to 550 185 to 255	83.4 to 119 39.9 to 55.1	100 130
Mineral Lard Cutting Oil: Federal Specification Grade 1		140 to 190 86 to 110	29.8 to 41 17.22 to 23	100 130
Federal Specification Grade 2		190 to 220 110 to 125	41 to 47.5 23 to 26.4	100 130
Petrolatum	.825	100 77	20.6 14.8	130 160
Turbine Lubricating Oil: Federal Specification (Penn Base)	.91 Average	400 to 440 185 to 205	86.6 to 95.2 39.9 to 44.3	100 130
VEGETABLE OILS: Castor Oil	.96 @ 68 F	1,200 to 1,500 450 to 600	259.8 to 324.7 97.4 to 129.9	100 130
China Wood Oil	.943	1,425 580	308.5 125.5	69 100
Cocoanut Oil	.925	140 to 148 76 to 80	29.8 to 31.6 14.69 to 15.7	100 130
Corn Oil	.924	135 54	28.7 8.59	130 212
Cotton Seed Oil	.88 to .925	176 100	37.9 20.6	100 130
Linseed Oil, Raw	.925 to .939	143 93	30.5 18.94	100 130
Olive Oil	.912 to .918	200 115	43.2 24.1	100 130
Palm Oil	.924	221 125	47.8 26.4	100 130
Peanut Oil	.920	195 112	42 23.4	100 130
Rape Seed Oil	.919	250 145	54.1 31	100 130
Rosin Oil	.980	1,500 600	324.7 129.9	100 130

*Unless otherwise noted.
**Depends on origin or percent and type of solvent.

TABLE A.5 (CONT.)   VISCOSITY OF COMMON LIQUIDS

Liquid	*Sp Gr at 60 F	VISCOSITY		
		SSU	Centistokes	At F
Rosin (Wood)	1.09 Avg.	500 to 20,000 1,000 to 50,000	108.2 to 4,400 216.4 to 11,000	200 190
Sesame Oil	.923	184 110	39.6 23	100 130
Soja Bean Oil	.927 to .98	165 96	35.4 19.64	100 130
Turpentine	.86 to .87	33 32.6	2.11 2.0	60 100
SUGAR, SYRUPS, MOLASSES, ETC. Corn Syrups	1.4 to 1.47	5,000 to 500,000 1,500 to 60,000	1,100 to 110,000 324.7 to 13,200	100 130
Glucose	1.35 to 1.44	35,000 to 100,000 4,000 to 11,000	7,700 to 22,000 880 to 2,420	100 150
Honey (Raw)		340	73.6	100
Molasses "A" (First)	1.40 to 1.46	1,300 to 23,000 700 to 8,000	281.1 to 5,070 151.5 to 1,760	100 130
Molasses "B" (Second)	1.43 to 1.48	6,400 to 60,000 3,000 to 15,000	1,410 to 13,200 660 to 3,300	100 130
Molasses "C" (Blackstrap or final)	1.46 to 1.49	17,000 to 250,000 6,000 to 75,000	2,630 to 5,500 1,320 to 16,500	100 130
Sucrose Solutions (Sugar Syrups): 60 Brix	1.29	230 92	49.7 18.7	70 100
62 Brix	1.30	310 111	67.1 23.2	70 100
64 Brix	1.31	440 148	95.2 31.6	70 100
66 Brix	1.326	650 195	140.7 42.0	70 100
68 Brix	1.338	1,000 275	216.4 59.5	70 100
70 Brix	1.35	1,650 400	364 86.6	70 100
72 Brix	1.36	2,700 640	595 138.6	70 100
74 Brix	1.376	5,500 1,100	1,210 238	70 100
76 Brix	1.39	10,000 2,000	2,200 440	70 100

*Unless otherwise noted.

Revised 1958. Copyright 1955 by the Hydraulic Institute.

TABLE A.5 (CONT.)   VISCOSITY OF COMMON LIQUIDS

Liquid	*Sp Gr at 60 F	VISCOSITY		
		SSU	Centistokes	At F
TARS:				
Tar-Coke Oven	1.12+	3,000 to 8,000 650 to 1,400	600 to 1,760 140.7 to 308	71 100
Tar-Gas House	1.16 to 1.30	15,000 to 300,000 2,000 to 20,000	3,300 to 66,000 440 to 4,400	70 100
Road Tar:				
Grade RT-2	1.07+	200 to 300 55 to 60	43.2 to 64.9 8.77 to 10.22	122 212
Grade RT-4	1.08+	400 to 700 65 to 75	86.6 to 154 11.63 to 14.28	122 212
Grade RT-6	1.09+	1,000 to 2,000 85 to 125	216.4 to 440 16.83 to 26.2	122 212
Grade RT-8	1.13+	3,000 to 8,000 150 to 225	660 to 1,760 31.8 to 48.3	122 212
Grade RT-10	1.14+	20,000 to 60,000 250 to 400	4,400 to 13,200 53.7 to 86.6	122 212
Grade RT-12	1.15+	114,000 to 456,000 500 to 800	25,000 to 75,000 108.2 to 173.2	122 212
Pine Tar	1.06	2,500 500	559 108.2	100 132
MISCELLANEOUS Corn Starch Solutions: 22 Baumé	1.18	150 130	32.1 27.5	70 100
24 Baumé	1.20	600 440	129.8 95.2	70 100
25 Baumé	1.21	1,400 800	303 173.2	70 100
Ink—Printers	1.00 to 1.38	2,500 to 10,000 1,100 to 3,000	550 to 2,200 238.1 to 660	100 130
Tallow	.918 Avg.	56	9.07	212
Milk	1.02 to 1.05		1.13	68
Varnish — Spar	.9	1,425 650	313 143	68 100
Water — Fresh	1.0		1.13 .55	60 130

*Unless otherwise noted.

Revised 1958. Copyright 1955 by the Hydraulic Institute.

TABLE A.6   VISCOSITY CONVERSION

The following table will give a comparison of various viscosity ratings so that if the viscosity is given in terms other than Saybolt Universal, it can be translated quickly by following horizontally to the Saybolt Universal column.

Seconds Saybolt Universal ssu	Kinematic Viscosity Centistokes *	Seconds Saybolt Furol ssf	Seconds Redwood I (Standard)	Seconds Redwood 2 (Admiralty)	Degrees Engler	Degrees Barbey	Seconds Parlin Cup #7	Seconds Parlin Cup #10	Seconds Parlin Cup #15	Seconds Parlin Cup #20	Seconds Ford Cup #3	Seconds Ford Cup #4
31	1.00	–	29	–	1.00	6200	–	–	–	–	–	–
35	2.56	–	32.1	–	1.16	2420	–	–	–	–	–	–
40	4.30	–	36.2	5.10	1.31	1440	–	–	–	–	–	–
50	7.40	–	44.3	5.83	1.58	838	–	–	–	–	–	–
60	10.3	–	52.3	6.77	1.88	618	–	–	–	–	–	–
70	13.1	12.95	60.9	7.60	2.17	483	–	–	–	–	–	–
80	15.7	13.70	69.2	8.44	2.45	404	–	–	–	–	–	–
90	18.2	14.44	77.6	9.30	2.73	348	–	–	–	–	–	–
100	20.6	15.24	85.6	10.12	3.02	307	–	–	–	–	–	–
150	32.1	19.30	128	14.48	4.48	195	–	–	–	–	–	–
200	43.2	23.5	170	18.90	5.92	144	40	–	–	–	–	–
250	54.0	28.0	212	23.45	7.35	114	46	–	–	–	–	–
300	65.0	32.5	254	28.0	8.79	95	52.5	15	6.0	3.0	30	20
400	87.60	41.9	338	37.1	11.70	70.8	66	21	7.2	3.2	42	28
500	110.0	51.6	423	46.2	14.60	56.4	79	25	7.8	3.4	50	34
600	132	61.4	508	55.4	17.50	47.0	92	30	8.5	3.6	58	40
700	154	71.1	592	64.6	20.45	40.3	106	35	9.0	3.9	67	45
800	176	81.0	677	73.8	23.35	35.2	120	39	9.8	4.1	74	50
900	198	91.0	762	83.0	26.30	31.3	135	41	10.7	4.3	82	57
1000	220	100.7	896	92.1	29.20	28.2	149	43	11.5	4.5	90	62
1500	330	150	1270	138.2	43.80	18.7	–	65	15.2	6.3	132	90
2000	440	200	1690	184.2	58.40	14.1	–	86	19.5	7.5	172	118
2500	550	250	2120	230	73.0	11.3	–	108	24	9	218	147
3000	660	300	2540	276	87.60	9.4	–	129	28.5	11	258	172
4000	880	400	3380	368	117.0	7.05	–	172	37	14	337	230
5000	1100	500	4230	461	146	5.64	–	215	47	18	425	290
6000	1320	600	5080	553	175	4.70	–	258	57	22	520	350
7000	1540	700	5920	645	204.5	4.03	–	300	67	25	600	410
8000	1760	800	6770	737	233.5	3.52	–	344	76	29	680	465
9000	1980	900	7620	829	263	3.13	–	387	86	32	780	520
10000	2200	1000	8460	921	292	2.82	–	430	96	35	850	575
15000	3300	1500	13700	–	438	2.50	–	650	147	53	1280	860
20000	4400	2000	18400	–	584	1.40	–	860	203	70	1715	1150

* Kinematic Viscosity (in centistokes) $= \dfrac{\text{Absolute viscosity (in centipoises)}}{\text{Specific Gravity}}$

Above 250 SSU, use the following approximate conversion:

$$\text{SSU} = \text{Centistokes} \times 4.62$$

Above the range of this table and within the range of the viscosimeter, multiply their rating by the following factors to convert to SSU:

Viscosimeter	Factor		Viscosimeter	Factor
Saybolt Furol	10.		Parlin cup #15	98.2
Redwood Standard	1.095		Parlin cup #20	187.0
Redwood Admiralty	10.87		Ford cup # 4	17.4
Engler - Degrees	34.5			

Copyright 1955 by the Hydraulic Institute.

TABLE A.6 (CONT.)   VISCOSITY CONVERSION

The following table will give a comparison of various viscosity ratings so that if the viscosity is given in terms other than Saybolt Universal, it can be translated quickly by following horizontally to the Saybolt Universal column.

Seconds Saybolt Universal ssu	Kinematic Viscosity Centistokes*	Approx. Seconds Mac Michael	Approx. Gardner Holt Bubble	Seconds Zahn Cup #1	Seconds Zahn Cup #2	Seconds Zahn Cup #3	Seconds Zahn Cup #4	Seconds Zahn Cup #5	Seconds Demmler Cup #1	Seconds Demmler Cup #10	Approx. Seconds Stormer 100 gm Load	Seconds Pratt and Lambert "F"
31	1.00	–	–	–	–	–	–	–	–	–	–	–
35	2.56	–	–	–	–	–	–	–	–	–	–	–
40	4.30	–	–	–	–	–	–	–	–	–	–	–
50	7.40	–	–	–	–	–	–	–	1.3	–	–	–
									2.3	–	2.6	–
60	10.3	–	–	–	–	–	–	–	3.2	–	3.6	–
70	13.1	–	–	–	–	–	–	–	4.1	–	4.6	–
80	15.7	–	–	–	–	–	–	–	4.9	–	5.5	–
90	18.2	–	–	–	–	–	–	–	5.7	–	6.4	–
100	20.6	125	–	38	18	–	–	–	6.5	–	7.3	–
150	32.1	145	–	47	20	–	–	–	10.0	1.0	11.3	–
200	43.2	165	A	54	23	–	–	–	13.5	1.4	15.2	–
250	54.0	198	A	62	26	–	–	–	16.9	1.7	19	–
300	65.0	225	B	73	29	–	–	–	20.4	2.0	23	–
400	87.0	270	C	90	37	–	–	–	27.4	2.7	31	7
500	110.0	320	D	–	46	–	–	–	34.5	3.5	39	8
600	132	370	F	–	55	–	–	–	41	4.1	46	9
700	154	420	G	–	63	22.5	–	–	48	4.8	54	9.5
800	176	470	–	–	72	24.5	–	–	55	5.5	62	10.8
900	198	515	H	–	80	27	18	–	62	6.2	70	11.9
1000	220	570	I	–	88	29	20	13	69	6.9	77	12.4
1500	330	805	M	–	–	40	28	18	103	10.3	116	16.8
2000	440	1070	Q	–	–	51	34	24	137	13.7	154	22
2500	550	1325	T	–	–	63	41	29	172	17.2	193	27.6
3000	660	1690	U	–	–	75	48	33	206	20.6	232	33.7
4000	880	2110	V	–	–	–	63	43	275	27.5	308	45
5000	1100	2635	W	–	–	–	77	50	344	34.4	385	55.8
6000	1320	3145	X	–	–	–	–	65	413	41.3	462	65.5
7000	1540	3670	–	–	–	–	–	75	481	48	540	77
8000	1760	4170	Y	–	–	–	–	86	550	55	618	89
9000	1980	4700	–	–	–	–	–	96	620	62	695	102
10000	2200	5220	Z	–	–	–	–	–	690	69	770	113
15000	3300	7720	Z2	–	–	–	–	–	1030	103	1160	172
20000	4400	10500	Z3	–	–	–	–	–	1370	137	1540	234

* Kinematic Viscosity (in centistokes) $= \dfrac{\text{Absolute viscosity (in centipoises)}}{\text{Specific Gravity}}$

Above 250 SSU, use the following approximate conversion:

$$\text{SSU} = \text{Centistokes} \times 4.62$$

Above the range of this table and within the range of the viscosimeter, multiply their rating by the following factors to convert to SSU:

Viscosimeter	Factor
Mac Michael	1.92 (approx.)
Demmler #1	14.6
Demmler #10	146.
Stormer	13. (approx.)

TABLE A.7   THEORETICAL DISCHARGE OF NOZZLES, IN US GALLONS PER MINUTE

Head Lbs.	Head Feet	Velocity of discharge, feet per second	1/16	⅛	3/16	¼	⅜	½	⅝	¾	⅞	1	1⅛	1¼	1⅜	1½	1¾	2	2¼	2½	2¾	3	3½	4	4½	5	5½	6
10	23.1	38.58	0.37	1.48	3.30	5.90	13.2	23.6	36.8	53.2	72.2	94.4	119	148	178	212	289	378	478	590	715	850	1160	1512	1912	2360	2860	3400
15	34.7	47.25	0.45	1.81	4.02	7.23	16.2	28.7	45.0	65.1	88.4	116	146	181	218	260	354	463	586	723	880	1040	1420	1852	2344	2800	3520	4160
20	46.2	54.55	0.52	2.09	4.66	8.35	18.7	33.4	52.0	75.3	102	134	169	209	252	300	409	534	676	835	1018	1200	1640	2136	2700	3340	4072	4800
25	57.8	60.99	0.58	2.33	5.23	9.33	20.9	37.2	58.2	84.1	114	149	189	233	282	336	457	597	756	933	1138	1350	1828	2390	3024	3730	4582	5400
30	69.3	66.82	0.64	2.56	5.71	10.2	22.8	40.9	63.7	92.2	125	164	207	256	309	368	501	654	828	1022	1240	1480	2000	2616	3312	4038	4960	5920
35	80.9	72.16	0.69	2.76	6.16	11.0	24.7	44.2	68.8	99.6	135	177	223	276	334	397	541	707	895	1104	1340	1590	2160	2828	3580	4416	5360	6360
40	92.4	77.14	0.74	2.95	6.60	11.8	26.4	47.2	73.6	106	144	189	239	295	357	425	578	755	956	1180	1430	1700	2320	3020	3824	4720	5720	6800
45	104.0	81.83	0.78	3.13	6.99	12.5	28.0	50.2	78.1	113	153	200	253	313	378	450	613	801	1014	1252	1520	1800	2450	3200	4056	5000	6080	7200
50	115.5	86.26	0.82	3.30	7.37	13.2	29.5	52.8	82.3	119	161	211	267	330	399	475	646	845	1069	1320	1600	1900	2580	3380	4276	5280	6400	7600
55	127.1	90.46	0.86	3.46	7.73	13.8	30.9	55.4	86.3	125	169	221	280	346	418	498	678	886	1122	1385	1680	2000	2720	3544	4488	5540	6720	8000
60	138.6	94.49	0.90	3.62	8.08	14.5	32.3	57.8	90.1	130	177	231	293	362	437	520	708	925	1171	1446	1755	2080	2840	3700	4684	5784	7020	8320
65	150.2	98.35	0.94	3.77	8.40	15.1	33.6	60.2	93.8	136	184	241	305	377	455	542	737	963	1219	1506	1830	2165	2950	3850	4876	6024	7320	8660
70	161.7	102.06	0.97	3.91	8.73	15.6	34.9	62.5	97.4	141	191	250	316	391	472	562	765	999	1265	1561	1895	2250	3060	4000	5060	6244	7580	9000
75	173.3	105.65	1.01	4.04	9.03	16.2	36.1	64.6	101	146	198	259	327	404	488	582	792	1034	1309	1616	1960	2330	3170	4136	5236	6464	7840	9320
80	184.8	109.11	1.04	4.18	9.33	16.7	37.8	66.6	104	150	204	267	338	418	504	601	818	1068	1352	1669	2030	2404	3280	4272	5400	6676	8120	9616
85	196.4	112.46	1.07	4.31	9.62	17.2	38.5	68.8	107	155	210	275	348	431	520	620	843	1101	1394	1720	2080	2480	3380	4400	5576	6880	8320	9920
90	207.9	115.72	1.10	4.43	9.89	17.7	39.6	70.8	110	160	217	283	358	443	535	637	867	1133	1434	1770	2150	2540	3470	4532	5736	7080	8600	10160
95	219.5	118.89	1.13	4.55	10.2	18.2	40.7	72.8	113	164	223	291	368	455	550	655	891	1164	1474	1820	2200	2620	3560	4656	5896	7280	8800	10480
100	231.1	121.98	1.16	4.67	10.4	18.7	41.7	74.6	116	168	228	299	378	467	564	672	914	1194	1512	1866	2260	2700	3650	4776	6048	7464	9040	10800
105	242.6	125.00	1.19	4.78	10.7	19.1	42.8	76.5	119	172	234	306	387	478	578	688	937	1224	1549	1912	2320	2760	3750	4896	6200	7648	9280	11040
110	254.2	127.94	1.22	4.90	10.9	19.6	43.8	78.3	122	177	239	313	396	490	591	705	959	1253	1586	1957	2380	2820	3840	5012	6344	7828	9520	11280
115	265.7	130.82	1.25	5.01	11.2	20.0	44.8	80.1	125	181	245	320	405	501	605	720	980	1281	1621	2002	2430	2880	3920	5124	6484	8008	9720	11520
120	277.3	133.63	1.27	5.12	11.4	20.4	45.7	81.8	127	184	250	327	414	512	618	736	1001	1308	1656	2044	2480	2950	4004	5232	6624	8176	9920	11800
125	288.8	136.38	1.30	5.22	11.7	20.9	46.7	83.5	130	188	255	334	422	522	630	751	1022	1335	1691	2086	2540	3000	4100	5340	6764	8344	10160	12000
130	300.4	139.08	1.33	5.32	11.9	21.3	47.6	85.1	133	192	260	341	431	532	643	766	1042	1362	1724	2128	2580	3070	4160	5448	6896	8512	10320	12280

The actual quantity discharged by a nozzle will be less than above table. A well tapered smooth nozzle may be assumed to give above 94 per cent of the values in the tables.

Index

484

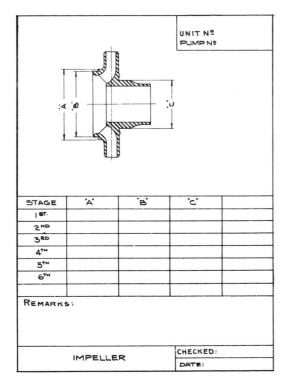

Fig. 26.5   Record card for impeller dimensions

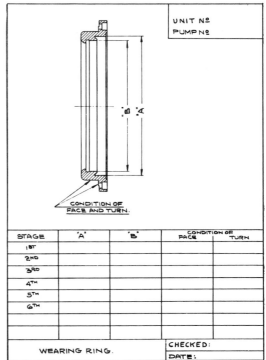

Fig. 26.6   Record card for wearing ring dimensions

sign and a great number of combinations are made for each size of casing, using different impeller sizes and designs. Without an identification number, the pump manufacturer would be at a loss to furnish correct repairs even though the size and type of the pump might be known.

Record of inspection and repairs

The working schedule of the semiannual and annual inspection program should be incorporated on maintenance cards, one for each pump in the installation. These cards should contain the identifying pump number, the date of the scheduled inspection, a complete record of all the items requiring individual inspection, and space for comments and observations of the inspecting personnel. Adequate maintenance does not stop with repair work on worn or damaged parts. A written record of the conditions of the parts to be repaired or replaced, of the rate and appearance of the wear, and of the method by which the re-

pair was carried out is as important as the repair job itself. These records can form the basis of preventive measures which will act to reduce both the frequency and cost of maintenance work. The type of inspection records and the extent of detail they may contain vary with the type of pump in question and availability of personnel. Typical examples of the records that can be kept of the impeller and wearing ring clearances of a multistage pump are shown in Fig. 26.5 and 26.6.

It is often advisable to take photographs of badly worn parts before they are repaired; photographs provide a more accurate and more graphic record of the damage than a description.

Complete records of maintenance and repair costs should always be kept for each individual pump, together with a record of its operating hours; study of these records may reveal whether a change in materials or design will be the most economical plan to follow.

27 *Operating Problems*

The operation of a centrifugal pump may be affected by hydraulic or mechanical troubles. Hydraulic troubles may cause a pump to fail to deliver water altogether; or, the pump may deliver an insufficient capacity, develop insufficient pressure, lose its prime after starting, or consume excessive power. Mechanical difficulties may appear at the stuffing boxes and bearings, or produce vibration, noise, or overheating of the pump.

It is important to note that there is often a definite connection between these two kinds of difficulties. For instance, increased wear at the running clearances must be classified as a mechanical trouble, but will result in a reduction of the net pump capacity—a hydraulic symptom—without necessarily causing a mechanical breakdown or even excessive vibration. As a result, it is of great advantage to classify symptoms and causes separately and to list a schedule of potentially contributory causes for each symptom (Table 27.1).

The cure for each trouble is almost always self-evident. The remainder of this chapter will be devoted to an analysis of some of the most important trouble spots —an analysis that may be helpful in diagnosing and preventing these troubles.

Pump noise

Pump noise often gives an experienced maintenance man a definite indication of the source of trouble. If a pump produces a crackling noise, the source of trouble is probably at the pump suction. This type of noise is usually associated with "cavitation" —the condition existing in flowing liquids when the pressure at any point falls below the vapor pressure of the liquid at the prevailing temperature. Some of the liquid flashes into vapor, and bubbles of the vapor are carried along with the liquid. If this happens in the suction area of a centrifugal pump or within the entrance of the impeller, the bubbles are carried into the impeller and undergo an increase in pressure and, therefore, condense. This process is accompanied by a violent collapse of the bubbles, possible pitting and erosion of the impeller vanes, and a definite crackling noise. Of course, the presence of vapor within the liquid pumped causes a reduction in the pump capacity. Cavitation, therefore, is a direct result of insufficient pressure at the pump suction—operation with insufficient net positive suction head (NPSH).

This diagnosis may be checked. For instance, throttling the pump discharge will